Chromatography: Concepts, Methods and Applications

Chromatography: Concepts, Methods and Applications

Edited by
Judah Carter

Larsen & Keller
www.larsen-keller.com

Chromatography: Concepts, Methods and Applications
Edited by Judah Carter
ISBN: 978-1-63549-066-4 (Hardback)

© 2017 Larsen & Keller

⊟ Larsen & Keller

Published by Larsen and Keller Education,
5 Penn Plaza,
19th Floor,
New York, NY 10001, USA

Cataloging-in-Publication Data

Chromatography : concepts, methods and applications / edited by Judah Carter.
    p. cm.
Includes bibliographical references and index.
ISBN 978-1-63549-066-4
1. Chromatographic analysis. 2. Chemistry, Analytic.
I. Carter, Judah.

QD79.C4 C47 2017
543.8--dc23

The publisher's policy is to use permanent paper from mills that operate a sustainable forestry policy. Furthermore, the publisher ensures that the text paper and cover boards used have met acceptable environmental accreditation standards.

Printed and bound in the United States of America.

For more information regarding Larsen and Keller Education and its products, please visit the publisher's website www.larsen-keller.com

Table of Contents

Preface

This book elucidates the concepts and innovative models around prospective developments with respect to chromatography. It describes in detail the various basic concepts of this field. Chromatography is the science and technique of using different laboratory methods to separate chemical mixtures. It is basically a method of purification. This textbook is a valuable compilation of topics, ranging from the basic to the most complete theories in the field of chromatography. Such selected concepts that redefine this subject have been presented in this text. Different approaches, evaluations and methodologies on chromatography have been included in this textbook. It will provide comprehensive knowledge to the students.

A detailed account of the significant topics covered in this book is provided below:

Chapter 1- The term used for the separation of mixtures is termed as chromatography. Chromatography is known to be either preparative or analytical. This chapter helps the reader to develop a better understanding on chromatography.

Chapter 2- The extraction of one material from another is elution. The measurement of the quality of the separation is chromatographic response function. Apart from elution and chromatographic response function, the essential concepts of chromatography are retardation factor, theoretical plate, Van Deemter equation etc. The information presented in this text elucidates the fundamental concepts of chromatography.

Chapter 3- Chromatography has particular methods and techniques. Some of these are micellar electrokinetic chromatography, affinity chromatography, size-exclusion chromatography, two-dimensional chromatography and ion chromatography. This chapter discusses the methods of chromatography in a critical manner providing key analysis to the subject matter.

Chapter 4- Liquid chromatography is a technique that is used in combining the capabilities of liquid chromatography with the capabilities of mass spectrometry. It's a very powerful technique that is found to be useful in a number of applications. The chapter concentrates on topics such as ion suppression in liquid chromatography–mass spectrometry, fast protein liquid chromatography, countercurrent chromatography, column chromatography etc. The section on liquid chromatography offers an insightful focus, keeping in mind the complex topic.

Chapter 5- Chromatography can be divided into preparative and analytic chromatography. Gas chromatography is a type of analytic chromatography which is used in separating compounds and then further analyzing them. The following content also focuses on inverse gas chromatography, unresolved complex mixture, comprehensive two-dimensional gas chromatography, gas chromatography-mass spectrometry, methanizers etc.

Chapter 6- The device that is used in liquid and gas chromatography to detect various components is known as chromatography detector. These detectors can also be classified into destructive detectors and non-destructive detectors. Some of the destructive detectors are flame ionization detectors and mass spectrometry whereas the non-destructive detectors explained are thermal conductivity detector and electron capture detector. The following chapter helps the readers in developing an in-depth understanding on the various detectors in chromatography.

Chapter 7- Chromatography software is a software that is used in collecting and analyzing chromatographic results. They are usually provided by manufacturers and help in analyzing data. Agilent ChemStation, empower and OpenChrom are the topics that have been explained in the section. The following chapter helps the readers in developing an in-depth understanding of chromatography software.

Chapter 8- Chromatography is an interdisciplinary subject. The allied fields associated with chromatography are analytical chemistry, biochemistry and chemical engineering. This chapter will provide a glimpse of the allied fields of chromatography.

Chapter 9- The history of chromatography explains to the readers the growth of chromatography over the years. Chromatography was initially used to separate plant pigments. New forms of chromatography have been developed in the recent past. The section on evolution of chromatography offers an insightful focus, keeping in mind the subject matter.

I would like to make a special mention of my publisher who considered me worthy of this opportunity and also supported me throughout the process. I would also like to thank the editing team at the back-end who extended their help whenever required.

Editor

Introduction to Chromatography

The term used for the separation of mixtures is termed as chromatography. Chromatography is known to be either preparative or analytical. This chapter helps the reader to develop a better understanding on chromatography.

Chromatography is the collective term for a set of laboratory techniques for the separation of mixtures. The mixture is dissolved in a fluid called the *mobile phase,* which carries it through a structure holding another material called the *stationary phase*. The various constituents of the mixture travel at different speeds, causing them to separate. The separation is based on differential partitioning between the mobile and stationary phases. Subtle differences in a compound's partition coefficient result in differential retention on the stationary phase and thus changing the separation.

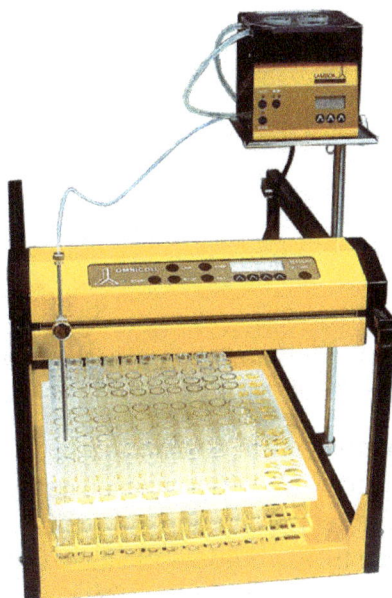

Automated fraction collector and sampler for chromatographic techniques

Chromatography may be preparative or analytical. The purpose of preparative chromatography is to separate the components of a mixture for more advanced use (and is thus a form of purification). Analytical chromatography is done normally with smaller amounts of material and is for measuring the relative proportions of analytes in a mixture. The two are not mutually exclusive.

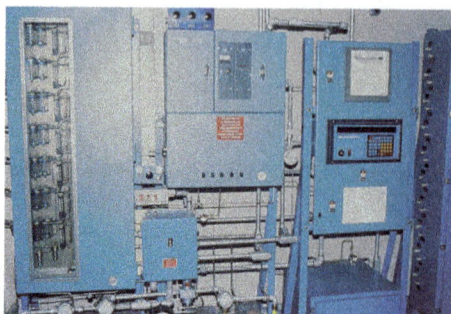

Pictured is a sophisticated gas chromatography system. This instrument records concentrations of acrylonitrile in the air at various points throughout the chemical laboratory.

History

Thin layer chromatography is used to separate components of a plant extract, illustrating the experiment with plant pigments that gave chromatography its name

Chromatography was first employed in Russia by the Italian-born scientist Mikhail Tsvet in 1900. He continued to work with chromatography in the first decade of the 20th century, primarily for the separation of plant pigments such as chlorophyll, carotenes, and xanthophylls. Since these components have different colors (green, orange, and yellow, respectively) they gave the technique its name. New types of chromatography developed during the 1930s and 1940s made the technique useful for many separation processes.

Chromatography technique developed substantially as a result of the work of Archer John Porter Martin and Richard Laurence Millington Synge during the 1940s and 1950s, for which they won a Nobel prize. They established the principles and basic techniques of partition chromatography, and their work encouraged the rapid development of several chromatographic methods: paper chromatography, gas chromatography, and what would become known as high performance liquid chromatography. Since then, the technology has advanced rapidly. Researchers found that the main principles of Tsvet's chromatography could be applied in many different ways, resulting in the different varieties of chromatography described below. Advances are continually improving the technical performance of chromatography, allowing the separation of increasingly similar molecules.

Chromatography Terms

- The analyte is the substance to be separated during chromatography. It is also normally what is needed from the mixture.

- Analytical chromatography is used to determine the existence and possibly also the concentration of analyte(s) in a sample.

- A bonded phase is a stationary phase that is covalently bonded to the support particles or to the inside wall of the column tubing.

- A chromatogram is the visual output of the chromatograph. In the case of an optimal separation, different peaks or patterns on the chromatogram correspond to different components of the separated mixture.

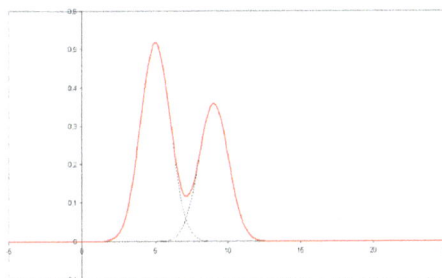

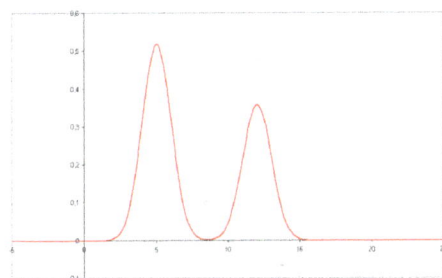

Plotted on the x-axis is the retention time and plotted on the y-axis a signal (for example obtained by a spectrophotometer, mass spectrometer or a variety of other detectors) corresponding to the response created by the analytes exiting the system. In the case of an optimal system the signal is proportional to the concentration of the specific analyte separated.

- A chromatograph is equipment that enables a sophisticated separation, e.g. gas chromatographic or liquid chromatographic separation.

- Chromatography is a physical method of separation that distributes components to separate between two phases, one stationary (stationary phase), the other (the mobile phase) moving in a definite direction.

- The eluate is the mobile phase leaving the column.

- The eluent is the solvent that carries the analyte.

- An eluotropic series is a list of solvents ranked according to their eluting power.

- An immobilized phase is a stationary phase that is immobilized on the support particles, or on the inner wall of the column tubing.

- The mobile phase is the phase that moves in a definite direction. It may be a liquid (LC and Capillary Electrochromatography (CEC)), a gas (GC), or a supercritical fluid (supercritical-fluid chromatography, SFC). The mobile phase consists of the sample being separated/analyzed and the solvent that moves the sample through the column. In the case of HPLC the mobile phase consists of a non-polar solvent(s) such as hexane in normal phase or a polar solvent such as methanol in reverse phase chromatography and the sample being separated. The mobile phase moves through the chromatography column (the stationary phase) where the sample interacts with the stationary phase and is separated.

- Preparative chromatography is used to purify sufficient quantities of a substance for further use, rather than analysis.

- The retention time is the characteristic time it takes for a particular analyte to pass through the system (from the column inlet to the detector) under set conditions.

- The sample is the matter analyzed in chromatography. It may consist of a single component or it may be a mixture of components. When the sample is treated in the course of an analysis, the phase or the phases containing the analytes of interest is/are referred to as the sample whereas everything out of interest separated from the sample before or in the course of the analysis is referred to as waste.

- The solute refers to the sample components in partition chromatography.

- The solvent refers to any substance capable of solubilizing another substance, and especially the liquid mobile phase in liquid chromatography.

- The stationary phase is the substance fixed in place for the chromatography procedure. Examples include the silica layer in thin layer chromatography

- The detector refers to the instrument used for qualitative and quantitative detection of analytes after separation.

Chromatography is based on the concept of partition coefficient. Any solute partitions between two immiscible solvents. When we make one solvent immobile (by adsorption on a solid support matrix) and another mobile it results in most common applications of chromatography. If the matrix support, or stationary phase, is polar (e.g. paper, silica etc.) it is forward phase chromatography, and if it is non-polar (C-18) it is reverse phase.

Techniques by Chromatographic Bed Shape

Column Chromatography

Column chromatography is a separation technique in which the stationary bed is within a tube. The particles of the solid stationary phase or the support coated with a liquid stationary phase may fill the whole inside volume of the tube (packed column) or be concentrated on or along the inside tube wall leaving an open, unrestricted path for the mobile phase in the middle part of the tube (open tubular column). Differences in rates of movement through the medium are calculated to different retention times of the sample.

In 1978, W. Clark Still introduced a modified version of column chromatography called flash column chromatography (flash). The technique is very similar to the traditional column chromatography, except for that the solvent is driven through the column by applying positive pressure. This allowed most separations to be performed in less than 20 minutes, with improved separations compared to the old method. Modern flash chromatography systems are sold as pre-packed plastic cartridges, and the solvent is pumped through the cartridge. Systems may also be linked with detectors and fraction collectors providing automation. The introduction of gradient pumps resulted in quicker separations and less solvent usage.

In expanded bed adsorption, a fluidized bed is used, rather than a solid phase made by a packed bed. This allows omission of initial clearing steps such as centrifugation and filtration, for culture broths or slurries of broken cells.

Phosphocellulose chromatography utilizes the binding affinity of many DNA-binding proteins for phosphocellulose. The stronger a protein's interaction with DNA, the higher the salt concentration needed to elute that protein.

Planar Chromatography

Planar chromatography is a separation technique in which the stationary phase is present as or on a plane. The plane can be a paper, serving as such or impregnated by a substance as the stationary bed (paper chromatography) or a layer of solid particles spread on a support such as a glass plate (thin layer chromatography). Different compounds in the sample mixture travel different distances according to how strongly they interact with the stationary phase as compared to the mobile phase. The specific Retention factor (R_f) of each chemical can be used to aid in the identification of an unknown substance.

Paper Chromatography

Paper chromatography is a technique that involves placing a small dot or line of sample solution onto a strip of *chromatography paper*. The paper is placed in a container with a shallow layer of solvent and sealed. As the solvent rises through the paper, it meets the sample mixture, which starts to travel up the paper with the solvent. This paper is made of cellulose, a polar substance, and the compounds within the mixture travel farther if they are non-polar. More polar substances bond with the cellulose paper more quickly, and therefore do not travel as far.

Thin Layer Chromatography (TLC)

Thin layer chromatography (TLC) is a widely employed laboratory technique and is similar to paper chromatography. However, instead of using a stationary phase of paper, it involves a stationary phase of a thin layer of adsorbent like silica gel, alumina, or cellulose on a flat, inert substrate. Compared to paper, it has the advantage of faster runs, better separations, and the choice between different adsorbents. For even better resolution and to allow for quantification, high-performance TLC can be used. An older popular use had been to differentiate chromosomes by observing distance in gel (separation of was a separate step).

Displacement Chromatography

The basic principle of displacement chromatography is: A molecule with a high affinity for the chromatography matrix (the displacer) competes effectively for binding sites, and thus displace all molecules with lesser affinities. There are distinct differences between displacement and elution chromatography. In elution mode, substances typically emerge from a column in narrow, Gaussian peaks. Wide separation of peaks, preferably to baseline, is desired for maximum purification. The speed at which any component of a mixture travels down the column in elution mode depends on many factors. But for two substances to travel at different speeds, and thereby be resolved, there must be substantial differences in some interaction between the biomolecules and the chromatography matrix. Operating parameters are adjusted to maximize the effect of this difference. In many cases, baseline separation of the peaks can be achieved only with gradient elution and low column loadings. Thus, two drawbacks to elution mode chromatography, especially at the preparative scale, are operational complexity, due to gradient solvent pumping, and low throughput, due to low column loadings. Displacement chromatography has advantages over elution chromatography in that components are resolved into consecutive zones of pure substances rather than "peaks". Because the process takes advantage of the nonlinearity of the isotherms, a larger column feed can be separated on a given column with the purified components recovered at significantly higher concentrations.

Techniques by Physical State of Mobile Phase

Gas Chromatography

Gas chromatography (GC), also sometimes known as gas-liquid chromatography, (GLC), is a separation technique in which the mobile phase is a gas. Gas chromatographic separation is always carried out in a column, which is typically "packed" or "capillary". Packed columns are the routine work horses of gas chromatography, being cheaper and easier to use and often giving adequate performance. Capillary columns generally give far superior resolution and although more expensive are becoming widely used, especially for complex mixtures. Both types of column are made from non-adsorbent and chemically inert materials. Stainless steel and glass are the usual materials for packed columns and quartz or fused silica for capillary columns.

Gas chromatography is based on a partition equilibrium of analyte between a solid or viscous liquid stationary phase (often a liquid silicone-based material) and a mobile gas (most often helium). The stationary phase is adhered to the inside of a small-diameter (commonly 0.53 – 0.18mm inside diameter) glass or fused-silica tube (a capillary column) or a solid matrix inside a larger metal tube (a packed column). It is widely used in analytical chemistry; though the high temperatures used in GC make it unsuitable for high molecular weight biopolymers or proteins (heat denatures them), frequently encountered in biochemistry, it is well suited for use in the petrochemical, environmental monitoring and remediation, and industrial chemical fields. It is also used extensively in chemistry research.

Liquid Chromatography

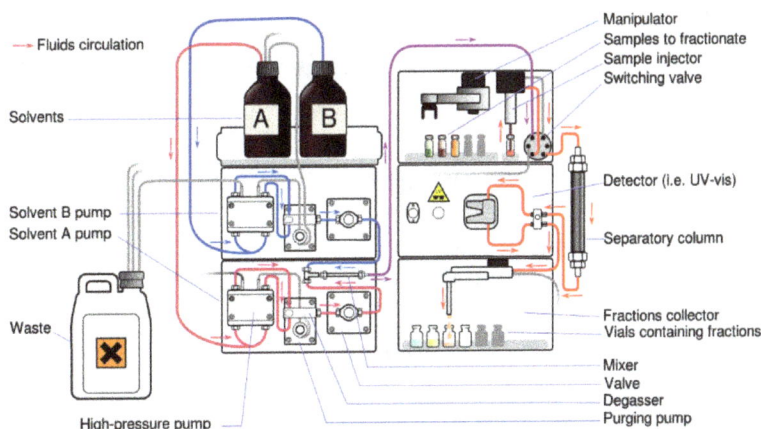

Preparative HPLC apparatus

Liquid chromatography (LC) is a separation technique in which the mobile phase is a liquid. It can be carried out either in a column or a plane. Present day liquid chromatography that generally utilizes very small packing particles and a relatively high pressure is referred to as high performance liquid chromatography (HPLC).

In HPLC the sample is forced by a liquid at high pressure (the mobile phase) through a column that is packed with a stationary phase composed of irregularly or spherically shaped particles, a porous monolithic layer, or a porous membrane. HPLC is historically divided into two different sub-classes based on the polarity of the mobile and stationary phases. Methods in which the stationary phase is more polar than the mobile phase (e.g., toluene as the mobile phase, silica as the stationary phase) are termed normal phase liquid chromatography (NPLC) and the opposite (e.g., water-methanol mixture as the mobile phase and C18 = octadecylsilyl as the stationary phase) is termed reversed phase liquid chromatography (RPLC).

Specific techniques under this broad heading are listed below.

Affinity Chromatography

Affinity chromatography is based on selective non-covalent interaction between an analyte and specific molecules. It is very specific, but not very robust. It is often used in biochemistry in the purification of proteins bound to tags. These fusion proteins are labeled with compounds such as His-tags, biotin or antigens, which bind to the stationary phase specifically. After purification, some of these tags are usually removed and the pure protein is obtained.

Affinity chromatography often utilizes a biomolecule's affinity for a metal (Zn, Cu, Fe, etc.). Columns are often manually prepared. Traditional affinity columns are used as a preparative step to flush out unwanted biomolecules.

However, HPLC techniques exist that do utilize affinity chromatography properties. Immobilized Metal Affinity Chromatography (IMAC) is useful to separate aforementioned molecules based on the relative affinity for the metal (I.e. Dionex IMAC). Often these columns can be loaded with different metals to create a column with a targeted affinity.

Supercritical Fluid Chromatography

Supercritical fluid chromatography is a separation technique in which the mobile phase is a fluid above and relatively close to its critical temperature and pressure.

Techniques by Separation Mechanism

Ion Exchange Chromatography

Ion exchange chromatography (usually referred to as ion chromatography) uses an ion exchange mechanism to separate analytes based on their respective charges. It is usually performed in columns but can also be useful in planar mode. Ion exchange chromatography uses a charged stationary phase to separate charged compounds including anions, cations, amino acids, peptides, and proteins. In conventional methods the stationary phase is an ion exchange resin that carries charged functional groups that interact with oppositely charged groups of the compound to retain. Ion exchange chromatography is commonly used to purify proteins using FPLC.

Size-exclusion Chromatography

Size-exclusion chromatography (SEC) is also known as gel permeation chromatography (GPC) or gel filtration chromatography and separates molecules according to their size (or more accurately according to their hydrodynamic diameter or hydrodynamic volume). Smaller molecules are able to enter the pores of the media and, therefore, molecules are trapped and removed from the flow of the mobile phase. The average residence time in the pores depends upon the effective size of the analyte molecules. However, molecules that are larger than the average pore size of the packing are excluded and thus suffer essentially no retention; such species are the first to be eluted. It is generally a low-resolution chromatography technique and thus it is often reserved for the final, "polishing" step of a purification. It is also useful for determining the tertiary structure and quaternary structure of purified proteins, especially since it can be carried out under native solution conditions.

Expanded Bed Adsorption Chromatographic Separation

An expanded bed chromatographic adsorption (EBA) column for a biochemical separation process comprises a pressure equalization liquid distributor having a self-cleaning function below a porous blocking sieve plate at the bottom of the expanded bed, an upper part nozzle assembly having a backflush cleaning function at the top of the expanded bed, a better distribution of the feedstock liquor added into the expanded bed ensuring that the fluid passed through the expanded bed layer displays a state of piston flow. The expanded bed layer displays a state of piston flow. The expanded bed chromatographic separation column has advantages of increasing the separation efficiency of the expanded bed.

Expanded-bed adsorption (EBA) chromatography is a convenient and effective technique for the capture of proteins directly from unclarified crude sample. In EBA chromatography, the settled bed is first expanded by upward flow of equilibration buffer. The crude feed, a mixture of soluble proteins, contaminants, cells, and cell debris, is then passed upward through the expanded bed. Target proteins are captured on the adsorbent, while particulates and contaminants pass through. A change to elution buffer while maintaining upward flow results in desorption of the target protein in expanded-bed mode. Alternatively, if the flow is reversed, the adsorbed particles will quickly settle and the proteins can be desorbed by an elution buffer. The mode used for elution (expanded-bed versus settled-bed) depends on the characteristics of the feed. After elution, the adsorbent is cleaned with a predefined cleaning-in-place (CIP) solution, with cleaning followed by either column regeneration (for further use) or storage.

Special Techniques

Reversed-phase Chromatography

Reversed-phase chromatography (RPC) is any liquid chromatography procedure in which the mobile phase is significantly more polar than the stationary phase. It is so named because in normal-phase liquid chromatography, the mobile phase is significantly less polar than the stationary phase. Hydrophobic molecules in the mobile phase tend to adsorb to the relatively hydrophobic stationary phase. Hydrophilic molecules in the mobile phase will tend to elute first. Separating columns typically comprise a C8 or C18 carbon-chain bonded to a silica particle substrate.

Hydrophobic Interaction Chromatography

Hydrophobic interactions between proteins and the chromatographic matrix can be exploited to purify proteins. In hydrophobic interaction chromatography the matrix material is lightly substituted with hydrophobic groups. These groups can range from methyl, ethyl, propyl, octyl, or phenyl groups. At high salt concentrations, non-polar sidechains on the surface on proteins "interact" with the hydrophobic groups; that is, both types of groups are excluded by the polar solvent (hydrophobic effects are augmented by increased ionic strength). Thus, the sample is applied to the column in a buffer which is highly polar. The eluant is typically an aqueous buffer with decreasing salt concentrations, increasing concentrations of detergent (which disrupts hydrophobic interactions), or changes in pH.

In general, Hydrophobic Interaction Chromatography (HIC) is advantageous if the sample is sensitive to pH change or harsh solvents typically used in other types of chromatography but not high salt concentrations. Commonly, it is the amount of salt in the buffer which is varied. In 2012, Müller and Franzreb described the effects of temperature on HIC using Bovine Serum Albumin (BSA) with four different types of hydrophobic resin. The study altered temperature as to effect the binding affinity of BSA onto the matrix. It was concluded that cycling temperature from 50 degrees to 10 degrees would not be adequate to effectively wash all BSA from the matrix but could be very effective if the column would only be used a few times. Using temperature to effect change allows labs to cut costs on buying salt and saves money.

If high salt concentrations along with temperature fluctuations want to be avoided you can use a more hydrophobic to compete with your sample to elute it. [source] This so-called salt independent method of HIC showed a direct isolation of Human Immunoglobulin G (IgG) from serum with satisfactory yield and used Beta-cyclodextrin as a competitor to displace IgG from the matrix. This largely opens up the possibility of using HIC with samples which are salt sensitive as we know high salt concentrations precipitate proteins.

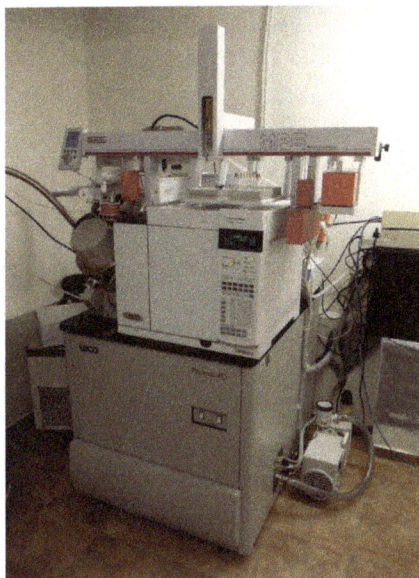

Two-dimensional chromatograph GCxGC-TOFMS at Chemical Faculty of GUTGdańsk, Poland, 2016

Two-dimensional Chromatography

In some cases, the chemistry within a given column can be insufficient to separate some analytes. It is possible to direct a series of unresolved peaks onto a second column with different physico-chemical (Chemical classification) properties. Since the mechanism of retention on this new solid support is different from the first dimensional separation, it can be possible to separate compounds that are indistinguishable by one-dimensional chromatography. The sample is spotted at one corner of a square plate,developed, air-dried, then rotated by 90° and usually redeveloped in a second solvent system.

Simulated Moving-bed Chromatography

The simulated moving bed (SMB) technique is a variant of high performance liquid chromatography; it is used to separate particles and/or chemical compounds that would be difficult or impossible to resolve otherwise. This increased separation is brought about by a valve-and-column arrangement that is used to lengthen the stationary phase indefinitely. In the moving bed technique of preparative chromatography the feed entry and the analyte recovery are simultaneous and continuous, but because of practical difficulties with a continuously moving bed, simulated moving bed technique was proposed. In the simulated moving bed technique instead of moving the bed, the sample inlet and the analyte exit positions are moved continuously, giving the impression of a moving bed. True moving bed chromatography (TMBC) is only a theoretical concept. Its simulation, SMBC is achieved by the use of a multiplicity of columns in series and a complex valve arrangement, which provides for sample and solvent feed, and also analyte and waste takeoff at appropriate locations of any column, whereby it allows switching at regular intervals the sample entry in one direction, the solvent entry in the opposite direction, whilst changing the analyte and waste takeoff positions appropriately as well.

Pyrolysis Gas Chromatography

Pyrolysis gas chromatography mass spectrometry is a method of chemical analysis in which the sample is heated to decomposition to produce smaller molecules that are separated by gas chromatography and detected using mass spectrometry.

Pyrolysis is the thermal decomposition of materials in an inert atmosphere or a vacuum. The sample is put into direct contact with a platinum wire, or placed in a quartz sample tube, and rapidly heated to 600–1000 °C. Depending on the application even higher temperatures are used. Three different heating techniques are used in actual pyrolyzers: Isothermal furnace, inductive heating (Curie Point filament), and resistive heating using platinum filaments. Large molecules cleave at their weakest points and produce smaller, more volatile fragments. These fragments can be separated by gas chromatography. Pyrolysis GC chromatograms are typically complex because a wide range of different decomposition products is formed. The data can either be used as fingerprint to prove material identity or the GC/MS data is used to identify individual fragments to obtain structural information. To increase the volatility of polar fragments, various methylating reagents can be added to a sample before pyrolysis.

Besides the usage of dedicated pyrolyzers, pyrolysis GC of solid and liquid samples can be performed directly inside Programmable Temperature Vaporizer (PTV) injectors that provide quick heating (up to 30 °C/s) and high maximum temperatures of 600–650 °C. This is sufficient for some pyrolysis applications. The main advantage is that no dedicated instrument has to be purchased and pyrolysis can be performed as part of routine GC analysis. In this case quartz GC inlet liners have to be used. Quantitative data can be acquired, and good results of derivatization inside the PTV injector are published as well.

Fast Protein Liquid Chromatography

Fast protein liquid chromatography (FPLC), is a form of liquid chromatography that is often used to analyze or purify mixtures of proteins. As in other forms of chromatography, separation is possible because the different components of a mixture have different affinities for two materials, a moving fluid (the "mobile phase") and a porous solid (the stationary phase). In FPLC the mobile phase is an aqueous solution, or "buffer". The buffer flow rate is controlled by a positive-displacement pump and is normally kept constant, while the composition of the buffer can be varied by drawing fluids in different proportions from two or more external reservoirs. The stationary phase is a resin composed of beads, usually of cross-linked agarose, packed into a cylindrical glass or plastic column. FPLC resins are available in a wide range of bead sizes and surface ligands depending on the application.

Countercurrent Chromatography

An example of a HPCCC system

Countercurrent chromatography (CCC) is a type of liquid-liquid chromatography, where both the stationary and mobile phases are liquids. The operating principle of CCC equipment requires a column consisting of an open tube coiled around a bobbin. The bobbin is rotated in a double-axis gyratory motion (a cardioid), which causes a variable gravity (G) field to act on the column during each rotation. This motion causes the column to see one partitioning step per revolution and components of the sample separate in the column due to their partitioning coefficient between the two immiscible liquid phases used. There are many types of CCC available today. These include HSCCC (High Speed CCC) and HPCCC (High Performance CCC). HPCCC is the latest and best performing version of the instrumentation available currently.

Chiral Chromatography

Chiral chromatography involves the separation of stereoisomers. In the case of enan-tiomers, these have no chemical or physical differences apart from being three-dimensional mirror images. Conventional chromatography or other separation processes are incapable of separating them. To enable chiral separations to take place, either the mobile phase or the stationary phase must themselves be made chiral, giving differing affinities between the analytes. Chiral chromatography HPLC columns (with a chiral stationary phase) in both normal and reversed phase are commercially available.

References

- McMurry, John (2011). Organic chemistry: with biological applications (2nd ed.). Belmont, CA: Brooks/Cole. p. 395. ISBN 9780495391470.

- Hostettmann, K; Marston, A; Hostettmann, M (1998). Preparative Chromatography Techniques Applications in Natural Product Isolation (Second ed.). Berlin, Heidelberg: Springer Berlin Heidelberg. p. 50. ISBN 9783662036310.

- Harwood, Laurence M.; Moody, Christopher J. (1989). Experimental organic chemistry: Principles and Practice (Illustrated ed.). WileyBlackwell. pp. 180–185. ISBN 978-0-632-02017-1.

- Anfinsen, Christian B.; Edsall, John Tileston; Richards, Frederic Middlebrook, eds. (1976). Advances in Protein Chemistry. pp. 6–7. ISBN 978-0-12-034230-3.

- Bailon, Pascal; Ehrlich, George K.; Fung, Wen-Jian and Berthold, Wolfgang (2000) An Overview of Affinity Chromatography, Humana Press. ISBN 978-0-89603-694-9.

- Ninfa, Alexander J., David P. Ballou, and Marilee Benore. Fundamental Laboratory Approaches for Biochemistry and Biotechnology. Hoboken, NJ: John Wiley, 2010.

Fundamental Concepts of Chromatography

The extraction of one material from another is elution. The measurement of the quality of the separation is chromatographic response function. Apart from elution and chromatographic response function, the essential concepts of chromatography are retardation factor, theoretical plate, Van Deemter equation etc. The information presented in this text elucidates the fundamental concepts of chromatography.

Elution

In analytical and organic chemistry, elution is the process of extracting one material from another by washing with a solvent; as in washing of loaded ion-exchange resins to remove captured ions.

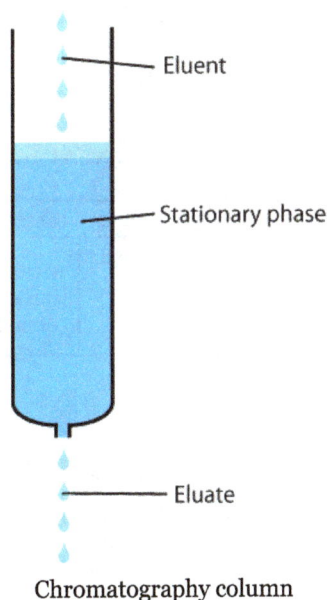

Chromatography column

In a liquid chromatography experiment, for example, an analyte is generally adsorbed, or "bound to", an adsorbent in a liquid chromatography column. The adsorbent, a solid phase (stationary phase), is a powder which is coated onto a solid support. Based on an adsorbent's composition, it can have varying affinities to "hold" onto other molecules—forming a thin film on its surface. Elution then is the process of removing analytes from the adsorbent by running a solvent, called an "eluent", past the adsorbent/analyte complex. As the solvent molecules "elute", or travel down through the chromatography column, they can either pass by the adsorbent/analyte complex or they can displace the analyte by binding to the adsorbent in its place. After the solvent molecules displace the analyte, the analyte can be carried out of the column for analysis. This is why as the mobile phase passes out of the column, it typically flows into a detector or is collected for compositional analysis.

Predicting and controlling the order of elution is a key aspect of column chromatographic methods.

Eluotropic Series

An eluotropic series is listing of various compounds in order of eluting power for a given adsorbent. The "eluting power" of a solvent is largely a measure of how well the solvent can "pull" an analyte off the adsorbent to which it is attached. This often happens when the eluent adsorbs onto the stationary phase, displacing the analyte. Such series are useful for determining necessary solvents needed for chromatography of chemical compounds. Normally such a series progresses from non-polar solvents, such as n-hexane, to polar solvents such as methanol or water. The order of solvents in an eluotropic series depends both on the stationary phase as well as on the compound used to determine the order.

Adsorption Strength (Least Strongly Adsorbed -> Most Strongly Adsorbed)								
Saturated hydrocarbons; alkyl halides	Unsaturated hydrocarbons; Alkenyl Halides	Aromatic Hydrocarbons; Aryl Halides	Polyhalogenated Hydrocarbons	Ethers	Esters	Aldehydes and Ketones	Alcohols	Acids and Bases (Amines)

Eluting Power (Least Eluting Power -> Greatest Eluting Power)												
Hexane or Pentane	Cyclohexane	Benzene	Dichloromethane	Chloroform	Ether (anhydrous)	Ethyl Acetate (anhydrous)	Acetone (anhydrous)	Ethanol	Methanol	Water	Pyridine	Acetic Acid

Eluent

The eluent or eluant is the "carrier" portion of the mobile phase. It moves the analytes through the chromatograph. In liquid chromatography, the eluent is the liquid solvent; in gas chromatography, it is the carrier gas.

Eluate

The eluate is the analyte material that emerges from the chromatograph. It specifically includes both the analytes and solutes passing through the column, while the eluent is only the carrier.

Elution Time and Elution Volume

The elution time of a solute is the time between the start of the separation (the time at which the solute enters the column) and the time at which the solute elutes. In the same way, the elution volume is the volume of eluent required to cause elution. Under standard conditions for a known mix of solutes in a certain technique, the elution volume may be enough information to identify solutes. For instance, a mixture of amino acids may be separated by ion-exchange chromatography. Under a particular set of conditions, the amino acids will elute in the same order and at the same elution volume.

Response Factor

Response factor, usually in chromatography and spectroscopy, is the ratio between a signal produced by an analyte, and the quantity of analyte which produces the signal. Ideally, and for easy computation, this ratio is unity (one). In real-world scenarios, this is often not the case.

Expression

The response factor f_i can be expressed on a molar, volume or mass basis. Where the true amount of sample and standard are equal:

$$f_i = \frac{A_i}{A_{st}} f_{st}$$

where A is the signal (e.g. peak area) and the subscript i indicates the sample and the subscript st indicates the standard. The response factor of the standard is assigned an arbitrary factor, for example 1 or 100. Response factor of sample/Response factor of standard=RRF

Chromatography

One of the main reasons to use response factors is to compensate for the irreproducibility of manual injections into a gas chromatograph (GC). Injection volumes for GC's can be 1 microliter (μL) or less and are difficult to reproduce. Differences in the volume of injected analyte leads to differences in the areas of the peaks in the chromatogram and any quantitative results are suspect.

To compensate for this error, a known amount of an internal standard (a second compound that does not interfere with the analysis of the primary analyte) is added to all solutions (standards and unknowns). This way if the injection volumes (and hence the peak areas) differ slightly, the ratio of the areas of the analyte and the internal standard will remain constant from one run to the next.

This comparison of runs also applies to solutions with different concentrations of the analyte. The area of the internal standard becomes the value to which all other areas are referenced. Below is the mathematical derivation and application of this method.

Consider an analysis of octane (C_8H_{18}) using nonane (C_9H_{20}) as the internal standard. The 3 chromatograms below are for 3 different samples.

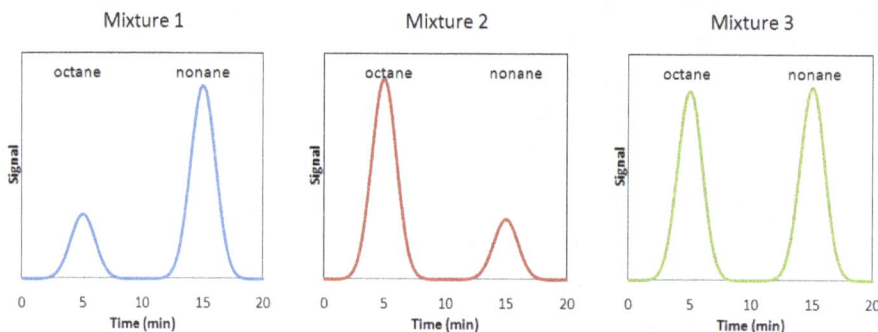

The amount of octane in each sample is different, but the amount of nonane is the same (in practice this is not a requirement). Due to scaling, the areas of the nonane peak appear to have different areas, but in reality the areas are identical. Therefore, the relative amounts of octane in each sample increases in the order of mixture 1 (least) < mixture 3 < mixture 2 (most).

This conclusion is reached because the ratio of the area of octane to that of nonane is the least in mixture 1 and the most, in mixture 2. Mixture 3 has an intermediate ratio. This ratio can be written as $Area_{octane} / Area_{nonane}$

In chromatography, the area of a peak is proportional to the number of moles times some constant of proportionality, $Area = k \times n$. The number of moles of compound is equal to the concentration (molarity, M) times the volume, $n = MV$. From these equations, the following derivation is made:

$$\frac{Area_{octane}}{Area_{nonane}} = \frac{k_{octane} \times M_{octane} \times V_{octane}}{k_{nonane} \times M_{nonane} \times V_{nonane}}$$

Since both compounds are in the same solution and are injected together, the volume terms are equal and cancel out. The above equation is then rearranged to solve for the ratio of the k's. This ratio is then called the response factor, F.

$$F = \frac{k_{octane}}{k_{nonane}} = \frac{Area_{octane} / M_{octane}}{Area_{nonane} / M_{nonane}}$$

The response factor, F, is equal to the ratios of the k's, which are constant. Therefore, F is constant. What this means is that regardless of the amounts of octane and nonane in solution, the ratio of the ratios of area to concentration will always yield a constant.

In practice, a solution containing known amounts of both octane and nonane is injected into a GC and a response factor, F, is calculated. Then a separate solution with an unknown amount of octane and a known amount of nonane is injected. The response factor is applied to the data from the second solution and the unknown concentration of the octane is found.

$$\left(\frac{Area_{octane} / M_{octane}}{Area_{nonane} / M_{nonane}} \right)_1 = F = \left(\frac{Area_{octane} / M_{octane}}{Area_{nonane} / M_{nonane}} \right)_2$$

This example deals with the analysis of octane and nonane, but can be applied to any two compounds.

Retardation Factor

In chromatography, the retardation factor (R) is the fraction of an analyte in the mobile phase of a chromatographic system. In planar chromatography in particular, the retardation factor R_f is defined as the ratio of the distance traveled by the center of a spot to the distance traveled by the solvent front. Ideally, the values for R_F are equivalent to the R values used in column chromatography.

Although the term retention factor is sometimes used synonymously with retardation factor in regard to planar chromatography the term is not defined in this context. However, in column chromatography, the retention factor or capacity factor (k) is defined as the ratio of time an analyte is retained in the stationary phase to the time it is retained in the mobile phase, which is inversely proportional to the retardation factor.

General Definition

In chromatography the retardation factor, R, is the fraction of the sample in the mobile phase at equilibrium, defined as:

$$R = \frac{\text{quantity of substance in the mobile phase}}{\text{total quantity of substance in the system}}$$

Planar Chromatography

The retardation factor, R_f, is commonly used in paper chromatography and thin layer chromatography for analyzing and comparing different substances. It can can be mathematically described by the following ratio:

$$R_f = \frac{\text{migration distance of substance}}{\text{migration distance of solvent front}}$$

An R_f value will always be in the range 0 to 1; if the substance moves, it can only move in the direction of the solvent flow, and cannot move faster than the solvent. For example, if particular substance in an unknown mixture travels 2.5 cm and the solvent front travels 5.0 cm, the retardation factor would be 0.5. One can choose a mobile phase with different characteristics (particularly polarity) in order to control how far the substance being investigated migrates.

An R_f value is characteristic for any given compound (provided that the same stationary and mobile phases are used). It can provide corroborative evidence as to the identity of a compound. If the identity of a compound is suspected but not yet proven, an authentic sample of the compound, or standard, is spotted and run on a TLC plate side by side (or on top of each other) with the compound in question. Note that this identity check must be performed on a single plate, because it is difficult to duplicate all the factors which influence R_f exactly from experiment to experiment.

Relationship with Retention Factor

In terms of retention factor (k), retardation factor (R) is defined as follows:

$$R = \frac{1}{k+1}$$

and conversely:

$$k = \frac{1-R}{R}$$

Chromatographic Response Function

Chromatographic response function, often abbreviated to CRF, is a coefficient which measures the quality of the separation in the result of a chromatography.

The CRF concept have been created during the development of separation optimization, to compare the quality of many simulated or real chromatographic separations. Many CRFs have been proposed and discussed.

In high performance liquid chromatography the CRF is calculated from various parameters of the peaks of solutes (like width, retention time, symmetry etc.) are considered into the calculation. In TLC the CRFs are based on the placement of the spots, measured as RF values.

CRFs Examples in thin Layer Chromatography

The CRFs in thin layer chromatography characterize the *equal-spreading* of the spots.

The ideal case, when the RF of the spots are uniformly distributed in <0,1> range (for example 0.25,0.5 and 0.75 for three solutes) should be characterized as the best situation possible.

The simplest criteria are ΔR_F and ΔR_F product (Wang et al., 1996). They are the smallest difference between sorted RF values, or product of such differences.

Another function is the multispot response function (MRF) as developed by De Spiegeleer et al. It is based also of differences product. This function always lies between 0 and 1. When two RF values are equal, it is equal to 0, when all RF values are equal-spread, it is equal to 1. The L and U values - upper and lower limit of RF - give possibility to avoid the band region.

$$MRF = \frac{(U - hR_{Fn})(hR_{F1} - L)\prod\limits_{i=1}^{n-1}(hR_{Fi+1} - hR_{Fi})}{[(U-L)/(n+1)]^{n+1}}$$

The last example of coefficient sensitive to minimal distance between spots is Retention distance (Komsta et al., 2007)

$$R_D = \left[(n+1)^{(n+1)}\prod\limits_{i=0}^{n}(R_{F(i+1)} - R_{Fi})\right]^{\frac{1}{n}}$$

The second group are criteria insensitive for minimal difference between RF values (if two compounds are not separated, such CRF functions will not indicate it). They are equal to zero in equal-spread state increase when situation is getting worse.

There are:

Separation response (Bayne et al., 1987)

$$D = \sqrt{\sum\limits_{i=1}^{n}\left(R_{Fi} - \frac{i-1}{n-1}\right)}$$

Performance index (Gocan et al., 1991)

$$I_p = \sqrt{\frac{\sum(\Delta hR_{Fi} - \Delta hR_{Fl})^2}{n(n+1)}}$$

Informational entropy (Gocan et al., 1991, second reference)

$$S_m = \sqrt{\frac{\sum(\Delta hR_{Fi} - \Delta hR_{Ft})^2}{n+1}}$$

Retention uniformity (Komsta et al., 2007)

$$R_U = 1 - \sqrt{\frac{6(n+1)}{n(2n+1)} \sum_{i=1}^{n} \left(R_{Fi} - \frac{i}{n+1} \right)^2}$$

In all above formulas, n is the number of compounds separated, $R_{f(1...n)}$ are the Retention factor of the compounds sorted in non-descending order, $R_{fo} = 0$ and $R_{f(n+1)} = 1$.

Theoretical Plate

A theoretical plate in many separation processes is a hypothetical zone or stage in which two phases, such as the liquid and vapor phases of a substance, establish an equilibrium with each other. Such equilibrium stages may also be referred to as an equilibrium stage, ideal stage, or a theoretical tray. The performance of many separation processes depends on having a series of equilibrium stages and is enhanced by providing more such stages. In other words, having more theoretical plates increases the efficiency of the separation process be it either a distillation, absorption, chromatographic, adsorption or similar process.

Applications

The concept of theoretical plates and trays or equilibrium stages is used in the design of many different types of separation.

Distillation Columns

The concept of theoretical plates in designing distillation processes has been discussed in many reference texts. Any physical device that provides good contact between the vapor and liquid phases present in industrial-scale distillation columns or laboratory-scale glassware distillation columns constitutes a "plate" or "tray". Since an actual, physical plate is rarely a 100% efficient equilibrium stage, the number of actual plates is more than the required theoretical plates.

$$N_a = \frac{N_t}{E}$$

where N_a is the number of actual, physical plates or trays, N_t is the number of theoretical plates or trays and E is the plate or tray efficiency.

So-called bubble-cap or valve-cap trays are examples of the vapor and liquid contact devices used in industrial distillation columns. Another example of vapor and liquid contact devices are the spikes in laboratory Vigreux fractionating columns.

The trays or plates used in industrial distillation columns are fabricated of circular steel plates and usually installed inside the column at intervals of about 60 to 75 cm (24 to 30 inches) up the height of the column. That spacing is chosen primarily for ease of installation and ease of access for future repair or maintenance.

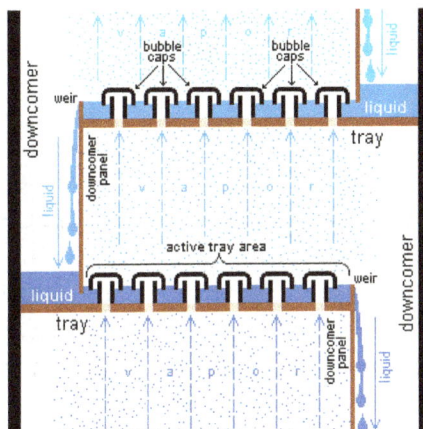

Typical bubble cap trays used in industrial distillation columns

An example of a very simple tray is a perforated tray. The desired contacting between vapor and liquid occurs as the vapor, flowing upwards through the perforations, comes into contact with the liquid flowing downwards through the perforations. In current modern practice, as shown in the adjacent diagram, better contacting is achieved by installing bubble-caps or valve caps at each perforation to promote the formation of vapor bubbles flowing through a thin layer of liquid maintained by a weir on each tray.

To design a distillation unit or a similar chemical process, the number of theoretical trays or plates (that is, hypothetical equilibrium stages), N_t, required in the process should be determined, taking into account a likely range of feedstock composition and the desired degree of separation of the components in the output fractions. In industrial continuous fractionating columns, N_t is determined by starting at either the top or bottom of the column and calculating material balances, heat balances and equilibrium flash vaporizations for each of the succession of equilibrium stages until the desired end product composition is achieved. The calculation process requires the availability of a great deal of vapor–liquid equilibrium data for the components present in the distillation feed, and the calculation procedure is very complex.

In an industrial distillation column, the N_t required to achieve a given separation also depends upon the amount of reflux used. Using more reflux decreases the number of plates required and using less reflux increases the number of plates required. Hence, the calculation of N_t is usually repeated at various reflux rates. N_t is then divided by the tray efficiency, E, to determine the actual number of trays or physical plates, N_a, needed in the separating column. The final design choice of the number of trays to be installed in an industrial distillation column is then selected based upon an economic balance between the cost of additional trays and the cost of using a higher reflux rate.

There is a very important distinction between the theoretical plate terminology used in discussing conventional distillation trays and the theoretical plate terminology used in the discussions below of packed bed distillation or absorption or in chromatography or other applications. The theoretical plate in conventional distillation trays has no "height". It is simply a hypothetical equilibrium stage. However, the theoretical plate in packed beds, chromatography and other applications is defined as having a height.

Distillation and Absorption Packed Beds

Distillation and absorption separation processes using packed beds for vapor and liquid contacting have an equivalent concept referred to as the plate height or the height equivalent to a theoretical plate (HETP).HETP arises from the same concept of equilibrium stages as does the theoretical plate and is numerically equal to the absorption bed length divided by the number of theoretical plates in the absorption bed (and in practice is measured in this way).

$$N_t = \frac{H}{\text{HETP}}$$

where N_t is the number of theoretical plates (also called the "plate count"), H is the total bed height and HETP is the height equivalent to a theoretical plate.

The material in packed beds can either be random dumped packing (1-3" wide) such as Raschig rings or structured sheet metal. Liquids tend to wet the surface of the packing and the vapors contact the wetted surface, where mass transfer occurs.

Chromatographic Processes

The theoretical plate concept was also adapted for chromatographic processes by Martin and Synge. The IUPAC's Gold Book provides a definition of the number of theoretical plates in a chromatography column.

The same equation applies in chromatography processes as for the packed bed processes, namely:

$$N_t = \frac{H}{\text{HETP}}$$

In chromatography, the HETP may also be calculated with the Van Deemter equation.

Other Applications

The concept of theoretical plates or trays applies to other processes as well, such as capillary electrophoresis and some types of adsorption.

Van Deemter Equation

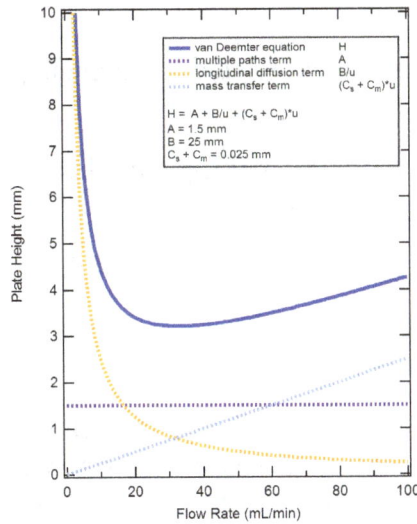

The Van Deemter equation in chromatography relates the variance per unit length of a separation column to the linear mobile phasevelocity by considering physical, kinetic, and thermodynamic properties of a separation. These properties include pathways within the column, diffusion (axial and longitudinal), and mass transferkinetics between stationary and mobile phases. In liquid chromatography, the mobile phase velocity is taken as the exit velocity, that is, the ratio of the flow rate in ml/second to the cross-sectional area of the 'column-exit flow path.' For a packed column, the cross-sectional area of the column exit flow path is usually taken as 0.6 times the cross-sectional area of the column. Alternatively, the linear velocity can be taken as the ratio of the column length to the dead time. If the mobile phase is a gas, then the pressure correction must be applied. The variance per unit length of the column is taken as the ratio of the column length to the column efficiency in theoretical plates. The Van Deemter equation is a hyperbolic function that predicts that there is an optimum velocity at which there will be the minimum variance per unit column length and, thence, a maximum efficiency. The Van Deemter equation was the result of the first application of rate theory to the chromatography elution process.

Van Deemter Equation

The van Deemter equation relates the resolving power (HETP, height equivalent to a theoretical plate) of a chromatographic column to the various flow and kinetic parameters which cause peak broadening, as follows:

$$HETP = A + \frac{B}{u} + (C_s + C_m) \cdot u$$

Where

- HETP = height equivalent to a theoretical plate, a measure of the resolving power of the column [m]

- A = Eddy-diffusion parameter, related to channeling through a non-ideal packing [m]

- B = diffusion coefficient of the eluting particles in the longitudinal direction, resulting in dispersion [m^2 s^{-1}]

- C = Resistance to mass transfer coefficient of the analyte between mobile and stationary phase [s]

- u = Linear Velocity [m s^{-1}]

In open tubularcapillaries, the A term will be zero as the lack of packing means channeling does not occur. In packed columns, however, multiple distinct routes ("channels") exist through the column packing, which results in band spreading. In the latter case, A will not be zero.

The form of the van Deemter equation is such that HETP achieves a minimum value at a particular flow velocity. At this flow rate, the resolving power of the column is maximized, although in practice, the elution time is likely to be impractical. Differentiating the van Deemter equation with respect to velocity, setting the resulting expression equal to zero, and solving for the optimum velocity yields the following:

$$u = \sqrt{\frac{B}{C}}$$

Plate Count

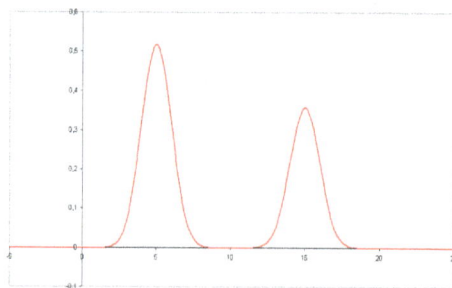

Two well resolved peaks in a chromatogram

The plate height given as:

$$H = \frac{L}{N}$$

with L the column length and : N the number of theoretical plates can be estimated from a chromatogram by analysis of the retention time t_R for each component and its standard deviation σ as a measure for peak width, provided that the elution curve represents a Gaussian curve.

In this case the plate count is given by:

$$N = \left(\frac{t_R}{\sigma} \right)^2$$

By using the more practical peak width at half height $W_{1/2}$ the equation is:

$$N = 8 \ln(2) \cdot \left(\frac{t_R}{W_{1/2}} \right)^2$$

or with the width at the base of the peak:

$$N = 16 \cdot \left(\frac{t_R}{W_{base}} \right)^2$$

Expanded Van Deemter

The Van Deemter equation can be further expanded to:

$$H = 2\lambda d_p + \frac{2\gamma D_m}{u} + \frac{\omega(d_p \text{ or } d_c)^2 u}{D_m} + \frac{R d_f^2 u}{D_s}$$

Where:

- H is plate height
- λ is particle shape (with regard to the packing)
- d_p is particle diameter
- γ, ω, and R are constants
- D_m is the diffusion coefficient of the mobile phase
- d_c is the capillary diameter
- d_f is the film thickness
- D_s is the diffusion coefficient of the stationary phase.
- u is the linear velocity

Rodrigues Equation

The Rodrigues equation is an equation used in chromatography to describe the efficiency of a bed of permeable (large-pore) particles. It is thus an extension of Van Deemter's equation. It was developed by Alirio E. Rodrigues *et al.*.

Equation

The equation is:

$$HETP = A + \frac{B}{u} + C \cdot f(\lambda) \cdot u$$

Where

- HETP is the height equivalent to a theoretical plate

- A = Eddy diffusion

- B = Longitudinal diffusion

- C = Resistance to mass transfer

- u = Flow rate

$$f(\lambda) = \frac{3}{\lambda} \left[\frac{1}{tanh(\lambda)} - \frac{1}{\lambda} \right]$$

- λ = Intraparticular Péclet number

Monolithic HPLC Column

A monolithic HPLC column, or monolithic column, is a column used in high-performance liquid chromatography (HPLC). The internal structure of the monolithic column is created in such a way that many channels form inside the column. The material inside the column which separates the channels can be porous and functionalized. In contrast, most HPLC configurations use particulate packed columns; in these configurations, tiny beads of an inert substance, typically a modified silica, are used inside the column.

Technology Overview

In analytical chromatography, the goal is to separate and uniquely identify each of the compounds in a substance. Alternatively, preparative scale chromatography is a method of purification of large batches of material in a production environment. The basic methods of separation in HPLC rely on a mobile phase (water, organic solvents, etc.) being passed through a stationary phase (particulate silica packings, monoliths, etc.) in a closed environment (column); the differences in reactivity among the solvent of interest and the mobile and stationary phases distinguish compounds from one another in a series of adsorption and desorption phenomena. The results are then visually displayed in a resulting chromatogram. Stationary phases are available in many varieties of packing styles as well as chemical structures and can be functionalized for added specificity. Monolithic-style columns, or monoliths, are one of many types of stationary phase structure.

Monoliths, in chromatographic terms, are porous rod structures characterized by mesopores and macropores. These pores provide monoliths with high permeability, a large number of channels, and a high surface area available for reactivity. The backbone of a monolithic column is composed of either an organic or inorganic substrate, and can easily be chemically altered for specific applications. Their unique structure gives them several physico-mechanical properties that enable them to perform competitively against traditionally packed columns.

Historically, the typical HPLC column consists of high-purity particulate silica compressed into stainless steel tubing. To decrease run times and increase selectivity, smaller diffusion distances have been pursued. To achieve smaller diffusion distances there has been a decrease in the particle sizes. However, as the particle size decreases, the backpressure (for a given column diameter and a given volumetric flow) increases proportionally. Pressure is inversely proportional to the square of the particle size; i.e., when particle size is halved, pressure increases by a factor of four. This is because as the particle sizes get smaller, the interstitial voids (the spaces between the particles) do as well, and it is harder to push the compounds through the smaller spaces. Modern HPLC systems are generally designed to withstand about 5,000 pounds per square inch (340 bar) of backpressure. This presents a significant limitation to the chromatographer.

Monoliths also have very short diffusion distances, while also providing multiple pathways for solute dispersion. Packed particle columns have pore connectivity values of about 1.5, while monoliths have values ranging from 6 to greater than 10. This means that, in a particulate column, a given analyte may diffuse into and out of the same pore, or enter through one pore and exit through a connected pore. By contrast, an analyte in a monolith is able to enter one channel and exit through any of 6 or more different venues. Little of the surface area in a monolith is inaccessible to compounds in the mobile phase. The high degree of interconnectivity in monoliths confers an advantage seen in the low backpressures and readily achievable high flow rates.

Monoliths are ideally suited for large molecules. As mentioned previously, particle sizes are decreasing in an attempt to achieve higher resolution and faster separations, which led to higher backpressures. When the smaller particle sizes are used to separate biomolecules, backpressures increase further because of the large molecule size. In monoliths, where backpressures are low and channel sizes are large, small molecule separations are less efficient. This is demonstrated by the dynamic binding capacities, a measure of how much sample can bind to the surface of the stationary phase. Dynamic binding capacities of monoliths for large molecules can be an order of ten times greater than that for particulate packings.

Monoliths exhibit no shear forces or eddying effects. High interconnectivity of the mesopores allows for multiple avenues of convective flow through the column. Mass transport of solutes through the column is relatively unaffected by flow rate. This is completely at odds to traditional particulate packings, whereby eddy effects and shear forces contribute greatly to the loss of resolution and capacity, as seen in the vanDeemter curve. Monoliths can, however, suffer from a different flow disadvantage: wall effects. Silica monoliths, especially, have a tendency to pull away from the sides of their column encasing. When this happens, the flow of the mobile phase occurs around the stationary phase as well as through it, decreasing resolution. Wall effects have been reduced greatly by advances in column construction.

Other advantages of monoliths conferred by their individual construction include greater column to column and batch to batch reproducibility. One technique of creating monolith columns is to polymerize the structure in situ. This involves filling the mold or column tubing with a mixture of monomers, a cross-linking agent, a free-radical initiator, and a porogenic solvent, then initiating the polymerization process under carefully controlled thermal or irradiating conditions. Monolithic in situ polymerization avoids the primary source of column to column variability, which is the packing procedure.

Additionally, packed particle columns must be maintained in a solvent environment and cannot be exposed to air during or after the packing procedure. If exposed to air, the pores dry out and no longer provide adequate surface area for reactivity; the column must be repacked or discarded. Further, because particle compression and packing uniformity are not relevant to monoliths, they exhibit greater mechanical robustness; if particulate columns are dropped, for example, the integrity of the column may be corrupted. Monolithic columns are more physically stable than their particulate counterparts.

Technology Development

The roots of liquid chromatography extend back over a century ago to 1900, when Russian botanist Mikhail Tsvet began experimenting with plant pigments in chlorophyll. He noted that, when a solvent was applied, distinct bands appeared that migrated at different rates along a stationary phase. For this new observation, he coined the term "chromatography," a colored picture. His first lecture on the subject was presented in 1903, but his most important contribution occurred three years later, in 1906, when the paper "Adsorption analysis and chromatographic method. Applications on the chemistry of chlorophyll," was published. Rivalry with a colleague who readily and vocally denounced his work meant that chromatographic analysis was shelved for almost 25 years. The great irony of the matter is that it was his rival's students who later took up the chromatography banner in their work with carotins.

Greatly unchanged from Tswett's time until the 1940s, normal phase chromatography was performed by passing a gravity-fed solvent through small glass tubes packed with pellicular adsorbent beads. It was in the 1940s, however, that there was a great revolution in gas chromatography (GC). Although GC was a wonderful technique for analyzing inorganic compounds, less than 20% of organic molecules are able to be separated using this technique. It was Richard Synge, who in 1952 won the Nobel Prize in Chemistry for his work with partition chromatography, who applied the theoretical knowledge gained from his work in GC to LC. From this revolution, the 1950s also saw the advent of paper chromatography, reversed-phase partition chromatography (RPC), and hydrophobic interaction chromatography (HIC). The first gels for use in LC were created using cross-linked dextrans (Sephadex) in an attempt to realize Synge's prediction that a unique single-piece stationary phase could provide an ideal chromatographic solution.

In the 1960s, polyacrylamide and agarose gels were created in a further attempt to create a single-piece stationary phase, but the purity of and stability of available components did not prove useful for implementation in the HPLC. In this decade, affinity chromatography was invented, an ultra-violet (UV) detector was used for the first time in conjunction with LC, and, most importantly, the modern HPLC was born. Csaba Horvath led the development of modern HPLC by piecing together laboratory equipment to suit his purposes. In 1968, Picker Nuclear Company marketed the first commercially available HPLC as a "Nucleic Acid Analyzer." The following year, the first international symposia on HPLC was held, and Kirkland at DuPont was able to functionalize controlled porosity pellicular particles for the first time.

The 1970s and 1980s witnessed a renewed interest in separations media with reduced interparticular void volumes. chromatography showed, for the first time, that chromatography media could support high flow rates without sacrificing resolution. Monoliths aptly fit into this new class of media, as they exhibit no void volume and can withstand flow rates up to 9mL/minute. Polymeric monoliths as they exist today were developed independently by three different labs in the late 1980s led by Hjerten, Svec, and Tennikova. Simultaneously, bioseparations became increasingly important, and monolith technologies proved beneficial in biotechnology separations.

Though industry focus in the 1980s was on biotechnology, focus in the 1990s shifted to process engineering. While mainstream chromatographers were using 3µm particulate columns, sub-2µm columns were in research phase. The smaller particles meant better resolution and shorter run times; there was also an associated increase in backpressure.

In order to withstand the pressure, a new field of chromatography came into being: UHPLC or UPLC- ultra high pressure liquid chromatography. The new instruments were able to endure pressures of up to 15,000 pounds per square inch (1,000 bar), as opposed to conventional machines, which, as previously state, can hold up to 5,000 pounds per square inch (340 bar). UPLC is an alternative solution to the same problems monolithic columns solve. Similarly to UPLC, monolith chromatography can help the bottom line by increasing sample throughput, but without the need to spend capital on new equipment.

In 1996, Nobuo Tanaka, at the Kyoto Institute of Technology, prepared silica monoliths using a colloidal suspension synthesis (aka "sol-gel") developed by a colleague. The process is different from that used in polymeric monoliths. Polymeric monoliths, as mentioned above, are created in situ, using a mixture of monomers and a porogen within the column tubing. Silica monoliths, on the other hand, are created in a mold, undergo a significant amount of shrinkage, and are then clad in a polymeric shrink tubing like PEEK (polyetheretherketone) to reduce wall effects. This method limits the size of columns that can be produced to less than 15 cm long, and though standard analytical inner diameters are readily achieved, there is currently a trend in developing nanoscalecapillary and prep scale silica monoliths.

Technology Life Cycle

Silica monoliths have only been commercially available since 2001, when Merck began their Chromolith campaign. The Chromolith technology was licensed from Tanaka's group at Kyoto Institute of Technology. The new product won the PittCon Editors' Gold Award for Best New Product, as well as an R&D 100 Award, both in 2001.

Individual monolith columns have a life cycle that generally exceeds that of its particulate competitors. When selecting an HPLC column supplier, column lifetime was second only to column-to-column reproducibility in importance to the purchaser. Chromolith columns, for example, have demonstrated reproducibility of 3,300 sample injections and 50,000 column volumes of mobile phase. Also important to the life cycle of the monolith is its increased mechanical robustness; polymeric monoliths are able to withstand pH ranges from 1 to 14, can endure elevated temperatures, and do not need to be handled delicately. "Monoliths are still teenagers," affirms Frantisec Svec, a leader in the field of novel stationary phases for LC.

Industry Evolution

Liquid chromatography as we know it today really got its start in 1969, when the first modern HPLC was designed and marketed as a nucleic acid analyzer. Columns throughout the 1970s were unreliable, pump flow rates were inconsistent, and many biologically active compounds escaped detection by UV and fluorescence detectors. Focus on purification methods in the '70s morphed into faster analyses in the 1980s, when computerized controls were integrated into HPLC equipment. Higher degrees of computerization then led to emphasis on more precise, faster, automated equipment in the 1990s. Atypical of many technologies of the '60s and '70s, the emphasis in improvements was not on "bigger and better," but on "smaller and better". At the same time the HPLC user-interface was improving, it was critical to be able to isolate hundreds of peptides or biomarkers from ever decreasing sample sizes.

Laboratory analytical instrumentation has only been recognized as a separate and distinct industry by NAICS and SIC since 1987. This market segmentation includes not only gas and liquid chromatography, but also mass spectrometry and spectrophotometric instruments. Since first recognized as a separate market, sales of analytical laboratory equipment increased from about $3.5 billion in 1987 to more than $26 billion in 2004. Revenues in the world liquid chromatography market, specifically, are expected to grow from $3.4 billion in 2007 to $4.7 billion in 2013, with a slight decrease in spending expected in 2008 and 2009 from the worldwide economic slump and decreased or stagnant spending. The pharmaceutical industry alone accounts for 35% of all the HPLC instruments in use. The main source of growth in LC stems from biosciences and pharmaceutical companies.

Technology Applications

In its earliest form, liquid chromatography was used to separate the pigments of chlorophyll by a Russian botanist. Decades later, other chemists used the procedure for the study of carotins. Liquid chromatography was then used for the isolation of small molecules and organic compounds like amino acids, and most recently has been used in peptide and DNA research. Monolith columns have been instrumental in advancing the field of biomolecular research.

In recent trade shows and international meetings for HPLC, interest in column monoliths and biomolecular applications has grown steadily, and this correlation is no coincidence. Monoliths have been shown to possess great potential in the "omics" fields-genomics, proteomics, metabolomics, and pharmacogenomics, among others. The reductionist approach to understanding the chemical pathways of the body and reactions to different stimuli, like drugs, are essential to new waves of healthcare like personalized medicine.

Pharmacogenomics studies how responses to pharmaceutical products differ in efficacy and toxicity based on variations in the patient's genome; it is a correlation of drug response to gene expression in a patient. Jeremy K. Nicholson of the Imperial College, London, used a postgenomic viewpoint to understand adverse drug reactions and the molecular basis of human disesase. His group studied gut microbial metabolic profiles and were able to see distinct differences in reactions to drug toxicity and metabolism even among various geographical distributions of the same race. Affinity monolith chromatography provides another approach to drug response measurements. David

Hage at the University of Nebraska binds ligands to monolithic supports and measures the equilibrium phenomena of binding interactions between drugs and serum proteins. A monolith-based approach at the University of Bologna, Italy, is currently in use for high-speed screening of drug candidates in the treatment of Alzheimer's. In 2003, Regnier and Liu of Purdue University described a multi-dimensional LC procedure for identifying single nucleotide polymorphisms (SNPs) in proteins. SNPs are alterations in the genetic code that can sometimes cause changes in protein conformation, as is the case with sickle cell anemia. Monoliths are particularly useful in these kinds of separations because of their superior mass transport capabilities, low backpressures coupled with faster flow rates, and relative ease of modification of the support surface.

Bioseparations on a production scale are enhanced by monolith column technologies as well. The fast separations and high resolving power of monoliths for large molecules means that real-time analysis on production fermentors is possible. Fermentation is well known for its use in making alcoholic beverages, but is also an essential step in the production of vaccines for rabies and other viruses. Real-time, on-line analysis is critical for monitoring of production conditions, and adjustments can be made if necessary. Boehringer Ingelheim Austria has validated a method with cGMP (commercial good manufacturing practices) for production of pharmaceutical-grade DNA plasmids. They are able to process 200L of fermentation broth on an 800mL monolith. At BIA Separations, processing time of the tomato mosaic virus decreased considerably from the standard five days of manually intensive work to equivalent purity and better recovery in only two hours with a monolith column. Other viruses have been purified on monoliths as well.

Another area of interest for HPLC is forensics. GC-MS (Gas Chromatography-Mass Spectroscopy) is generally considered the gold standard for forensic analysis. It is used in conjunction with online databases for rapid analysis of compounds in tests for blood alcohol, cause of death, street drugs, and food analysis, especially in poisoning cases. Analysis of buprenorphine, a heroin substitute, demonstrated the potential utility of multidimensional LC as a low-level detection method. HPLC methods can measure this compound at 40 ng/mL, compared to GC-MS at 0.5 ng/mL, but LC-MS-MS can detect buprenorphine at levels as low as 0.02 ng/mL. The sensitivity of multidimensional LC is therefore 2000 times greater than that of conventional HPLC.

Industry Applications

The liquid chromatography marketplace is incredibly diverse. Five to ten firms are consistently market leaders, yet nearly half of the market is made up of small, fragmented companies. This section of the report will focus on the roles that a few companies have had in bringing monolith column technologies to the commercial market.

In 1998, start-up biotechnology company BIA Separations of Ljubljana, Slovenia, came into being. The technology was originally developed by Tatiana Tennikova and Frantisek Svec during a collaboration between their respective institutes. The patent for these columns was acquired by BIA Separations and Ales Podgornik and Milos Barut developed the first commercially available monolith column in the form of a short disc encapsulated in a plastic housing. Trademarked CIM, BIA Separations has since introduced full lines of reversed-phase, normal-phase, ion-exchange, and affinity polymeric monoliths. Ales Podgornik and Janez Jancar then went on to develop large scale tube monolithic columns for industrial use. The largest column currently available is 8L. In May 2008, LC instrumentation powerhouse Agilent technologies agreed to market BIA Separations' analytical columns based on monolith technology. Agilent's commercialized the columns with strong and weak ion exchange phases and Protein A in September 2008 when they unveiled their new Bio-Monolith product line at the BioProcess International conference.

While BIA Separations was the first to commercially market polymeric monoliths, Merck KGaA was the first company to market silica monoliths. In 1996, Tanaka and co-workers at the Kyoto Institute of Technology published extensive work on silica monolith technologies. Merck was later issued a license from Kyoto Institute of Technology to develop and produce the silica monoliths. Promptly thereafter, in 2001, Merck introduced its Chromolith line of monolithic HPLC columns at analytical instrumentation trade show PittCon. Initially, says Karin Cabrera, senior scientist at Merck, the high flow rate was the selling point for the Chromolith line. Based on customer feedback, though, Merck soon learned that the columns were more stable and longer-lived than particle-packed columns. The columns were the recipients of various new product awards. Difficulties in production of the silica monoliths and tight patent protection have precluded attempts by other companies at developing a similar product. It has been noted that there are more patents concerning how to encapsulate the silica rod than there are on the manufacture of the silica itself.

Historically, Merck has been known for its superior chemical products, and, in liquid chromatography, for the purity and reliability of its particulate silica. Merck is not known for its LC columns. Five years after the introduction of its Chromolith line, Merck made a very strategic marketing decision. They granted a worldwide sublicense of the technology to a small (less than $100M in sales), innovative company well known for its cutting-edge column technology: Phenomenex. This was a superior strategic move for two reasons. As mentioned above, Merck is not well known for its column manufacturing. Furthermore, having more than one silica monolith manufacturer serves to better validate the technology. Having sublicensed the technology from Merck, Phenomenex introduced its Onyx product line in January 2005.

On the other side of monolith technologies are the polymerics. Unlike the inorganic silica columns, the polymer monoliths are made of an organic polymer base. Dionex, traditionally known for its ion chromatography capabilities, has led this side of the field. In the 1990s, Dionex first acquired a license for the polymeric monolith technology developed by leading monolithic chromatography researcher Frantisec Svec while he was at Cornell University. In 2000, they acquired LC Packings, whose competencies were in LC column packings. LC Packings/Dionex revealed their first monolithic capillary column at the Montreux LC-MS Conference. Earlier that year, another company, Isco, introduced a polystyrene divinylbenzene (PS-DVB) monolith column under the brand SWIFT. In January 2005, Dionex was sold the rights to Teledyne Isco's SWIFT media products, intellectual property, technology, and related assets. Though the core competencies of Dionex have traditionally been in ion chromatography, through strategic acquisitions and technology transfers, it has quickly established itself as the primary producer of polymeric monoliths.

Economic Impact

Though the many advances of HPLC and monoliths are highly visible within the confines of the analytical and pharmaceutical industries, it is unlikely that general society is aware of these developments. Currently, consumers may witness technology developments in the analytical sciences industry in the form of a broader array of available pharmaceutical products of higher purity, advanced forensic testing in criminal trials, better environmental monitoring, and faster returns on clinical tests. In the future, presumably, this may not be the case. As medicine becomes more individualized over time, consumer awareness that something is improving their quality of care seems more likely. The further thought that monoliths or HPLC are involved is unlikely to concern the general public, however.

There are two main cost drivers behind technological change in this industry. Though many different analytical areas use LC, including food and beverage industries, forensics labs, and clinical testing facilities, the largest impetus toward technology developments comes from the research and development and production arms of the pharmaceutical industry. The areas in which high-throughput monolithic column technologies are likely to have the largest economic impact are R&D and downstream processing.

From the Research and Development field comes the desire for more resolved, faster separations from smaller sample quantities. The only phase of drug development under direct control of a pharmaceutical company is the R&D stage. The goal of analytical work is to obtain as much information as possible from the sample. At this stage, high-throughput and analysis of tiny sample quantities are critical. Pharmaceutical companies are looking for tools that will better enable them to measure and predict the efficacy of candidate drugs in shorter times and with less expensive clinical trials. To this end, nano-scale separations, highly automated HPLC equipment, and multi-dimensional chromatography have become influential.

The prevailing method to increase the sensitivity of analytical methods has been multi-dimensional chromatography. This practice uses other analysis techniques in conjunction with liquid chromatography. For example, mass spectrometry (MS) has very much gained in popularity as an on-line analytical technique following HPLC. It is limited, however, in that MS, like nuclear magnetic resonance spectroscopy (NMR) or electrospray ionization techniques (ESI), is only feasible when using very small quantities of solute and solvent; LC-MS is used with nano or capillary scale techniques, but cannot be used in prep-scale. Another tactic for increasing selectivity in multi-dimensional chromatography is to use two columns with different selectivity orthogonally; ie... linking an ion exchange column to a C18 endcapped column. In 2007, Karger reported that, through multi-dimensional chromatography and other techniques, starting with only about 12,000 cells containing 1-4µg of protein, he was able to identify 1867 unique proteins. Of those, Karger can isolate 4 that may be of interest as cervical cancer markers. Today, liquid chromatographers using multi-dimensional LC can isolate compounds at the femtomole (10^{-15} mole) and attomole (10^{-18} mole) levels.

After a drug has been approved by the U.S. Food and Drug Administration (FDA), the emphasis at a pharmaceutical company is on getting a product to market. This is where prep or process scale chromatography has a role. In contrast to analytical analysis, preparatory scale chromatography focuses on isolation and purity of compounds. There is a trade-off between the degree of purity of compound and the amount of time required to achieve that purity. Unfortunately, many of the preparatory or process scale solutions used by pharmaceutical companies are proprietary, due to difficulties in patenting a process. Hence, there is not a great deal of literature available. However, some attempts to address the problems of prep scale chromatography include monoliths and simulated moving beds.

A comparison of immunoglobulin protein capture on a conventional column and a monolithic column yields some economically interesting results. If processing times are equivalent, process volumes of IgG, an antibody, are 3,120L for conventional columns versus 5,538L for monolithic columns. This represents a 78% increase in process volume efficiency, while at the same time only a tenth of the media waste volume is generated. Not only is the monolith column more economically prudent when considering the value of product processing times, but, at the same time, less media is used, representing a significant reduction in variable costs.

Separation Process

A separation process is a method to achieve any phenomenon that converts a mixture of chemical substance into two or more distinct product mixtures, which may be referred to as mixture. at least one of which is enriched in one or more of the mixture's constituents. In some cases, a separation may fully divide the mixture into its pure constituents. Separations differ in chemical properties or physical properties such as size, shape, mass, density, or chemical affinity, between the constituents of a mixture. They are often classified according to the particular differences they use to achieve separation. Usually there is only physical movement and no substantial chemical modification. If no single difference can be used to accomplish a desired separation, multiple operations will often be performed in combination to achieve the desired end.

With a few exceptions, elements or compounds are naturally found in an impure state. Often these impure raw materials must be separated into their purified components before they can be put to productive use, making separation techniques essential for the modern industrial economy. In some cases, these separations require total purification, as in the electrolysis refining of bauxite ore for aluminum metal, but a good example of an incomplete separation technique is oil refining. Crude oil occurs naturally as a mixture of various hydrocarbons and impurities. The refining process splits this mixture into other, more valuable mixtures such as natural gas, gasoline and chemical feedstocks, none of which are pure substances, but each of which must be separated from the raw crude. In both of these cases, a series of separations is necessary to obtain the desired end products. In the case of oil refining, crude is subjected to a long series of individual distillation steps, each of which produces a different product or intermediate.

Separators are used for the separation of liquid. In fact, they are vertically supported centrifuges and are built with flying bearings. They work with centrifugal force. A separator belongs to the continues sedimentation centrifuges. Both the purified liquid and the liquid separated from each other are continuously discharged by using a pump (under pressure) or pressure free. The solid material can be discharged discontinuously (chamber drum, solid walled disc drum), pseudo continuously (self-cleaning disc drum) or continuously (nozzle drum). The drum is the centerpiece of the separator, in which the separation process takes place. There are two types of drums: the chamber drum (known as chamber separators) and the disc drum (known as disc separators). The power transmission on the spindle and thereby on the drum can take place by using one of the three drive motors: helical gears, a belt drive or direct drive, via a special motor. The sealing of the separators is differentiated into four types: open, semi-closed, hydro-hermetic (sealing of the product space) or fully hermetic (absolute airtight).

The purpose of a separation may be *analytical*,can be used as a lie components in the original mixture without any attempt to save the fractions, or may be *preparative*, i.e. to "prepare" fractions or samples of the components that can be saved. The separation can be done on a small scale, effectively a laboratory scale for analytical or preparative purposes, or on a large scale, effectively an industrial scale for preparative purposes, or on some intermediate scale.

List of Separation Techniques

- Adsorption, adhesion of atoms, ions or molecules of gas, liquid, or dissolved solids to a surface

- Capillary electrophoresis

- Centrifugation and cyclonic separation, separates based on density differences

- Chromatography separates dissolved substances by different interaction with (i.e., travel through) a material

- Crystallization

- Decantation

- Demister (vapor), removes liquid droplets from gas streams

- Distillation, used for mixtures of liquids with different boiling points

- Drying, removes liquid from a solid by vaporisation

- Electrophoresis, separates organic molecules based on their different interaction with a gel under an electric potential (i.e., different travel)

- Electrostatic separation, works on the principle of corona discharge, where two plates are placed close together and high voltage is applied. This high voltage is used to separate the ionized particles.

- Elutriation

- Evaporation

- Extraction

    o Leaching

    o Liquid-liquid extraction

    o Solid phase extraction

- Field flow fractionation

- Flotation

    o Dissolved air flotation, removes suspended solids non-selectively from slurry by bubbles that are generated by air coming out of solution

    o Froth flotation, recovers valuable, hydrophobic solids by attachment to air bubbles generated by mechanical agitation of an air-slurry mixture, which float, and are recovered

    o Deinking, separating hydrophobic ink particles from hydrophilic paper pulp in paper recycling

- Flocculation, separates a solid from a liquid in a colloid, by use of a flocculant, which promotes the solid clumping into flocs

- Filtration – Mesh, bag and paper filters are used to remove large particulates suspended in fluids (e.g., fly ash) while membrane processes including microfiltration, ultrafiltration, nanofiltration, reverse osmosis, dialysis (biochemistry) utilising synthetic membranes, separates micrometre-sized or smaller species

- Fractional distillation

- Fractional freezing

- Oil-water separation, gravimetrically separates suspended oil droplets from waste water in oil refineries, petrochemical and chemical plants, natural gas processing plants and similar industries

- Magnetic separation

- Precipitation

- Recrystallization

- Scrubbing, separation of particulates (solids) or gases from a gas stream using liquid.

- Sedimentation, separates using vocal density pressure differences

  o Gravity separation

- Sieving

- Stripping

- Sublimation

- Vapor-liquid separation, separates by gravity, based on the Souders-Brown equation

- Winnowing

- Zone refining

Partition Coefficient

In the physical sciences, a partition-coefficient (P) or distribution-coefficient (D) is the ratio of concentrations of a compound in a mixture of two immiscible phases at equilibrium. This ratio is therefore a measure of the difference in solubility of the compound in these two phases. The partition-coefficient generally refers to the concentration ratio of un-ionized species of compound whereas the distribution-coefficient refers to the concentration ratio of all species of the compound (ionized plus un-ionized).

In the chemical and pharmaceutical sciences, both phases usually are solvents. Most commonly, one of the solvents is water while the second is hydrophobic such as 1-octanol. Hence the partition coefficient measures how hydrophilic ("water-loving") or hydrophobic ("water-fearing") a chemical substance is. Partition coefficients are useful in estimating the distribution of drugs within the body. Hydrophobic drugs with high octanol/water partition coefficients are mainly distributed to hydrophobic areas such as lipid bilayers of cells. Conversely hydrophilic drugs (low octanol/water partition coefficients) are found primarily in aqueous regions such as blood serum.

If one of the solvents is a gas and the other a liquid, a gas/liquid partition coefficient can be determined. For example, the blood/gas partition coefficient of a general anesthetic measures how easily the anesthetic passes from gas to blood. Partition coefficients can also be defined when one of the phases is solid, for instance, when one phase is a molten metal and the second is a solid metal, or when both phases are solids. The partitioning of a substance into a solid results in a solid solution.

Partition coefficients can be measured experimentally in various ways (by shake-flask, HPLC, etc.) or estimated via calculation based on a variety of methods (fragment-based, atom-based, etc.).

Nomenclature

Despite formal recommendation to the contrary, the term *partition coefficient* remains the predominantly used term in the scientific literature.

In contrast, the IUPAC recommends that the title term no longer be used, rather, that it be replaced with more specific terms. For example, *partition constant*, defined as $(K_D)_A = [A]_{org} / [A]_{aq}$, where K_D is the process equilibrium constant, [A] represents the concentration of solute A being tested, and "org" and "aq" refer to the organic and aqueous phases, respectively. The IUPAC further recommends "partition ratio" for cases where transfer activity coefficients can be determined, and "distribution ratio" for the ratio of total analytical concentrations of a solute between phases, regardless of chemical form.

Partition Coefficient and Log P

The partition coefficient, abbreviated *P*, is defined as a particular ratio of the concentrations of a solute between the two solvents (a biphase of liquid phases), specifically for un-ionized solutes, and the logarithm of the ratio is thus log *P*. When one of the solvents is water and the other is a non-polar solvent, then the log *P* value is a measure of lipophilicity or hydrophobicity. The defined precedent is for the lipophilic and hydrophilic phase types to always be in the numerator and denominator, respectively; for example, in a biphasic system of *n*-octanol (hereafter simply "octanol") and water:

$$P_{oct/wat} = \log(\frac{[\text{solute}]_{octanol}^{\text{un-ionized}}}{[\text{solute}]_{water}^{\text{un-ionized}}})$$

To a first approximation, the non-polar phase in such experiments is usually dominated by the un-ionized form of the solute, which is electrically neutral, though this may not be true for the aqueous phase. To measure the *partition coefficient of ionizable solutes*, the pH of the aqueous phase is adjusted such that the predominant form of the compound in solution is the un-ionized, or its measurement at another pH of interest requires consideration of all species, un-ionized and ionized.

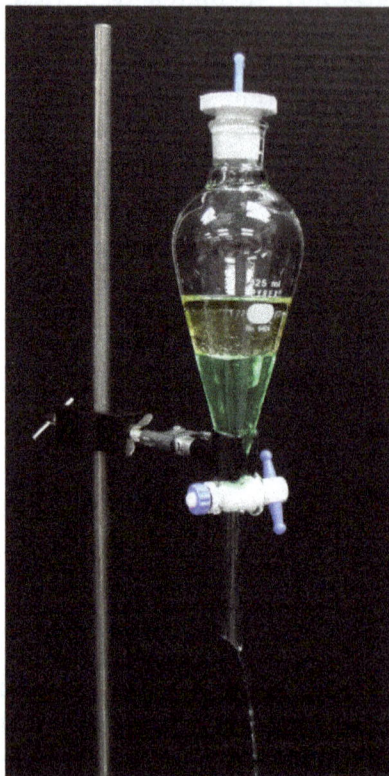

An equilibrium of dissolved substance distributed between a hydrophobic phase and a hydrophilic phase is established in special glassware such as this separatory funnel that allows shaking and sampling from which the log *P* is determined. Here, the green substance has a greater solubility in the lower layer than in the upper layer.

A corresponding partition coefficient for ionizable compounds, abbreviated log P^{I}, is derived for cases where there are dominant ionized forms of the molecule, such that one must consider partition of all forms, ionized and un-ionized, between the two phases (as well as the interaction of the two equilibria, partition and ionization). M is used to indicate the number of ionized forms; for the I-th form (I = 1, 2, ... , M) the logarithm of the corresponding partition coefficient, log $P^{\mathrm{I}}_{\mathrm{oct/wat}}$, is defined in the same manner as for the un-ionized form. For instance, for an octanol-water partition, it is:

$$\log P^{\mathrm{I}}_{\mathrm{oct/wat}} = \log\left(\frac{[\text{solute}]^{\mathrm{I}}_{\text{octanol}}}{[\text{solute}]^{\mathrm{I}}_{\text{water}}}\right)$$

To distinguish between this and the standard, un-ionized, partition coefficient, the un-ionized is often assigned the symbol $\log P^o$, such that the indexed $\log P^I_{oct/wat}$ expression for ionized solutes becomes simply an extension of this, into the range of values $I > o$.

Distribution Coefficient and Log D

The distribution coefficient, $\log D$, is the ratio of the sum of the concentrations of all forms of the compound (ionized plus un-ionized) in each of the two phases, one essentially always aqueous; as such, it depends on the pH of the aqueous phase, and $\log D = \log P$ for non-ionizable compounds at any pH. For measurements of distribution coefficients, the pH of the aqueous phase is buffered to a specific value such that the pH is not significantly perturbed by the introduction of the compound. The value of each $\log D$ is then determined as the logarithm of a ratio—of the sum of the experimentally measured concentrations of the solute's various forms in one solvent, to the sum of such concentrations of its forms in the other solvent; it can be expressed as:

$$\log D_{oct/wat} = \log\left(\frac{\left[\text{solute}\right]^{ionized}_{octanol} + \left[\text{solute}\right]^{un\text{-}ionized}_{octanol}}{\left[\text{solute}\right]^{ionized}_{water} + \left[\text{solute}\right]^{un\text{-}ionized}_{water}}\right)$$

In the above formula, the superscripts "ionized" each indicate the sum of concentrations of all ionized species in their respective phases. In addition, since $\log D$ is pH-dependent, the pH at which the $\log D$ was measured must be specified. In areas such as drug discovery—areas involving partition phenomena in biological systems such as the human body—the $\log D$ at the physiologic pH, 7.4, is of particular interest.

It is often convenient to express the $\log D$ in terms of P^I, defined above (which includes P^o as state $I = o$), thus covering both un-ionized and ionized species. For example, in octanol-water:

$$\log D_{oct/wat} = \log\left(\sum_{I=0}^{M} f^I P^I_{oct/wat}\right)$$

which sums the individual partition coefficients (not their logarithms), and where indicates the pH-dependent mole fraction of the I-th form (of the solute) in the aqueous phase, and other variables are defined as previously.

Example Partition Coefficient Data

The values in the following table are from the Dortmund Data Bank. They are sorted by the partition coefficient, smallest to largest (acetamide being hydrophilic, and 2,2',4,4',5-pentachlorobiphenyl lipophilic), and are presented with the temperature at which they were measured (which impacts the values).

Values for other compounds may be found in a variety of available reviews and monographs.Critical discussions of the challenges of measurement of log P, and related computation of its estimated values, appear in several reviews.

Applications

Pharmacology

A drug's distribution coefficient strongly affects how easily the drug can reach its intended target in the body, how strong an effect it will have once it reaches its target, and how long it will remain in the body in an active form. Hence, the log P of a molecule is one criterion used in decision-making by medicinal chemists in pre-clinical drug discovery, for example, in the assessment of druglikeness of drug candidates. Likewise, it is used to calculate lipophilic efficiency in evaluating the quality of research compounds, where the efficiency for a compound is defined as its potency, via measured values of pIC_{50} or pEC_{50}, minus its value of log P.

Pharmacokinetics

Drug permeability in brain capillaries (y-axis) as a function of partition coefficient (x-axis).

In the context of pharmacokinetics (what the body does to a drug), the distribution coefficient has a strong influence on ADME properties of the drug. Hence the hydrophobicity of a compound (as measured by its distribution coefficient) is a major determinant of how drug-like it is. More specifically, for a drug to be orally absorbed, it normally must first pass through lipid bilayers in the intestinal epithelium (a process known as transcellular transport). For efficient transport, the drug must be hydrophobic enough to partition into the lipid bilayer, but not so hydrophobic, that once it is in the bilayer, it will not partition out again. Likewise, hydrophobicity plays a major role in determining where drugs are distributed within the body after absorption and as a consequence in how rapidly they are metabolized and excreted.

Pharmacodynamics

In the context of pharmacodynamics (what a drug does to the body), the hydrophobic effect is the major driving force for the binding of drugs to their receptor targets.

On the other hand, hydrophobic drugs tend to be more toxic because they, in general, are retained longer, have a wider distribution within the body (e.g., intracellular), are somewhat less selective in their binding to proteins, and finally are often extensively metabolized. In some cases the metabolites may be chemically reactive. Hence it is advisable to make the drug as hydrophilic as possible while it still retains adequate binding affinity to the therapeutic protein target. For cases where a drug reaches its target locations through passive mechanisms (i.e., diffusion through membranes), the ideal distribution coefficient for the drug is typically intermediate in value (neither too lipophilic, nor too hydrophilic); in cases where molecules reach their targets otherwise, no such generalization applies.

Environmental Science

The hydrophobicity of a compound can give scientists an indication of how easily a compound might be taken up in groundwater to pollute waterways, and its toxicity to animals and aquatic life. Partition coefficient can also used to predict the mobility of radionuclides in groundwater. In the field of hydrogeology, the octanol-water partition coefficient, or K_{ow}, is used to predict and model the migration of dissolved hydrophobic organic compounds in soil and groundwater.

Agrochemical Research

Hydrophobic insecticides and herbicides tend to be more active. Hydrophobic agrochemicals in general have longer half lives and therefore display increased risk of adverse environmental impact.

Metallurgy

In metallurgy, the partition coefficient is an important factor in determining how different impurities are distributed between molten and solidified metal. It is a critical parameter for purification using zone melting, and determines how effectively an impurity can be removed using directional solidification, described by the Scheil equation.

Consumer Product Development

Many other industries take into account distribution coefficients for example in the formulation of make-up, topical ointments, dyes, hair colors and many other consumer products.

Measurement

A number of methods of measuring distribution coefficients have been developed, including the shake-flask, reverse phase HPLC, and pH-metric techniques.

Shake Flask-type

The classical and most reliable method of log P determination is the *shake-flask method*, which consists of dissolving some of the solute in question in a volume of octanol and water, then measuring the concentration of the solute in each solvent. The most common method of measuring the distribution of the solute is by UV/VIS spectroscopy.

HPLC-based

A faster method of log P determination makes use of high-performance liquid chromatography. The log P of a solute can be determined by correlating its retention time with similar compounds with known log P values.

An advantage of this method is that it is fast (5–20 minutes per sample); a disadvantage is that the solute's chemical structure must be known beforehand. Moreover, since the value of log P is determined by linear regression, several compounds with similar structures must have known log P values, and extrapolation from one chemical class to another—applying a regression equation derived from one chemical class to a second one—may not be reliable, since each chemical classes will have its characteristic regression parameters.

pH-Metric

The pH-metric set of techniques determine lipophilicity pH profiles directly from a single acid-base titration in a two-phase water-organic solvent system. Hence, a single experiment can be used to measure the logs of the partition coefficient (log P) giving the distribution of molecules that are primarily neutral in charge, as well as the distribution coefficient (log D) of all forms of the molecule over a pH range, e.g., between 2 and 12. The method does, however, require the separate determination of the pK_a value(s) of the substance.

Electrochemical

Polarized liquid interfaces have been used to examine the thermodynamics and kinetics of the transfer of charged species from one phase to another. Two main methods exist. The first is ITIES, Interfaces between two immiscible electrolyte ssolutions. The second is droplet experiments. Here a reaction at a triple interface between a conductive solid, droplets of a redox active liquid phase and an electrolyte solution have been used to determine the energy required to transfer a charged species across the interface.

Prediction

There are many situations where prediction of partition coefficients prior to experimental measurement is useful. For example, tens of thousands of industrially man-ufactured chemicals are in common use, but only a small fraction have undergone rigorous toxicological evaluation. Hence there is a need to prioritize the remainder for testing. QSAR equations which in turn are based on calculated partition coef-ficients can be used to provide toxicity estimates. Calculated partition coefficients are also widely used in drug discovery to optimize screening libraries and to predict druglikeness of designed drug candidates before they are synthesized. As discussed in more detail below, estimates of partition coefficients can be made using a variety of methods including fragment-based, atom-based, and knowledge-based that rely solely on knowledge of the structure of the chemical. Other prediction methods rely on other experimental measurements such as solubility. The methods also differ in accuracy and whether they can be applied to all molecules, or only ones similar to molecules already studied.

Atom-based

Standard approaches of this type, using atomic contributions, have been named by those formulating them with a prefix letter: AlogP, XlogP, MlogP, etc. A con-ventional method for predicting log P through this type of method is to param-eterize the distribution coefficient contributions of various atoms to the over-all molecular partition coefficient, which produces a parametric model. This parametric model can be estimated using constrained least-squares estimation, using a training set of compounds with experimentally measured partition co-efficients. In order to get reasonable correlations, the most common elements contained in drugs (hydrogen, carbon, oxygen, sulfur, nitrogen, and halogens) are divided into several different atom types depending on the environment of the atom within the molecule. While this method is generally the least accurate, the advantage is that it is the most general, being able to provide at least a rough estimate for a wide variety of molecules.

Fragment-based

The most common of these uses a group contribution method, and is termed ClogP. It has been shown that the log P of a compound can be determined by the sum of its non-overlapping molecular fragments (defined as one or more atoms covalently bound to each other within the molecule). Fragmentary log P values have been determined in a statistical method analogous to the atomic methods (least squares fitting to a training set). In addition, Hammett type corrections are included to account of electronic and steric effects. This meth-od in general gives better results than atomic based methods, but cannot be used to predict partition coefficients for molecules containing unusual func-tional groups for which the method has not yet been parameterized (most likely because of the lack of experimental data for molecules containing such functional groups).

Knowledged-based

A typical data mining based prediction uses support vector machines, decision trees, or neural networks. This method is usually very successful for calculating log P values when used with compounds that have similar chemical structures and known log P values. Molecule mining approaches apply a similarity matrix based prediction or an automatic fragmentation scheme into molecular sub-structures. Furthermore, there exist also approaches using maximum common subgraph searches or molecule kernels.

Log D from log P and pK_a

For cases where the molecule is un-ionized:

- $$\log D \cong \log P$$

For other cases, estimation of log D at a given pH, from log P and the known mole fraction of the un-ionized form, f^0, in the case where partition of ionized forms into non-polar phase can be neglected, can be formulated as:

- $$\log D \cong \log P + \log\left(f^0\right)..$$

The following approximate expressions are valid only for monoprotic acids and bases:

- $$\log D_{acids} \cong \log P + \log\left[\frac{1}{(1+10^{pH-pK_a})}\right], \text{ and}$$

- $$\log D_{bases} \cong \log P + \log\left[\frac{1}{(1+10^{pK_a-pH})}\right]$$

Further approximations for when the compound is largely ionized:

- for acids with $\left(\text{pH} - \text{p}K_a\right) > 1, \log D_{acids} \cong \log P + \text{p}K_a - \text{pH}$, and

- for bases with $\left(\text{p}K_a - \text{pH}\right) > 1, \log D_{bases} \cong \log P - \text{p}K_a + \text{pH}$.

For prediction of pK_a, which in turn can be used to estimate log D, Hammett type equations have frequently been applied.

Log P from Log S

If the solubility of an organic compound is known or predicted in both water and 1-octanol, then log P can be estimated as:

- $$\log P = \log S_o - \log S_w$$

There are a variety of approaches to predict solubilities, and so log S.

Binding Selectivity

Binding selectivity is defined with respect to the binding of ligands to a substrate forming a complex. A selectivity coefficient is the equilibrium constant for the reaction of displacement by one ligand of another ligand in a complex with the substrate. Binding selectivity is of major importance in biochemistry and in chemical separation processes.

Selectivity coefficient

The concept of selectivity is used to quantify the extent to which a given substrate, A, binds two different ligands, B and C. The simplest case is where the complexes formed have 1:1 stoichiometry. Then, the two interactions may be characterized by equilibrium constants K_{AB} and K_{AC}.

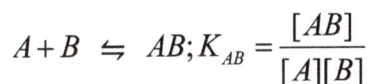

$$A + B \leftrightarrows AB; K_{AB} = \frac{[AB]}{[A][B]}$$

$$A + C \leftrightarrows AC; K_{AC} = \frac{[AC]}{[A][C]}$$

[..] represents a concentration. A selectivity coefficient is defined as the ratio of the two equilibrium constants.

$$K_{B,C} = \frac{K_{AC}}{K_{AB}}$$

The selectivity coefficient is in fact the equilibrium constant for the displacement reaction

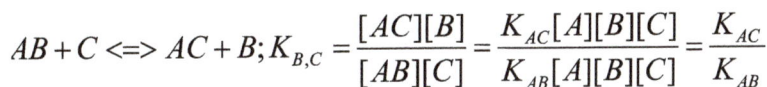

$$AB + C <=> AC + B; K_{B,C} = \frac{[AC][B]}{[AB][C]} = \frac{K_{AC}[A][B][C]}{K_{AB}[A][B][C]} = \frac{K_{AC}}{K_{AB}}$$

It is easy to show that the same definition applies to complexes of a different stoichiometry, $A_p B_q$ and $A_p C_q$. The greater the selectivity coefficient, the more the ligand C will displace the ligand B from the complex formed with the substrate A. An alternative interpretation is that the greater the selectivity coefficient, the lower the concentration of C that is needed to displace B from AB. Selectivity coefficients are determined experimentally by measuring the two equilibrium constants, K_{AB} and K_{AC}.

Applications

Biochemistry

In biochemistry the substrate is known as a receptor. A receptor is a protein molecule, embedded in either the plasma membrane or the cytoplasm of a cell, to which one or more specific kinds of signalling molecules may bind. A ligand may be a peptide or another small molecule, such as a neurotransmitter, a hormone, a pharmaceutical drug, or a toxin. The specificity of a receptor is determined by its spatial geometry and the way it binds to the ligand through non-covalent interactions, such as hydrogen bonding or Van der Waals forces.

If a receptor can be isolated a synthetic drug can be developed either to stimulate the receptor, an agonist or to block it, an antagonist. The stomach ulcer drug cimetidine was developed as an H_2 antagonist by chemically engineering the molecule for maximum specificity to an isolated tissue containing the receptor. The further use of quantitative structure-activity relationships (QSAR) led to the development of other agents such as ranitidine.

It is important to note that "selectivity" when referring to a drug is relative and not absolute. For example, in a higher dose, a specific drug molecule may also bind to other receptors than those said to be "selective".

Chelation Therapy

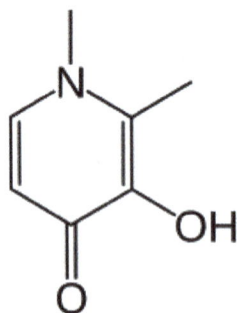

Deferiprone

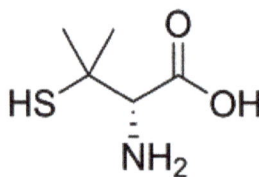

Penicillamine

Chelation therapy is a form of medical treatment in which a chelating ligand is used to selectively remove a metal from the body. When the metal exists as a divalent ion, such as with lead, Pb^{2+} or mercury, Hg^{2+} selectivity against calcium, Ca^{2+} and mag-nesium, Mg^{2+}, is essential in order that the treatment does not remove essential metals.

Selectivity is determined by various factors. In the case of iron overload, which may occur in individuals with β-thalessemia who have received blood transfusions, the target metal ion is in the +3 oxidation state and so forms stronger complexes than the divalent ions. It also forms stronger complexes with oxygen-donor ligands than with nitrogen-donor ligands. deferoxamine, a naturally occurring siderophore produced by the actinobacter *Streptomyces pilosus* and was used initially as a chelation therapy agent. Synthetic siderophores such as deferiprone and deferasirox have been developed, using the known structure of deferoxamine as a starting point.

Chelation occurs with the two oxygen atoms.

Wilson's disease is caused by a defect in copper metabolism which results in accumulation of copper metal in various organs of the body. The target ion in this case is divalent, Cu^{2+}. This ion is classified as borderline in the scheme of Ahrland, Chatt and Davies. This means that it forms roughly equally strong complexes with ligands whose donor atoms are N, O or F as with ligands whose donor atoms are P, S or Cl. Penicillamine, which contains nitrogen and sulphur donor atoms, is used as this type of ligand binds more strongly to copper ions than to calcium and magnesium ions.

Treatment of poisoning by heavy metals such as lead and mercury is more problematical, because the ligands used do not have high specificity relative to calcium. For example, EDTA may be administered as a calcium salt to reduce the removal of calcium from bone together with the heavy metal.

Chromatography

In column chromatography a mixture of substances is dissolved in a mobile phase and passed over a stationary phase in a column. A selectivity factor is defined as the ratio of distribution coefficients, which describe the equilibrium distribution of an analyte between the stationary phase and the mobile phase. The selectivity factor is equal to the selectivity coefficient with the added assumption that the activity of the stationary phase, the substrate in this case, is equal to 1, the standard assumption for a pure phase. The resolution of a chromatographic column, R_S is related to the selectivity factor by:

$$R_S = \frac{\sqrt{N}}{4}\left(\frac{\alpha-1}{\alpha}\right)\left(\frac{k_B}{1+k_B}\right)$$

where α is selectivity factor, N is the number of theoretical plates k_A and k_B are the retention factors of the two analytes. Retention factors are proportional to distribution coefficients. In practice substances with a selectivity factor very close to 1 can be separated. This is particularly true in gas-liquid chromatography where column lengths up to 60 m are possible, providing a very large number of theoretical plates.

In ion-exchange chromatography the selectivity coefficient is defined in a slightly different way

Solvent Extraction

Solvent extraction is used to extract individual lanthanoid elements from the mixtures found in nature in ores such as monazite. In one process, the metal ions in aqueous solution are made to form complexes with tributylphosphate (TBP), which are extracted into an organic solvent such as kerosene. Complete separation is effected by using a countercurrent exchange method. A number of cells are arranged as a cascade. After equilibration, the aqueous component of each cell is transferred to the previous cell and the organic component is transferred to the next cell, which initially contains only water. In this way the metal ion with the most stable complex passes down the cascade in the organic phase and the metal with the least stable complex passes up the cascade in the aqueous phase.

If solubility in the organic phase is not an issue, a selectivity coefficient is equal to the ratio of the stability constants of the TBP complexes of two metal ions. For lanthanoid elements which are adjacent in the periodic table this ratio is not much greater than 1, so many cells are needed in the cascade.

Chemical Sensors

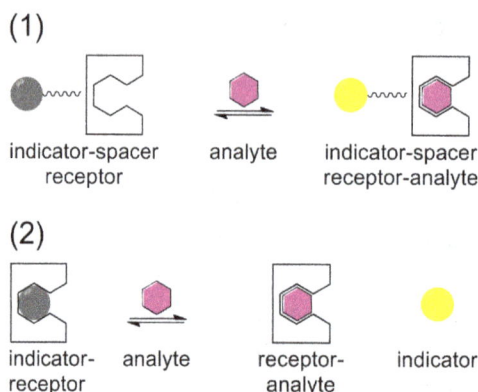

Types of Chemosensors. (1.) Indicator-spacer-receptor (ISR) (2.) Indicator-Displacement Assay (IDA)

A potentiometric selectivity coefficient defines the ability of an ion-selective electrode to distinguish one particular ion from others. The selectivity coefficient, $K_{B,C}$ is evaluated by means of the emf response of the ion-selective electrode in mixed solutions of the primary ion, B, and interfering ion, C (fixed interference method) or less desirably, in separate solutions of B and C (separate solution method). For example, a potassium ion-selective membrane electrode utilizes the naturally occurring macrocyclic antibiotic valinomycin. In this case the cavity in the macrocyclic ring is just the right size to encapsulate the potassium ion, but too large to bind the sodium ion, the most likely interference, strongly.

Chemical sensors, are being developed for specific target molecules and ions in which the target (guest) form a complex with a sensor (host). The sensor is designed to be an excellent match in terms of the size and shape of the target in order to provide for the maximum binding selectivity. An indicator is associated with the sensor which undergoes a change when the target forms a complex with the sensor . The indicator change is usually a colour change (gray to yellow in the illustration) seen in absorbance or, with greater sensitivity, luminescence. The indicator may be attached to the sensor via a spacer, in the ISR arrangement, or it may be displaced from the sensor, IDA arrangement.

References

- Foreman, J.C.; Johansen, T., eds. (2003). Textbook of receptor pharmacology (2nd. ed.). Boca Raton, Fla.: CRC Press. ISBN 0-8493-1029-6.

- Walker, M.; Shah, H.H. (1997). Everything you should know about chelation therapy (4th ed.). New Canaan, Conn.: Keats Pub. ISBN 0-87983-730-6.

- Skoog, D.A; West, D.M.; Holler, J.F.; Crouch, S.R. (2004). Fundamentals of Analytical Chemistry (8th ed.). Thomson Brooks/Cole. ISBN 0-03-035523-0. Section 30E

- Rydberg, J.; Musikas, C; Choppin, G.R., eds. (2004). Solvent Extraction Principles and Practice ((2nd. ed.). Boca Raton, Fla.: CRC Press. ISBN 0-8247-5063-2.

- Florinel-Gabriel Bănică, Chemical Sensors and Biosensors: Fundamentals and Applications, John Wiley and Sons, Chichester, 2012, Print ISBN 978-0-470-71066-1

Methods and Techniques of Chromatography

Chromatography has particular methods and techniques. Some of these are micellar electrokinetic chromatography, affinity chromatography, size-exclusion chromatography, two-dimensional chromatography and ion chromatography. This chapter discusses the methods of chromatography in a critical manner providing key analysis to the subject matter.

Capillary Electrochromatography

Capillary electrochromatography (CEC) is a chromatographic technique in which the mobile phase is driven through the chromatographic bed by electroosmosis. Capillary electrochromatography is a combination of two analytical techniques, High performance liquid chromatography and capillary electrophoresis. Capillary electrophoresis aims to separate analytes on the basis of their mass to charge ratio by passing a high voltage across ends of a capillary tube, which is filled with the analyte. High performance liquid chromatography separates analytes by passing them, under high pressure, through a column filled with stationary phase. The interactions between the analytes and the stationary phase and mobile phase lead to the separation of the analytes. In capillary electrochromatography capillaries, packed with HPLC stationary phase, are subjected to a high voltage. Separation is achieved by electrophoretic migration of solutes and differential partitioning.

Principle

Capillary electrochromatography (CEC) combines the principles used in HPLC and CE. The mobile phase is driven across the chromatographic bed using electroosmosis instead of pressure (as in HPLC). Electroosmosis is the motion of liquid induced by an applied potential across a porous material, capillary tube, membrane or any other fluid conduit. Electroosmotic flow is caused by the Coulomb force induced by an electric field on net mobile electric charge in a solution. Under alkaline conditions, the surface silanol groups of the fused silica will become ionised leading to a negatively charged surface. This surface will have a layer of positively charged ions in close proximity which are relatively immobilised. This layer of ions is called the Stern layer. The thickness of the double layer is given by the formula:

$$\delta = [\varepsilon r \, \varepsilon o \, RT / 2cF^2]^{1/2}$$

Where:

εr = dielectric constant or relative permittivity of medium

εo = permittivity of a vacuum

R = universal gas constant

T = absolute temperature

c = molar concentration

F = Faraday constant

When an electric field is applied to the fluid (usually via electrodes placed at inlets and outlets), the net charge in the electrical double layer is induced to move by the resulting Coulomb force. The resulting flow is termed electroosmotic flow. In CEC positive ions of the electrolyte added along with the analyte accumulate in the electrical double layer of the particles of the column packing on application of an electric field they move towards the cathode and drag the liquid mobile phase with them.

The relationship between the linear velocity u of the liquid in the capillary and the applied electric field is given by the Smoluchowski equation as follows:

$$u = \varepsilon r \varepsilon o \zeta E \eta$$

Where:

εr = Dielectric constant or relative permittivity of medium

εo = Permittivity of a vacuum

ζ = Potential across the Stern layer (Zeta Potential)

E = Electric field strength

η = Viscosity of the solvent

Separation of components in CEC is based on interactions between the stationary phase and differential electrophoretic migration of solutes.

Instrumentation and Working

The basic components of an capillary electrochromatograph are a sample vial, source and destination vials, a packed capillary, electrodes, a high voltage power supply, a detector, and a data output and handling device. The source vial, destination vial and capillary are filled with an electrolyte such as an aqueous buffer solution. The capillary is packed with stationary phase. To introduce the sample, the capillary inlet is placed into a vial containing the sample and then returned to the source vial (sample is introduced into the capillary via capillary action, pressure, or siphoning). The migration of the analytes is then initiated by an electric field that is applied between the source and destination vials and is supplied to the electrodes by the high-voltage power supply. The analytes separate as they migrate due to their electrophoretic mobility, and are detected near the outlet end of the capillary. The output of the detector is sent to a data output and handling device such as an integrator or computer. The data is then displayed as an electropherogram, which reports detector response as a function of time. Separated chemical compounds appear as peaks with different migration times in an electropherogram.

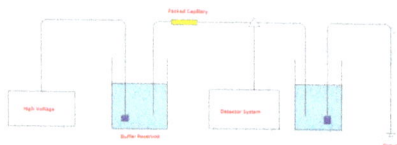

Figure 1. Diagram of Mechanism of Capillary Electrochromatography

Advantages

Avoiding the use of pressure to introduce the mobile phase into the column, results in a number of important advantages. Firstly, the pressure driven flow rate across a column depends directly on the square of the particle diameter and inversely on the length of the column. This restricts the length of the column and size of the particle, particle size is seldom less than 3 micrometer and the length of the column is restricted to 25 cm. Electrically driven flow rate is independent of length of column and size .A second advantage of using electroosmosis to pass the mobile phase into the column is the plug like flow velocity profile of EOF reduces the solute dispersion in the column, increasing column efficiency.

Capillary Electrophoresis

Capillary electrophoresis (CE) is a family of electrokinetic separation methods performed in submillimeter diameter capillaries and in micro- and nanofluidic channels. Very often, CE refers to capillary zone electrophoresis (CZE), but other electrophoretic techniques including capillary gel electrophoresis (CGE), capillary isoelectric focusing (CIEF), capillary isotachophoresis and micellar electrokinetic chromatography (MEKC) belong also to this class of methods. In CE methods, analytes migrate through electrolyte solutions under the influence of an electric field. Analytes can be separated according to ionic mobility and/or partitioning into an alternate phase via Non-covalent interactions. Additionally, analytes may be concentrated or "focused" by means of gradients in conductivity and pH.

Instrumentation

The instrumentation needed to perform capillary electrophoresis is relatively simple. A basic schematic of a capillary electrophoresis system is shown in *figure 1*. The system's main components are a sample vial, source and destination vials, a capillary, electrodes, a high voltage power supply, a detector, and a data output and handling device. The source vial, destination vial and capillary are filled with an electrolyte such as an aqueous buffer solution. To introduce the sample, the capillary inlet is placed into a vial containing the sample. Sample is introduced into the capillary via capillary action, pressure, siphoning, or electrokinetically, and the capillary is then returned to the source vial. The migration of the analytes is initiated by an electric field that is applied between the source and destination vials and is supplied to the electrodes by the high-voltage power supply. In the most common mode of CE, all ions, positive or negative, are pulled through the capillary in the same direction by electroosmotic flow. The analytes separate as they migrate due to their electrophoretic mobility, and are detected near the outlet end of the capillary. The output of the detector is sent to a data output and handling device such as an integrator or computer. The data is then displayed as an electropherogram, which reports detector response as a function of time. Separated chemical compounds appear as peaks with different retention times in an electropherogram. Capillary electrophoresis was first combined with mass spectrometry by Richard D. Smith and coworkers, and provides extremely high sensitivity for the analysis of very small sample sizes. Despite the very small sample sizes (typically only a few nanoliters of liquid are introduced into the capillary), high sensitivity and sharp peaks are achieved in part due to injection strategies that result in concentration of analytes into a narrow zone near the inlet of the capillary. This is achieved in either pressure or electrokinetic injections simply by suspending the sample in a buffer of lower conductivity (*e.g.* lower salt concentration) than the running buffer. A process called field-amplified sample stacking (a form of isotachophoresis) results in concentration of analyte in a narrow zone at the boundary between the low-conductivity sample and the higher-conductivity running buffer.

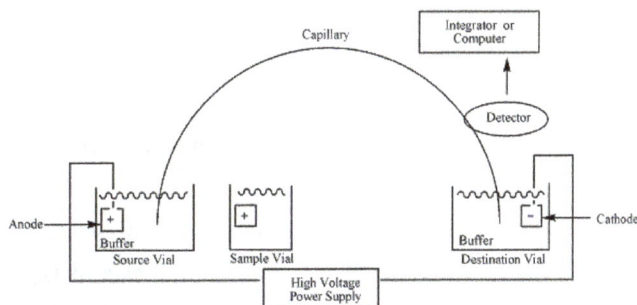

Figure 1: Diagram of capillary electrophoresis system

To achieve greater sample throughput, instruments with arrays of capillaries are used to analyze many samples simultaneously. Such capillary array electrophoresis (CAE) instruments with 16 or 96 capillaries are used for medium- to high-throughput capillary DNA sequencing, and the inlet ends of the capillaries are arrayed spatially to accept samples directly from SBS-standard footprint 96-well plates. Certain aspects of the instrumentation (such as detection) are necessarily more complex than for a single-capillary system, but the fundamental principles of design and operation are similar to those shown in Figure 1.

Detection

Separation by capillary electrophoresis can be detected by several detection devices. The majority of commercial systems use UV or UV-Vis absorbance as their primary mode of detection. In these systems, a section of the capillary itself is used as the detection cell. The use of on-tube detection enables detection of separated analytes with no loss of resolution. In general, capillaries used in capillary electrophoresis are coated with a polymer (frequently polyimide or Teflon) for increased flexibility. The portion of the capillary used for UV detection, however, must be optically transparent. For polyimide-coated capillaries, a segment of the coating is typically burned or scraped off to provide a bare window several millimeters long. This bare section of capillary can break easily, and capillaries with transparent coatings are available to increase the stability of the cell window. The path length of the detection cell in capillary electrophoresis (~ 50 micrometers) is far less than that of a traditional UV cell (~ 1 cm). According to the Beer-Lambert law, the sensitivity of the detector is proportional to the path length of the cell. To improve the sensitivity, the path length can be increased, though this results in a loss of resolution. The capillary tube itself can be expanded at the detection point, creating a "bubble cell" with a longer path length or additional tubing can be added at the detection point as shown in *figure 2*. Both of these methods, however, will decrease the resolution of the separation. Post-column detection utilizing a sheath flow configuration has also been described.

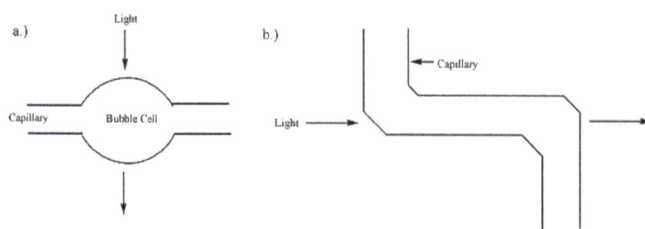

Figure 2: Techniques for increasing the pathlength of the capillary: a.) a bubble cell and b.) a z-cell (additional tubing).

Fluorescence detection can also be used in capillary electrophoresis for samples that naturally fluoresce or are chemically modified to contain fluorescent tags. This mode of detection offers high sensitivity and improved selectivity for these samples, but cannot be utilized for samples that do not fluoresce. Numerous labeling strategies are used to create fluorescent derivatives or conjugates of non-fluorescent molecules, including proteins and DNA. The set-up for fluorescence detection in a capillary electrophoresis system can be complicated. The method requires that the light beam be focused on the capillary, which can be difficult for many light sources. Laser-induced fluorescence has been used in CE systems with detection limits as low as 10^{-18} to 10^{-21} mol. The sensitivity of the technique is attributed to the high intensity of the incident light and the ability to accurately focus the light on the capillary. Multi-color fluorescence detection can be achieved by including multiple dichroic mirrors and bandpass filters to separate the fluorescence emission amongst multiple detectors (*e.g.*, photomultiplier tubes), or by using a prism or grating to project spectrally resolved fluorescence emission onto a position-sensitive detector such as a CCD array. CE systems with 4- and 5-color LIF detection systems are used routinely for capillary DNA sequencing and genotyping ("DNA fingerprinting") applications.

In order to obtain the identity of sample components, capillary electrophoresis can be directly coupled with mass spectrometers or Surface Enhanced Raman Spectroscopy (SERS). In most systems, the capillary outlet is introduced into an ion source that utilizes electrospray ionization (ESI). The resulting ions are then analyzed by the mass spectrometer. This set-up requires volatile buffer solutions, which will affect the range of separation modes that can be employed and the degree of resolution that can be achieved. The measurement and analysis are mostly done with a specialized gel analysis software.

For CE-SERS, capillary electrophoresis eluants can be deposited onto a SERS-active substrate. Analyte retention times can be translated into spatial distance by moving the SERS-active substrate at a constant rate during capillary electrophoresis. This allows the subsequent spectroscopic technique to be applied to specific eluants for identification with high sensitivity. SERS-active substrates can be chosen that do not interfere with the spectrum of the analytes.

Modes of Separation

The separation of compounds by capillary electrophoresis is dependent on the differential migration of analytes in an applied electric field. The electrophoretic migration velocity (u_p) of an analyte toward the electrode of opposite charge is:

$$u_p = \mu_p E$$

The electrophoretic mobility can be determined experimentally from the migration time and the field strength:

$$\mu_p = \left(\frac{L}{t_r}\right)\left(\frac{L_t}{V}\right)$$

where L is the distance from the inlet to the detection point, t_r is the time required for the analyte to reach the detection point (migration time), V is the applied voltage (field strength), and L_t is the total length of the capillary. Since only charged ions are affected by the electric field, neutral analytes are poorly separated by capillary electrophoresis.

The velocity of migration of an analyte in capillary electrophoresis will also depend upon the rate of electroosmotic flow (EOF) of the buffer solution. In a typical system, the electroosmotic flow is directed toward the negatively charged cathode so that the buffer flows through the capillary from the source vial to the destination vial. Separated by differing electrophoretic mobilities, analytes migrate toward the electrode of opposite charge. As a result, negatively charged analytes are attracted to the positively charged anode, counter to the EOF, while positively charged analytes are attracted to the cathode, in agreement with the EOF as depicted in *figure 3*.

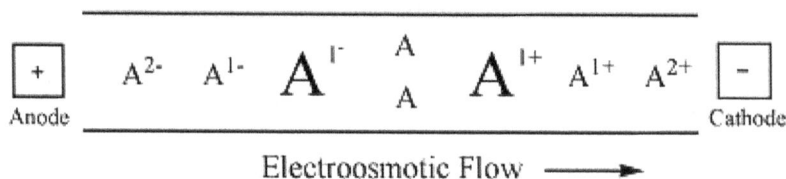

Figure 3: Diagram of the separation of charged and neutral analytes (A) according to their respective electrophoretic and electroosmotic flow mobilities

The velocity of the electroosmotic flow, u_o can be written as:

$$u_o = \mu_o E$$

where μ_o is the electroosmotic mobility, which is defined as:

$$\mu_o = \frac{\epsilon \zeta}{\eta}$$

where ζ is the zeta potential of the capillary wall, and $\dot{o}$ is the relative permittivity of the buffer solution. Experimentally, the electroosmotic mobility can be determined by measuring the retention time of a neutral analyte. The velocity (u) of an analyte in an electric field can then be defined as:

$$u_p + u_o = (\mu_p + \mu_o)E$$

Since the electroosmotic flow of the buffer solution is generally greater than that of the electrophoretic mobility of the analytes, all analytes are carried along with the buffer solution toward the cathode. Even small, triply charged anions can be redirected to the cathode by the relatively powerful EOF of the buffer solution. Negatively charged analytes are retained longer in the capilliary due to their conflicting electrophoretic mobilities. The order of migration seen by the detector is shown in *figure 3*: small multiply charged cations migrate quickly and small multiply charged anions are retained strongly.

Electroosmotic flow is observed when an electric field is applied to a solution in a capillary that has fixed charges on its interior wall. Charge is accumulated on the inner surface of a capillary when a buffer solution is placed inside the capillary. In a fused-silica capillary, silanol (Si-OH) groups attached to the interior wall of the capillary are ionized to negatively charged silanoate (Si-O⁻) groups at pH values greater than three. The ionization of the capillary wall can be enhanced by first running a basic solution, such as NaOH or KOH through the capillary prior to introducing the buffer solution. Attracted to the negatively charged silanoate groups, the positively charged cations of the buffer solution will form two inner layers of cations (called the diffuse double layer or the electrical double layer) on the capillary wall as shown in *figure 4*. The first layer is referred to as the fixed layer because it is held tightly to the silanoate groups. The outer layer, called the mobile layer, is farther from the silanoate groups. The mobile cation layer is pulled in the direction of the negatively charged cathode when an electric field is applied. Since these cations are solvated, the bulk buffer solution migrates with the mobile layer, causing the electroosmotic flow of the buffer solution. Other capillaries including Teflon capillaries also exhibit electroosmotic flow. The EOF of these capillaries is probably the result of adsorption of the electrically charged ions of the buffer onto the capillary walls. The rate of EOF is dependent on the field strength and the charge density of the capillary wall. The wall's charge density is proportional to the pH of the buffer solution. The electroosmotic flow will increase with pH until all of the available silanols lining the wall of the capillary are fully ionized.

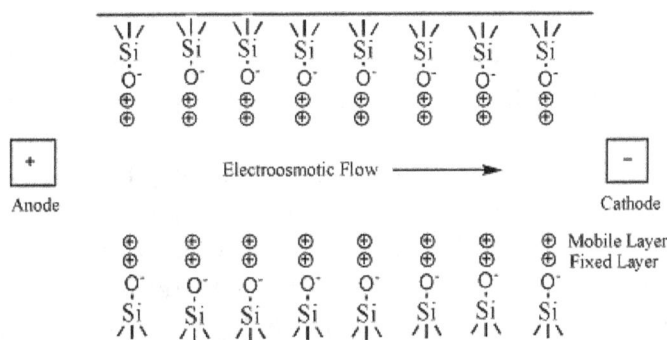

Figure 4: Depiction of the interior of a fused-silica gel capillary in the presence of a buffer solution.

In certain situations where strong electroosomotic flow toward the cathode is undesirable, the inner surface of the capillary can be coated with polymers, surfactants, or small molecules to reduce electroosmosis to very low levels, restoring the normal direction of migration (anions towar the anode, cations toward the cathode). CE instrumentation typically includes power supplies with reversible polarity, allowing the same instrument to be used in "normal" mode (with EOF and detection near the cathodic end of the capillary) and "reverse" mode (with EOF suppressed or reversed, and detection near the anodic end of the capillary). One of the most common approaches to suppressing EOF, reported by Stellan Hjertén in 1985, is to create a covalently attached layer of linear polyacrylamide. The silica surface of the capillary is first modified with a silane reagent bearing a polymerizable vinyl group (*e.g.* 3-methacryloxypropyltrimethoxysilane), followed by introduction of acrylamide monomer and a free radical initiator. The acrylamide is polymerized *in situ*, forming long linear chains, some of which are covalently attached to the wall-bound silane reagent. Numerous other strategies for covalent modification of capillary surfaces exist. Dynamic or adsorbed coatings (which can include polymers or small molecules) are also common. For example, in capillary sequencing of DNA, the sieving polymer (typically polydimethylacrylamide) suppresses electroosmotic flow to very low levels. A variety of dynamic capillary coating agents are commercially available to modify, suppress, or reverse the direction of electroosmotic flow. Besides modulating electroosmotic flow, capillary wall coatings can also serve the purpose of reducing interactions between "sticky" analytes (such as proteins) and the capillary wall. Such wall-analyte interactions, if severe, manifest as reduced peak efficiency, asymmetric (tailing) peaks, or even complete loss of analyte to the capillary wall.

Efficiency and Resolution

The number of theoretical plates, or separation efficiency, in capillary electrophoresis is given by:

$$N = \frac{\mu V}{2D_m}$$

where N is the number of theoretical plates, μ is the apparent mobility in the separation medium and D_m is the diffusion coefficient of the analyte. According to this equation, the efficiency of separation is only limited by diffusion and is proportional to the strength of the electric field, although practical considerations limit the strength of the electric field to several hundred volts per centimeter. Application of very high potentials (>20-30 kV) may lead to arcing or breakdown of the capillary. Further, application of strong electric fields leads to resistive heating (Joule heating) of the buffer in the capillary. At sufficiently high field strengths, this heating is strong enough that radial temperature gradients can develop within the capillary. Since electrophoretic mobility of ions is generally temperature-dependent (due to both temperature-dependent ionization and solvent viscosity effects), a non-uniform temperature profile results in variation of electrophoretic mobility across the capillary, and a loss of resolution. The onset of significant Joule heating can be determined by constructing an "Ohm's Law plot", wherein the current through the capillary is measured as a function of applied potential. At low fields, the current is proportional to the applied potential (Ohm's Law), whereas at higher fields the current deviates from the straight line as heating results in decreased resistance of the buffer. The best resolution is typically obtained at the maximum field strength for which Joule heating is insignificant (*i.e.* near the boundary between the linear and nonlinear regimes of the Ohm's Law plot). Generally capillaries of smaller inner diameter support use of higher field strengths, due to improved heat dissipation and smaller thermal gradients relative to larger capillaries, but with the drawbacks of lower sensitivity in absorbance detection due to shorter path length, and greater difficulty in introducing buffer and sample into the capillary (small capillaries require greater pressure and/ or longer times to force fluids through the capillary).

The efficiency of capillary electrophoresis separations is typically much higher than the efficiency of other separation techniques like HPLC. Unlike HPLC, in capillary electrophoresis there is no mass transfer between phases. In addition, the flow profile in EOF-driven systems is flat, rather than the rounded laminar flow profile characteristic of the pressure-driven flow in chromatography columns as shown in *figure 5*. As a result, EOF does not significantly contribute to band broadening as in pressure-driven chromatography. Capillary electrophoresis separations can have several hundred thousand theoretical plates.

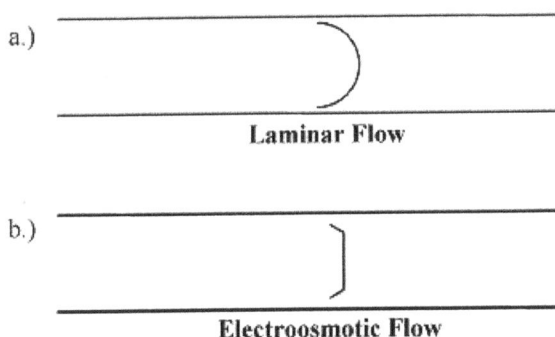

a.)

Laminar Flow

b.)

Electroosmotic Flow

Figure 5: Flow profiles of laminar and electroosmotic flow.

The resolution (R_s) of capillary electrophoresis separations can be written as:

$$R_s = \frac{1}{4}\left(\frac{\Delta\mu_p\sqrt{N}}{\mu_p+\mu_o}\right)$$

According to this equation, maximum resolution is reached when the electrophoretic and electroosmotic mobilities are similar in magnitude and opposite in sign. In addition, it can be seen that high resolution requires lower velocity and, correspondingly, increased analysis time.

Besides diffusion and Joule heating (discussed above), factors that may decrease the resolution in capillary electrophoresis from the theoretical limits in the above equation include, but are not limited to, the finite widths of the injection plug and detection window; interactions between the analyte and the capillary wall; instrumental non-idealities such as a slight difference in height of the fluid reservoirs leading to siphoning; irregularities in the electric field due to, *e.g.*, imperfectly cut capillary ends; depletion of buffering capacity in the reservoirs; and electrodispersion (when an analyte has higher conductivity than the background electrolyte). Identifying and minimizing the numerous sources of band broadening is key to successful method development in capillary electrophoresis, with the objective of approaching as close as possible to the ideal of diffusion-limited resolution.

Applications

Capillary electrophoresis may be used for the simultaneous determination of the ions NH_4^+, Na^+, K^+, Mg^{2+} and Ca^{2+} in saliva.

Micellar Electrokinetic Chromatography

Micellar electrokinetic chromatography (MEKC), is a chromatography technique, used in analytical chemistry. It is a modification of capillary electrophoresis (CE), where the samples are separated by differential partitioning between micelles (pseudo-stationary phase) and a surrounding aqueous buffer solution (mobile phase).

The basic set-up and detection methods used for MEKC are the same as those used in CE. The difference is that the solution contains a surfactant at a concentration that is greater than the critical micelle concentration (CMC). Above this concentration, surfactant monomers are in equilibrium with micelles.

In most applications, MEKC is performed in open capillaries under alkaline conditions to generate a strong electroosmotic flow. Sodium dodecyl sulfate (SDS) is the most commonly used surfactant in MEKC applications. The anionic character of the sulfate groups of SDS cause the surfactant and micelles to have electrophoretic mobility that is counter to the direction of the strong electroosmotic flow. As a result, the surfactant monomers and micelles migrate quite slowly, though their net movement is still toward the cathode. During a MEKC separation, analytes distribute themselves between the hydrophobic interior of the micelle and hydrophilic buffer solution as shown in *figure 1*.

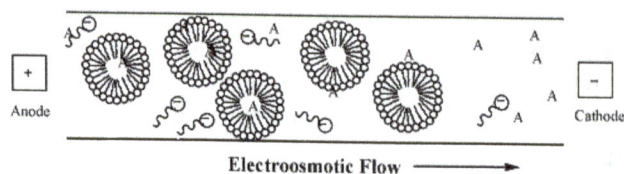

Figure 1: Depiction of the distribution of analytes (A) based on their hydrophobicity.

Analytes that are insoluble in the interior of micelles should migrate at the electroosmotic flow velocity, u_o, and be detected at the retention time of the buffer, t_M. Analytes that solubilize completely within the micelles (analytes that are highly hydrophobic) should migrate at the micelle velocity, u_c, and elute at the final elution time, t_c.

Theory

The micelle velocity is defined by:

$$u_c = u_p + u_o$$

where u_p is the electrophoretic velocity of a micelle.

The retention time of a given sample should depend on the capacity factor, k^1:

$$k^1 = \frac{n_c}{n_w}$$

where n_c is the total number of moles of solute in the micelle and n_w is the total moles in the aqueous phase. The retention time of a solute should then be within the range:

$$t_M \le t_r \le t_c$$

Charged analytes have a more complex interaction in the capillary because they exhibit electrophoretic mobility, engage in electrostatic interactions with the micelle, and participate in hydrophobic partitioning.

The fraction of the sample in the aqueous phase, R, is given by:

$$R = \frac{u_s - u_c}{u_o - u_c}$$

where u_s is the migration velocity of the solute. The value R can also be expressed in terms of the capacity factor:

$$R = \frac{1}{1 + k^1}$$

Using the relationship between velocity, tube length from the injection end to the detector cell (L), and retention time, $u_o = L/t_M$, $u_c = L/t_c$ and $u_s = L/t_r$, a relationship between the capacity factor and retention times can be formulated:

$$k^1 = \frac{t_r - t_M}{t_M(1 - (t_r/t_c))}$$

The extra term enclosed in parenthesis accounts for the partial mobility of the hydrophobic phase in MEKC. This equation resembles an expression derived for k^1 in conventional packed bed chromatography:

$$k = \frac{t_r - t_M}{t_M}$$

A rearrangement of the previous equation can be used to write an expression for the retention factor:

$$t_r = \left(\frac{1+k^1}{1+(t_M/t_c)k^1} \right) t_M$$

From this equation it can be seen that all analytes that partition strongly into the micellar phase (where k^1 is essentially ∞) migrate at the same time, t_c. In conventional chromatography, separation of similar compounds can be improved by gradient elution. In MEKC, however, techniques must be used to extend the elution range to separate strongly retained analytes.

Elution ranges can be extended by several techniques including the use of organic modifiers, cyclodextrins, and mixed micelle systems. Short-chain alcohols or acetonitrile can be used as organic modifiers that decrease t_M and k^1 to improve the resolution of analytes that co-elute with the micellar phase. These agents, however, may alter the level of the EOF. Cyclodextrins are cyclic polysaccharides that form inclusion complexes that can cause competitive hydrophobic partitioning of the analyte. Since analyte-cyclodextrin complexes are neutral, they will migrate toward the cathode at a higher velocity than that of the negatively charged micelles. Mixed micelle systems, such as the one formed by combining SDS with the neutral surfactant Brij-35, can also be used to alter the selectivity of MEKC.

History

In 1984, the Terabe group reported a technique that enabled capillary electrophoresis (CE) instrumentation to be used in the separation of neutral (as well as ionic) species.

Applications of MEKC

The simplicity and efficiency of MEKC have made it an attractive technique for a variety of applications. Further improvements can be made to the selectivity of MEKC by adding chiral selectors or chiral surfactants to the system. Unfortunately, this technique is not suitable for protein analysis because proteins are generally too large to partition into a surfactant micelle and tend to bind to surfactant monomers to form SDS-protein complexes.

Recent applications of MEKC include the analysis of uncharged pesticides, essential and branched-chain amino acids in nutraceutical products, hydrocarbon and alcohol contents of the marjoram herb.

MEKC has also been targeted for its potential to be used in combinatorial chemical analysis. The advent of combinatorial chemistry has enabled medicinal chemists to synthesize and identify large numbers of potential drugs in relatively short periods of time. Small sample and solvent requirements and the high resolving power of MEKC have enabled this technique to be used to quickly analyze a large number of compounds with good resolution.

Traditional methods of analysis, like high-performance liquid chromatography (HPLC), can be used to identify the purity of a combinatorial library, but assays need to be rapid with good resolution for all components to provide useful information for the chemist. The introduction of surfactant to traditional capillary electrophoresis instrumentation has dramatically expanded the scope of analytes that can be separated by capillary electrophoresis.

MEKC can also be used in routine quality control of antibiotics in pharmaceuticals or feedstuffs.

Size-exclusion Chromatography

Size-exclusion chromatography (SEC), also known as molecular sieve chromatography, is a chromatographic method in which molecules in solution are separated by their size, and in some cases molecular weight. It is usually applied to large molecules or macromolecular complexes such as proteins and industrial polymers. Typically, when an aqueous solution is used to transport the sample through the column, the technique is known as gel-filtration chromatography, versus the name gel permeation chromatography, which is used when an organic solvent is used as a mobile phase. SEC is a widely used polymer characterization method because of its ability to provide good molar mass distribution (Mw) results for polymers.

Applications

The main application of gel-filtration chromatography is the fractionation of proteins and other water-soluble polymers, while gel permeation chromatography is used to analyze the molecular weight distribution of organic-soluble polymers. Either technique should not be confused with gel electrophoresis, where an electric field is used to "pull" or "push" molecules through the gel depending on their electrical charges.

Advantages

The advantages of this method include good separation of large molecules from the small molecules with a minimal volume of eluate, and that various solutions can be applied without interfering with the filtration process, all while preserving the biological activity of the particles to be separated. The technique is generally combined with others that further separate molecules by other characteristics, such as acidity, basicity, charge, and affinity for certain compounds. With size exclusion chromatography, there are short and well-defined separation times and narrow bands, which lead to good sensitivity. There is also no sample loss because solutes do not interact with the stationary phase.

The other advantage to this experimental method is that in certain cases it is feasible to determine the approximate molecular weight of a compound. The shape and size of the compound (eluent) determine how the compound interacts with the gel (stationary phase). In order to determine the approximate molecular weight, the elution volumes of compounds with their corresponding molecular weights are obtained and then a plot of "K_{av}" vs "log(Mw)" is made, where $K_{av} = (V_e - V_o)/(V_t - V_o)$ and Mw is the molecular mass. This plot acts as a calibration curve, which is used to approximate the desired compound's molecular weight. The V_e component represents the volume at which the intermediate molecules elute such as molecules that have partial access to the beads of the column. In addition, V_t is the sum of the total volume between the beads and the volume within the beads. The V_o component represents the volume at which the larger molecules elute, which elute in the beginning. Disadvantages are, for example, that only a limited number of bands can be accommodated because the time scale of the chromatogram is short, and, in general, there has to be a 10% difference in molecular mass to have a good resolution

Discovery

The technique was invented by Grant Henry Lathe and Colin R Ruthven, working at Queen Charlotte's Hospital, London. They later received the John Scott Award for this invention. While Lathe and Ruthven used starch gels as the matrix, Jerker Porath and Per Flodin later introduced dextran gels; other gels with size fractionation properties include agarose and polyacrylamide. A short review of these developments has appeared.

There were also attempts to fractionate synthetic high polymers; however, it was not until 1964, when J. C. Moore of the Dow Chemical Company published his work on the preparation of gel permeation chromatography (GPC) columns based on cross-linked polystyrene with controlled pore size, that a rapid increase of research activity in this field began. It was recognized almost immediately that with proper calibration, GPC was capable to provide molar mass and molar mass distribution information for synthetic polymers. Because the latter information was difficult to obtain by other methods, GPC came rapidly into extensive use.

Theory and Method

SEC is used primarily for the analysis of large molecules such as proteins or polymers. SEC works by trapping smaller molecules in the pores of the adsorbent materials adsorption ("stationary phases"). The larger molecules simply pass by the pores because those molecules are too large to enter the pores. Larger molecules therefore flow through the column more quickly than smaller molecules, that is, the smaller the molecule, the longer the retention time.

One requirement for SEC is that the analyte does not interact with the surface of the stationary phases, with differences in elution time between analytes ideally being based solely on the solute volume the analytes can enter, rather than chemical or electrostatic interactions with the stationary phases. Thus, a small molecule that can penetrate every region of the stationary phase pore system can enter a total volume equal to the sum of the entire pore volume and the interparticle volume. This small molecule will elute late (after the molecule has penetrated all of the pore- and interparticle volume—approximately 80% of the column volume). At the other extreme, a very large molecule that cannot penetrate any the smaller pores can enter only the interparticle volume (~35% of the column volume) and will elute earlier when this volume of mobile phase has passed through the column. The underlying principle of SEC is that particles of different sizes will elute (filter) through a stationary phase at different rates. This results in the separation of a solution of particles based on size. Provided that all the particles are loaded simultaneously or near-simultaneously, particles of the same size should elute together.

However, as there are various measures of the size of a macromolecule (for instance, the radius of gyration and the hydrodynamic radius), a fundamental problem in the theory of SEC has been the choice of a proper molecular size parameter by which molecules of different kinds are separated. Experimentally, Benoit and co-workers found an excellent correlation between elution volume and a dynamically based molecular size, the hydrodynamic volume, for several different chain architecture and chemical compositions. The observed correlation based on the hydrodynamic volume became accepted as the basis of universal SEC calibration.

Still, the use of the hydrodynamic volume, a size based on dynamical properties, in the interpretation of SEC data is not fully understood. This is because SEC is typically run under low flow rate conditions where hydrodynamic factor should have little effect on the separation. In fact, both theory and computer simulations assume a thermodynamic separation principle: the separation process is determined by the equilibrium distribution (partitioning) of solute macromolecules between two phases --- a dilute bulk solution phase located at the interstitial space and confined solution phases within the pores of column packing material. Based on this theory, it has been shown that the relevant size parameter to the partitioning of polymers in pores is the mean span dimension (mean maximal projection onto a line). Although this issue has not been fully resolved, it is likely that the mean span dimension and the hydrodynamic volume are strongly correlated.

Each size exclusion column has a range of molecular weights that can be separated. The exclusion limit defines the molecular weight at the upper end of the column 'working' range and is where molecules are too large to be trapped in the stationary phase. The lower end of the range is defined by the permeation limit, which defines the molecular weight of a molecules that is small to penetrate all pores of the stationary phase. All molecules below this molecular mass are so small that they elute as a single band

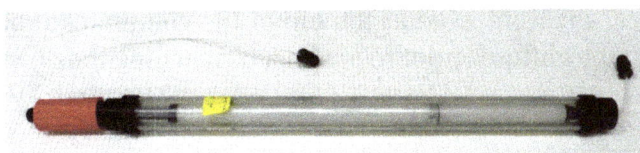

A size exclusion column.

This is usually achieved with an apparatus called a column, which consists of a hollow tube tightly packed with extremely small porous polymer beads designed to have pores of different sizes. These pores may be depressions on the surface or channels through the bead. As the solution travels down the column some particles enter into the pores. Larger particles cannot enter into as many pores. The larger the particles, the faster the elution.

The filtered solution that is collected at the end is known as the eluate. The void volume includes any particles too large to enter the medium, and the solvent volume is known as the column volume.

Factors Affecting Filtration

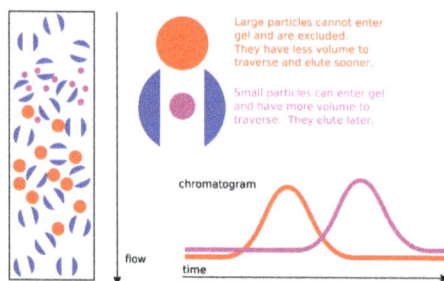

Large particles cannot enter gel and are excluded. They have less volume to traverse and elute sooner.

Small particles can enter gel and have more volume to traverse. They elute later.

chromatogram

flow

time

A cartoon illustrating the theory behind size exclusion chromatography

In real-life situations, particles in solution do not have a fixed size, resulting in the probability that a particle that would otherwise be hampered by a pore passing right by it. Also, the stationary-phase particles are not ideally defined; both particles and pores may vary in size. Elution curves, therefore, resemble Gaussian distributions. The stationary phase may also interact in undesirable ways with a particle and influence retention times, though great care is taken by column manufacturers to use stationary phases that are inert and minimize this issue.

Like other forms of chromatography, increasing the column length will enhance the resolution, and increasing the column diameter increases the capacity of the column. Proper column packing is important to maximize resolution: An overpacked column can collapse the pores in the beads, resulting in a loss of resolution. An underpacked column can reduce the relative surface area of the stationary phase accessible to smaller species, resulting in those species spending less time trapped in pores. Unlike affinity chromatography techniques, a solvent head at the top of the column can drastically diminish resolution as the sample diffuses prior to loading, broadening the downstream elution.

Analysis

In simple manual columns, the eluent is collected in constant volumes, known as fractions. The more similar the particles are in size the more likely they will be in the same fraction and not detected separately. More advanced columns overcome this problem by constantly monitoring the eluent.

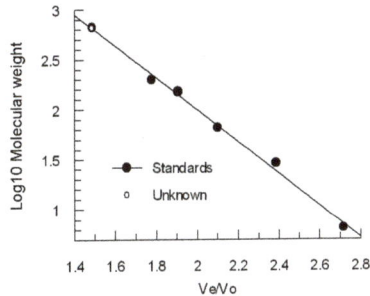

Standardization of a size exclusion column.

The collected fractions are often examined by spectroscopic techniques to determine the concentration of the particles eluted. Common spectroscopy detection techniques are refractive index (RI) and ultraviolet (UV). When eluting spectroscopically similar species (such as during biological purification), other techniques may be necessary to identify the contents of each fraction. It is also possible to analyse the eluent flow continuously with RI, LALLS, Multi-Angle Laser Light Scattering MALS, UV, and/or viscosity measurements.

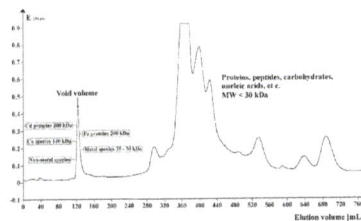

SEC Chromatogram of a biological sample.

The elution volume (Ve) decreases roughly linear with the logarithm of the molecular hydrodynamic volume. Columns are often calibrated using 4-5 standard samples (e.g., folded proteins of known molecular weight), and a sample containing a very large molecule such as thyroglobulin to determine the void volume. (Blue dextran is not recommended for Vo determination because it is heterogeneous and may give variable results) The elution volumes of the standards are divided by the elution volume of the thyroglobulin (Ve/Vo) and plotted against the log of the standards' molecular weights.

Applications

Biochemical Applications

In general, SEC is considered a low resolution chromatography as it does not discern similar species very well, and is therefore often reserved for the final "polishing" step of a purification. The technique can determine the quaternary structure of purified proteins that have slow exchange times, since it can be carried out under native solution conditions, preserving macromolecular interactions. SEC can also assay protein tertiary structure, as it measures the hydrodynamic volume (not molecular weight), allowing folded and unfolded versions of the same protein to be distinguished. For example, the apparent hydrodynamic radius of a typical protein domain might be 14 Å and 36 Å for the folded and unfolded forms, respectively. SEC allows the separation of these two forms, as the folded form will elute much later due to its smaller size.

Polymer Synthesis

SEC can be used as a measure of both the size and the polydispersity of a synthesised polymer, that is, the ability to be able to find the distribution of the sizes of polymer molecules. If standards of a known size are run previously, then a calibration curve can be created to determine the sizes of polymer molecules of interest in the solvent chosen for analysis (often THF). In alternative fashion, techniques such as light scattering and/or viscometry can be used online with SEC to yield absolute molecular weights that do not rely on calibration with standards of known molecular weight. Due to the difference in size of two polymers with identical molecular weights, the absolute determination methods are, in general, more desirable. A typical SEC system can quickly (in about half an hour) give polymer chemists information on the size and polydispersity of the sample. The preparative SEC can be used for polymer fractionation on an analytical scale.

Drawback

In SEC, mass is not measured so much as the hydrodynamic volume of the polymer molecules, that is, how much space a particular polymer molecule takes up when it is in solution. However, the approximate molecular weight can be calculated from SEC data because the exact relationship between molecular weight and hydrodynamic volume for polystyrene can be found. For this, polystyrene is used as a standard. But the relationship between hydrodynamic volume and molecular weight is not the same for all polymers, so only an approximate measurement can be obtained. Another drawback is the possibility of interaction between the stationary phase and the analyte. Any interaction leads to a later elution time and thus mimics a smaller analyte size.

When performing this method, the bands of the eluting molecules may be broadened. This can occur by turbulence caused by the flow of the mobile phase molecules passing through the molecules of the stationary phase. In addition, molecular thermal diffusion and friction between the molecules of the glass walls and the molecules of the eluent contribute to the broadening of the bands. Besides broadening, the bands also overlap with each other. As a result, the eluent usually gets considerably diluted. A few precautions can be taken to prevent the likelihood of the bands broadening. For instance, one can apply the sample in a narrow, highly concentrated band on the top of the column. The more concentrated the eluent is, the more efficient the procedure would be. However, it is not always possible to concentrate the eluent, which can be considered as one more disadvantage.

Absolute Size-exclusion Chromatography

Absolute size-exclusion chromatography (ASEC) is a technique that couples a dynamic light scattering (DLS) instrument to a size exclusion chromatography system for absolute size measurements of proteins and macromolecules as they elute from the chromatography system.

The definition of absolute used here is that it does not require calibration to obtain hydrodynamic size, often referred to as hydrodynamic diameter (D_H in units of nm). The sizes of the macromolecules are measured as they elute into the flow cell of the DLS instrument from the size exclusion column set. It should be noted that the hydrodynamic size of the molecules or particles are measured and not their molecular weights. For proteins a Mark-Houwink type of calculation can be used to estimate the molecular weight from the hydrodynamic size.

A big advantage of DLS coupled with SEC is the ability to obtain enhanced DLS resolution. Batch DLS is quick and simple and provides a direct measure of the average size but the baseline resolution of DLS is 3 to 1 in diameter. Using SEC, the proteins and protein oligomers are separated, allowing oligomeric resolution. Aggregation studies can also be done using ASEC although the aggregate concentration may not be calculated, the size of the aggregate will be measured only to be limited by the maximum size eluting from the SEC columns.

Limitations of ASEC include flow-rate, concentration, and precision. Because a correlation function requires anywhere from 3–7 seconds to properly build, a limited number of data points can be collected across the peak.

Gel Permeation Chromatography

Gel permeation chromatography (GPC) is a type of size exclusion chromatography (SEC), that separates analytes on the basis of size. The technique is often used for the analysis of polymers. As a technique, SEC was first developed in 1955 by Lathe and Ruthven. The term gel permeation chromatography can be traced back to J.C. Moore of the Dow Chemical Company who investigated the technique in 1964 and the proprietary column technology was licensed to Waters Corporation, who subsequently commercialized this technology in 1964. GPC systems and consumables are now also available from a number of manufacturers. It is often necessary to separate polymers, both to analyze them as well as to purify the desired product.

When characterizing polymers, it is important to consider the polydispersity index

(PDI) as well the molecular weight. Polymers can be characterized by a variety of definitions for molecular weight including the number average molecular weight (M_n), the weight average molecular weight (M_w) , the size average molecular weight (M_z), or the viscosity molecular weight (M_v). GPC allows for the determination of PDI as well as M_v and based on other data, the M_n, M_w, and M_z can be determined.

How GPC Works

GPC separates based on the size or hydrodynamic volume (radius of gyration) of the analytes. This differs from other separation techniques which depend upon chemical or physical interactions to separate analytes. Separation occurs via the use of porous beads packed in a column.

Schematic of pore vs. analyte size

The smaller analytes can enter the pores more easily and therefore spend more time in these pores, increasing their retention time. These smaller molecules spend more time in the column and therefore will elute last. Conversely, larger analytes spend little if any time in the pores and are eluted quickly. All columns have a range of molecular weights that can be separated.

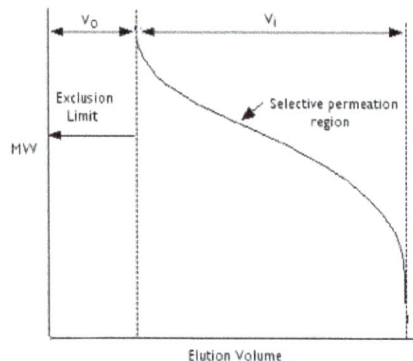

Range of molecular weights that can be separated for each packing material

If an analyte is either too large or too small, it will be either not retained or completely retained, respectively. Analytes that are not retained are eluted with the free volume outside of the particles (V_o), while analytes that are completely retained are eluted with volume of solvent held in the pores (V_i). The total volume can be considered by the following equation, where V_g is the volume of the polymer gel and V_t is the total volume:

As can be inferred, there is a limited range of molecular weights that can be separated by each column and therefore the size of the pores for the packing should be chosen according to the range of molecular weight of analytes to be separated. For polymer separations the pore sizes should be on the order of the polymers being analyzed. If a sample has a broad molecular weight range it may be necessary to use several GPC columns in tandem with one another to fully resolve the sample.

Application

GPC is often used to determine the relative molecular weight of polymer samples as well as the distribution of molecular weights. What GPC truly measures is the molecular volume and shape function as defined by the intrinsic viscosity. If comparable standards are used, this relative data can be used to determine molecular weights within ± 5% accuracy. Polystyrene standards with PDI of less than 1.2 are typically used to calibrate the GPC. Unfortunately, polystyrene tends to be a very linear polymer and therefore as a standard it is only useful to compare it to other polymers that are known to be linear and of relatively the same size.

Material and Methods

Instrumentation

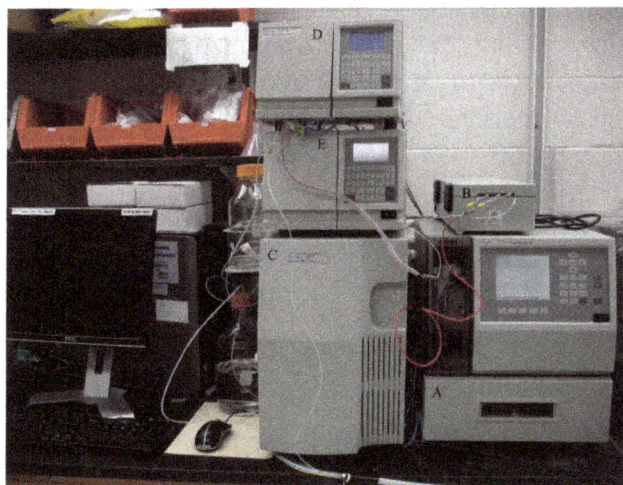

A typical GPC instrument including: A. Autosampler, B. Column, C. Pump, D. RI detector, E. UV-vis detector

Inside of an autosampler for running several samples without user interaction, e.g. overnight

Gel permeation chromatography is conducted almost exclusively in chromatography columns. The experimental design is not much different from other techniques of liquid chromatography. Samples are dissolved in an appropriate solvent, in the case of GPC these tend to be organic solvents and after filtering the solution it is injected onto a column. The separation of multi-component mixture takes place in the column. The constant supply of fresh eluent to the column is accomplished by the use of a pump. Since most analytes are not visible to the naked eye a detector is needed. Often multiple detectors are used to gain additional information about the polymer sample. The availability of a detector makes the fractionation convenient and accurate.

Gel

Gels are used as stationary phase for GPC. The pore size of a gel must be carefully controlled in order to be able to apply the gel to a given separation. Other desirable properties of the gel forming agent are the absence of ionizing groups and, in a given solvent, low affinity for the substances to be separated. Commercial gels like PLgel, Sephadex, Bio-Gel (cross-linked polyacrylamide), agarose gel and Styragel are often used based on different separation requirements.

Column

The column used for GPC is filled with a microporous packing material. The column is filled with the gel.

Eluent

The eluent (mobile phase) should be a good solvent for the polymer, should permit high detector response from the polymer and should wet the packing surface. The most common eluents in for polymers that dissolve at room temperature GPC are tetrahydrofuran (THF), *o*-dichlorobenzene and trichlorobenzene at 130–150 °C for crystalline polyalkynes and *m*-cresol and *o*-chlorophenol at 90 °C for crystalline condensation polymers such as polyamides and polyesters.

Pump

There are two types of pumps available for uniform delivery of relatively small liquid volumes for GPC: piston or peristaltic pumps.

Detector

In GPC, the concentration by weight of polymer in the eluting solvent may be monitored continuously with a detector. There are many detector types available and they can be divided into two main categories. The first is concentration sensitive detectors which includes UV absorption, differential refractometer (DRI) or refractive index (RI) detectors, infrared (IR) absorption and density detectors. The second category is molecular weight sensitive detectors, which include low angle light scattering detectors (LALLS) and multi angle light scattering (MALLS). The resulting chromatogram is therefore a weight distribution of the polymer as a function of retention volume.

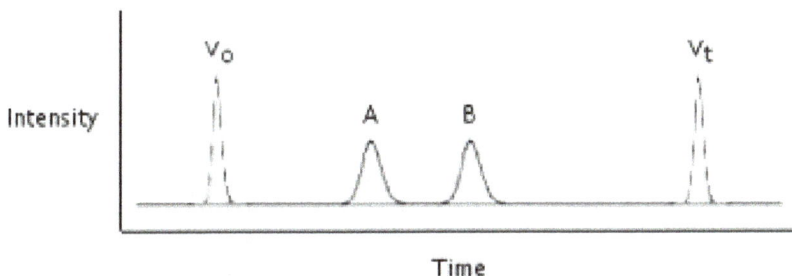

GPC Chromatogram; V_o = no retention, V_t = complete retention, A and B = partial retention

The most sensitive detector is the differential UV photometer and the most common detector is the differential refractometer (DRI). When characterizing copolymer, it is necessary to have two detectors in series. For accurate determinations of copolymer composition at least two of those detectors should be concentration detectors. The determination of most copolymer compositions is done using UV and RI detectors, although other combinations can be used.

Data Analysis

Gel permeation chromatography (GPC) has become the most widely used technique for analyzing polymer samples in order to determine their molecular weights and weight distributions. Examples of GPC chromatograms of polystyrene samples with their molecular weights and PDIs are shown on the left.

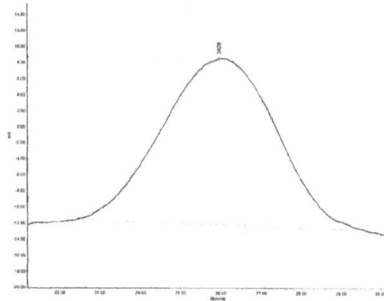

GPC Separation of Anionically Synthesized Polystyrene; M_n=3,000 g/mol, PDI=1.32

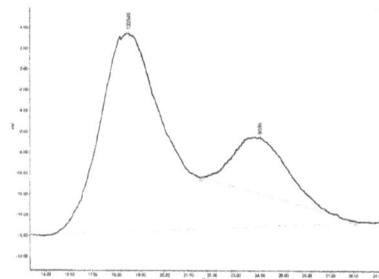

GPC Separation of Free-Radical Synthesized Polystyrene; M_n=24,000 g/mol, PDI=4.96

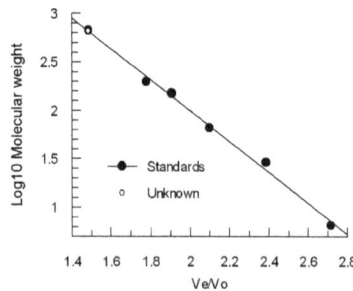

Standardization of a size exclusion column.

Benoit and co-workers proposed that the hydrodynamic volume, V_η, which is proportional to the product of [η] and M, where [η] is the intrinsic viscosity of the polymer in the SEC eluent, may be used as the universal calibration parameter. If the Mark–Houwink–Sakurada constants K and α are known, a plot of log [η]M versus elution volume (or elution time) for a particular solvent, column and instrument provides a universal calibration curve which can be used for any polymer in that solvent. By determining the retention volumes (or times) of monodisperse polymer standards (e.g. solutions of monodispersed polystyrene in THF), a calibration curve can be obtained by plotting the logarithm of the molecular weight versus the retention time or volume. Once the calibration curve is obtained, the gel permeation chromatogram of any other polymer can be obtained in the same solvent and the molecular weights (usually M_n and M_w) and the complete molecular weight distribution for the polymer can be determined. A typical calibration curve is shown to the right and the molecular weight from an unknown sample can be obtained from the calibration curve.

Advantages of GPC

As a separation technique GPC has many advantages. First of all, it has a well-defined separation time due to the fact that there is a final elution volume for all unretained analytes. Additionally, GPC can provide narrow bands, although this aspect of GPC is more difficult for polymer samples that have broad ranges of molecular weights present. Finally, since the analytes do not interact chemically or physically with the column, there is a lower chance for analyte loss to occur. For investigating the properties of polymer samples in particular, GPC can be very advantageous. GPC provides a more convenient method of determining the molecular weights of polymers. In fact most samples can be thoroughly analyzed in an hour or less. Other methods used in the past were fractional extraction and fractional precipitation. As these processes were quite labor-intensive molecular weights and mass distributions typically were not analyzed. Therefore, GPC has allowed for the quick and relatively easy estimation of molecular weights and distribution for polymer samples

Disadvantages of GPC

There are disadvantages to GPC, however. First, there is a limited number of peaks that can be resolved within the short time scale of the GPC run. Also, as a technique GPC requires around at least a 10% difference in molecular weight for a reasonable resolution of peaks to occur. In regards to polymers, the molecular masses of most of the chains will be too close for the GPC separation to show anything more than broad peaks. Another disadvantage of GPC for polymers is that filtrations must be performed before using the instrument to prevent dust and other particulates from ruining the columns and interfering with the detectors. Although useful for protecting the instrument, there is the possibility of the pre-filtration of the sample removing higher molecular weight sample before it can be loaded on the column. Another possibility to overcome these issues is the separation by field-flow fractionation (FFF).

Orthogonal Methods

Field-flow fractionation (FFF) can be considered as an alternative to GPC, especially when particles or high molar mass polymers cause clogging of the column, shear degradation is an issue or agglomeration takes place but cannot be made visible. FFF is separation in an open flow channel without having a static phase involved so no interactions occur. With one field-flow fractionation version, thermal field-flow fractionation, separation of polymers having the same size but different chemical compositions is possible.

Affinity Chromatography

Affinity chromatography is a method of separating biochemical mixtures based on a highly specific interaction such as that between antigen and antibody, enzyme and substrate, or receptor and ligand.

Uses

Affinity chromatography can be used to:

- Purify and concentrate a substance from a mixture into a buffering solution

- Reduce the amount of a substance in a mixture

- Discern what biological compounds bind to a particular substance

- Purify and concentrate an enzyme solution.

Principle

The stationary phase is typically a gel matrix, often of agarose; a linear sugar molecule derived from algae. Usually the starting point is an undefined heterogeneous group of molecules in solution, such as a cell lysate, growth medium or blood serum. The molecule of interest will have a well known and defined property, and can be exploited during the affinity purification process. The process itself can be thought of as an entrapment, with the target molecule becoming trapped on a solid or stationary phase or medium. The other molecules in the mobile phase will not become trapped as they do not possess this property. The stationary phase can then be removed from the mixture, washed and the target molecule released from the entrapment in a process known as elution. Possibly the most common use of affinity chromatography is for the purification of recombinant proteins.

Batch and Column Setups

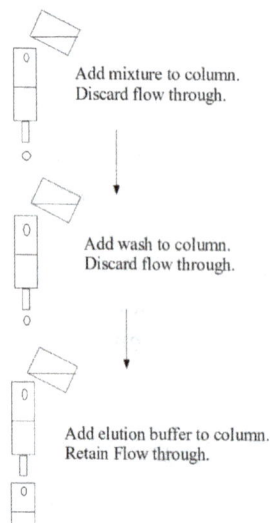

Add mixture to column.
Discard flow through.

Add wash to column.
Discard flow through.

Add elution buffer to column.
Retain Flow through.

Column chromatography

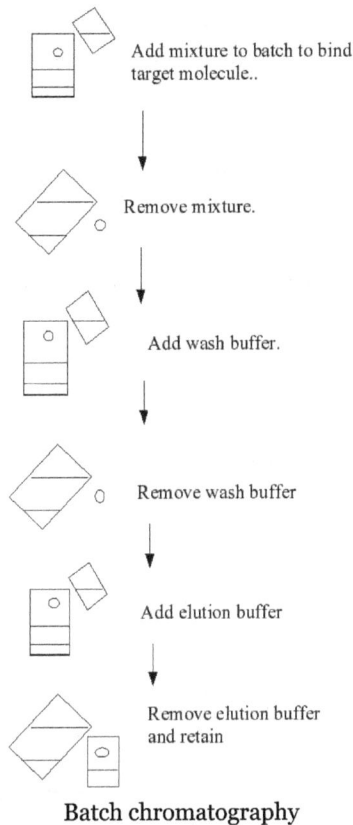

Add mixture to batch to bind target molecule..

Remove mixture.

Add wash buffer.

Remove wash buffer

Add elution buffer

Remove elution buffer and retain

Batch chromatography

Binding to the solid phase may be achieved by column chromatography whereby the solid medium is packed onto a column, the initial mixture run through the column to allow setting, a wash buffer run through the column and the elution buffer subsequently applied to the column and collected. These steps are usually done at ambient pressure. Alternatively, binding may be achieved using a batch treatment, for example, by adding the initial mixture to the solid phase in a vessel, mixing, separating the solid phase, removing the liquid phase, washing, re-centrifuging, adding the elution buffer, re-centrifuging and removing the elute.

Sometimes a hybrid method is employed such that the binding is done by the batch method, but the solid phase with the target molecule bound is packed onto a column and washing and elution are done on the column.

A third method, expanded bed absorption, which combines the advantages of the two methods mentioned above, has also been developed. The solid phase particles are placed in a column where liquid phase is pumped in from the bottom and exits at the top. The gravity of the particles ensure that the solid phase does not exit the column with the liquid phase.

Affinity columns can be eluted by changing salt concentrations, pH, pI, charge and ionic strength directly or through a gradient to resolve the particles of interest.

More recently, setups employing more than one column in series have been developed. The advantage compared to single column setups is that the resin material can be fully loaded, since non-binding product is directly passed on to a consecutive column with fresh column material. The resin costs per amount of produced product can thus be drastically be reduced. Since one column can always be eluted and regenerated while the other column is loaded, already two columns are sufficient to make full use of the advantages. Additional columns can give additional flexibility for elution and regeneration times, at the cost of additional equipment and resin costs.

Specific Uses

Affinity chromatography can be used in a number of applications, including nucleic acid purification, protein purification from cell free extracts, and purification from blood.

Various Affinity Media

Many different affinity media exist for a variety of possible uses. Briefly, they are (generalized):

- Activated/Functionalized – Works as a functional spacer, support matrix, and eliminates handling of toxic reagents.

- Amino Acid – Used with a variety of serum proteins, proteins, peptides, and enzymes, as well as rRNA and dsDNA.

- Avidin Biotin – Used in the purification process of biotin/avidin and their derivatives.

- Carbohydrate Bonding – Most often used with glycoproteins or any other carbohydrate-containing substance.

- Carbohydrate – Used with lectins, glycoproteins, or any other carbohydrate metabolite protein.

- Dye Ligand – This media is nonspecific, but mimics biological substrates and proteins.

- Glutathione – Useful for separation of GST tagged recombinant proteins.

- Heparin – This media is a generalized affinity ligand, and it is most useful for separation of plasma coagulation proteins, along with nucleic acid enzymes and lipases.

- Hydrophobic Interaction – Most commonly used to target free carboxyl groups and proteins.

- Immunoaffinity – Detailed below, this method utilizes antigens' and antibodies' high specificity to separate.

- Immobilized Metal Affinity Chromatography – Detailed further below, this method uses interactions between metal ions and proteins (usually specially tagged) to separate.

- Nucleotide/Coenzyme – Works to separate dehydrogenases, kinases, and trans-aminases.

- Nucleic Acid – Functions to trap mRNA, DNA, rRNA, and other nucleic acids/oligonucleotides.

- Protein A/G – This method is used to purify immunoglobulins.

- Speciality – Designed for a specific class or type of protein/coenzyme, this type of media will only work to separate a specific protein or coenzyme.

Immunoaffinity

Another use for the procedure is the affinity purification of antibodies from blood serum. If serum is known to contain antibodies against a specific antigen (for example if the serum comes from an organism immunized against the antigen concerned) then it can be used for the affinity purification of that antigen. This is also known as Immunoaffinity Chromatography. For example if an organism is immunised against a GST-fusion protein it will produce antibodies against the fusion-protein, and possibly antibodies against the GST tag as well. The protein can then be covalently coupled to a solid support such as agarose and used as an affinity ligand in purifications of antibody from immune serum.

For thoroughness the GST protein and the GST-fusion protein can each be coupled separately. The serum is initially allowed to bind to the GST affinity matrix. This will remove antibodies against the GST part of the fusion protein. The serum is then separated from the solid support and allowed to bind to the GST-fusion protein matrix. This allows any antibodies that recognize the antigen to be captured on the solid support. Elution of the antibodies of interest is most often achieved using a low pH buffer such as glycine pH 2.8. The eluate is collected into a neutral tris or phosphate buffer, to neutralize the low pH elution buffer and halt any degradation of the antibody's activity. This is a nice example as affinity purification is used to purify the initial GST-fusion protein, to remove the undesirable anti-GST antibodies from the serum and to purify the target antibody.

A simplified strategy is often employed to purify antibodies generated against peptide antigens. When the peptide antigens are produced synthetically, a terminal cysteine residue is added at either the N- or C-terminus of the peptide. This cysteine residue contains a sulfhydryl functional group which allows the peptide to be easily conjugated to a carrier protein (e.g. Keyhole limpet hemocyanin (KLH)). The same cysteine-containing peptide is also immobilized onto an agarose resin through the cysteine residue and is then used to purify the antibody.

Most monoclonal antibodies have been purified using affinity chromatography based on immunoglobulin-specific Protein A or Protein G, derived from bacteria.

Immobilized Metal Ion Affinity Chromatography

Immobilized metal ion affinity chromatography (IMAC) is based on the specific coordinate covalent bond of amino acids, particularly histidine, to metals. This technique works by allowing proteins with an affinity for metal ions to be retained in a column containing immobilized metal ions, such as cobalt, nickel, copper for the purification of histidine containing proteins or peptides, iron, zinc or gallium for the purification of phosphorylated proteins or peptides. Many naturally occurring proteins do not have an affinity for metal ions, therefore recombinant DNA technology can be used to introduce such a protein tag into the relevant gene. Methods used to elute the protein of interest include changing the pH, or adding a competitive molecule, such as imidazole.

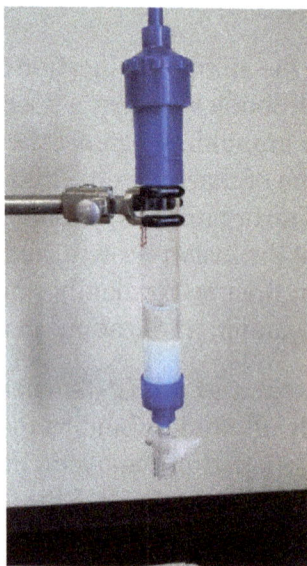

A chromatography column containing nickel-agarose beads used for purification of proteins with histidine tags

Recombinant Proteins

Possibly the most common use of affinity chromatography is for the purification of recombinant proteins. Proteins with a known affinity are protein tagged in order to aid their purification. The protein may have been genetically modified so as to allow it to be selected for affinity binding; this is known as a fusion protein. Tags include glutathione-S-transferase (GST), hexahistidine (His), and maltose binding protein (MBP). Histidine tags have an affinity for nickel or cobalt ions which have been immobilized by forming coordinate covalent bonds with a chelator incorporated in the stationary phase. For elution, an excess amount of a compound able to act as a metal ion ligand, such as imidazole, is used. GST has an affinity for glutathione which is commercially available immobilized as glutathione agarose. During elution, excess glutathione is used to displace the tagged protein.

Lectins

Lectin affinity chromatography is a form of affinity chromatography where lectins are used to separate components within the sample. Lectins, such as Concanavalin A are proteins which can bind specific carbohydrate (sugar) molecules. The most common application is to separate glycoproteins from non-glycosylated proteins, or one glycoform from another glycoform.

Ion Chromatography

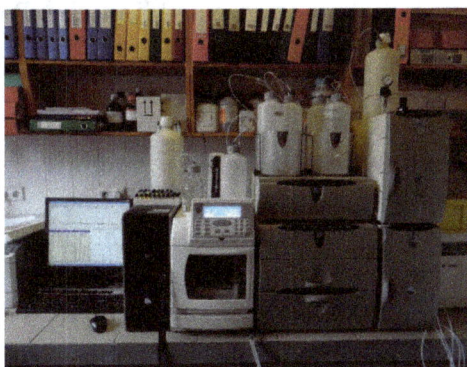

Dionex ICS-3000 ion chromatography system

Ion chromatography (or ion-exchange chromatography) is a chromatography process that separates ions and polar molecules based on their affinity to the ion exchanger. It works on almost any kind of charged molecule—including large proteins, small nucleotides, and amino acids. It is often used in protein purification, water analysis, and quality control.The water-soluble and charged molecules such as proteins, amino acids, and peptides bind to moieties which are oppositely charged by forming ionic bonds to the insoluble stationary phase. The equilibrated stationary phase consists of an ionizable functional group where the targeted molecules of a mixture to be separated and quantified can bind while passing through the column. This method applies the idea of the interaction between molecules and the stationary phase which are charged oppositely to each other. Cation exchange chromatography is used when the desired molecules to separate are cations; and anion exchange chromatography is used to separate anions, meaning that the beads in the column contain positively charged functional groups to attract the anions. The bound molecules then can be eluted and collected using an eluant which contains anions and cations by running higher concentration of ions through the column or changing pH of the column. One of the primary advantages for the use of ion chromatography is only one interaction involved during the separation as opposed to other separation techniques; therefore, ion chromatography may have higher matrix tolerance. However, there are also disadvantages involved when performing ion-exchange chromatography, such as constant evolution with the technique which leads to the inconsistency from column to column.

History

Table: History of ion exchange and ion chromatography, the analytical technique based on ion exchange

ca. 1850	Soil as an ion exchanger for Mg2+, Ca2+ and NH4+	Thomson & Way
1935	Sulfonated and aminated condensation polymers (phenol/formaldehyde)	Adams, Homes
1942	Sulfonated PS/DVB resin as cation exchanger (Manhattan Project)	d'Alelio
1947	Aminated PS/DVB resin as anion exchanger	McBurney
1953	Ion exclusion chromatography	Wheaton, Baumann
1957	Macroporous ion exchangers	Corte, Meyer, Kunin et al.
1959	Basic Theoretical principles	Helfferich
1967-70	Pellicular ion exchangers	Horvath, Kirkland
1975	Ion exchange chromatography with conductivity detection using a "stripper"	Small, Stevens, Bauman
1979	Conductivity detection without a "stripper"	Gjerde, Fritz, Schmuckler
1976-80	Ion pair chromatography	Waters, Bidlingmeier, Horvath et al.

Ion exchange chromatography

Ion chromatography has advanced through the accumulation of knowledge over a course of many years. Starting from 1947, Spedding and Powell used displacement ion-exchange chromatography for the separation of the rare earths. Additionally, they showed the ion-exchange separation of 14N and 15N isotopes in ammonia. At the start of the 1950s, Kraus and Nelson demonstrated the use of many analytical methods for metal ions dependent on their separation of their chloride, fluoride, nitrate or sulfate complexes by anion chromatography. Automatic in-line detection was progressively introduced from 1960 to 1980 as well as novel chromatographic methods for metal ion separations. A groundbreaking method by Small, Stevens and Bauman at Dow Chemical Co. unfolded the creation of the modern ion chromatography. Anions and cations could now be separated efficiently by a system of suppressed conductivity detection. In 1979, a method for anion chromatography with non-suppressed conductivity detection was introduced by Gjerde et al. Following it in 1980, was a similar method for cation chromatography.

As a result, a period of extreme competition began within the IC market, with supporters for both suppressed and nonsuppressed conductivity detection. This competition led to fast growth of new forms and the fast evolution of IC. A challenge that needs to be overcome in the future development of IC is the preparation of highly efficient monolithic ion-exchange columns and overcoming this challenge would be of great importance to the development of IC.

The boom of Ion exchange chromatography primarily began between 1935-1950 during World War II and it was through the "Manhattan project" that applications and IC were significantly extended. Ion chromatography was originally introduced by two English researchers, agricultural Sir Thompson and chemist J T Way. The works of Thompson and Way involved the action of water-soluble fertilizer salts, ammonium sulfate and potassium chloride. These salts could not easily be extracted from the ground due to the rain. They performed ion methods to treat clays with the salts, resulting in the extraction of ammonia in addition to the release of calcium. It was in the fifties and sixties that theoretical models were developed for IC for further understanding and it was not until the seventies that continuous detectors were utilized, paving the path for the development from low-pressure to high-performance chromatography. Not until 1975 was "ion chromatography" established as a name in reference to the techniques, and was thereafter used as a name for marketing purposes. Today IC is important for investigating aqueous systems, such as drinking water. It is a popular method for analyzing anionic elements or complexes that help solve environmentally relevant problems. Likewise, it also has great uses in the semiconductor industry.

Because of the abundant separating columns, elution systems, and detectors available, chromatography has developed into the main method for ion analysis.

When this technique was initially developed, it was primarily used for water treatment. Since 1935, ion exchange chromatography rapidly manifested into one of the most heavily leveraged techniques, with its principles often being applied to majority of fields of chemistry, including distillation, adsorption, and filtration.

Principle

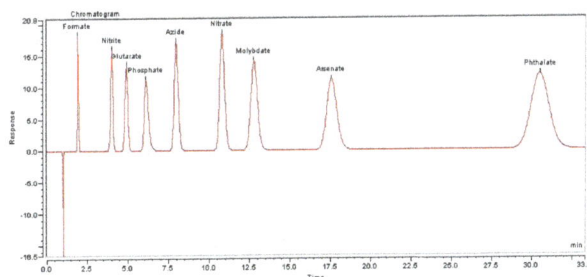

Ion Chromatography

Ion-exchange chromatography separates molecules based on their respective charged groups. Ion-exchange chromatography retains analyte molecules on the column based on coulombic (ionic) interactions. Essentially, molecules undergo electrostatic interactions with opposite charges on the stationary phase matrix. The stationary phase consists of an immobile matrix that contains charged ionizable functional groups or ligands. The stationary phase surface displays ionic functional groups (R-X) that interact with analyte ions of opposite charge. To achieve electroneutrality, these inert charges couple with exchangeable counterions in the solution. Ionizable molecules that are to be purified compete with these exchangeable counterions for binding to the immobilized charges on the stationary phase. These ionizable molecules are retained or eluted based on their charge. Initially, molecules that do not bind or bind weakly to the stationary phase are first to wash away. Altered conditions are needed for the elution of the molecules that bind to the stationary phase. The concentration of the exchangeable counterions, which competes with the molecules for binding, can be increased or the ph can be changed. A change in ph changes the charge on the particular molecules and, therefore, alters binding. The molecules then start eluting out based on the changes in their charges from the adjustments. Further such adjustments can be used to release the protein of interest. Additionally, concentration of counterions can be gradually varied to separate ionized molecules. This type of elution is called gradient elution. On the other hand, step elution can be used in which the concentration of counterions are varied in one step. This type of chromatography is further subdivided into cation exchange chromatography and anion-exchange chromatography. Positively charged molecules bind to anion exchange resins while negatively charged molecules bind to cation exchange resins. The ionic compound consisting of the cationic species M+ and the anionic species B- can be retained by the stationary phase.

Cation exchange chromatography retains positively charged cations because the stationary phase displays a negatively charged functional group:

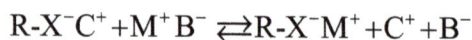

$$R\text{-}X^-C^+ + M^+B^- \rightleftharpoons R\text{-}X^-M^+ + C^+ + B^-$$

Anion exchange chromatography retains anions using positively charged functional group:

$$R\text{-}X^+A^- + M^+B^- \rightleftharpoons R\text{-}X^+B^- + M^+ + A^-$$

Note that the ion strength of either C+ or A- in the mobile phase can be adjusted to shift the equilibrium position, thus retention time.

The ion chromatogram shows a typical chromatogram obtained with an anion exchange column.

Procedure

It is possible to perform ion exchange chromatography in bulk, on thin layers of medium such as glass or plastic plates coated with a layer of the desired stationary phase, or in chromatography columns. Thin layer chromatography or column chromatography share similarities in that they both act within the same governing principles; there is constant and frequent exchange of molecules as the mobile phase travels along the stationary phase. It is not imperative to add the sample in minute volumes as the predetermined conditions for the exchange column have been chosen so that there will be strong interaction between the mobile and stationary phases. Furthermore, the mechanism of the elution process will cause a compartmentalization of the differing molecules based on their respective chemical characteristics. This phenomenon is due to an increase in salt concentrations at or near the top of the column, thereby displacing the molecules at that position, while molecules bound lower are released at a later point when the higher salt concentration reaches that area. These principles are the reasons that ion exchange chromatography is an excellent candidate for initial chromatography steps in a complex purification procedure as it can quickly yield small volumes of target molecules regardless of a greater starting volume.

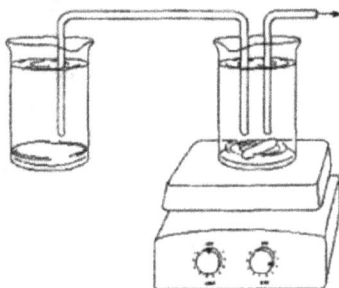

Chamber (left) contains high salt concentration. Stirred chamber (right) contains low salt concentration. Gradual stirring causes the formation of a salt gradient as salt travel from high to low concentrations.

Comparatively simple devices are often used to apply counterions of increasing gradient to a chromatography column. Counterions such as Copper(II) are chosen most often for effectively separating peptides and amino acids through complex formation.

A simple device can be used to create a salt gradient. Elution buffer is consistently being drawn from the chamber into the mixing chamber, thereby altering its buffer concentration. Generally, the buffer placed into the chamber is usually of high initial concentration, whereas the buffer placed into the stirred chamber is usually of low concentration. As the high concentration buffer from the left chamber is mixed and drawn into the column, the buffer concentration of the stirred column gradually increase. Altering the shapes of the stirred chamber, as well as of the limit buffer, allows for the production of concave, linear, or convex gradients of counterion.

A multitude of different mediums are used for the stationary phase. Among the most common immobilized charged groups used are trimethylaminoethyl (TAM), triethyl-aminoethyl (TEAE), diethyl-2-hydroxypropylaminoethyl (QAE), aminoethyl (AE), diethylaminoethyl (DEAE), sulpho (S), sulphomethyl (SM), sulphopropyl (SP), carboxy (C), and carboxymethyl (CM).

Successful column packing is an important aspect of ion chromatography. Stability and efficiency of a final column depends on packing methods, solvent used, and factors that affect mechanical properties of the column. In contrast to early inefficient dry- packing methods, wet slurry packing, in which particles that are suspended in an appropriate solvent are delivered into a column under pressure, shows significant improvement. Three different approaches can be employed in performing wet slurry packing: the balanced density method (solvent's density is about that of porous silica particles), the high viscosity method (a solvent of high viscosity is used), and the low viscosity slurry method (performed with low viscosity solvents).

Polystyrene is used as a medium for ion- exchange. It is made from the polymerization of styrene with the use of divinylbenzene and benzoyl peroxide. Such exchangers form hydrophobic interactions with proteins which can be irreversible. Due to this property, polystyrene ion exchangers are not suitable for protein separation. They are used on the other hand for the separation of small molecules in amino acid separation and removal of salt from water. Polystyrene ion exchangers with large pores can be used for the separation of protein but must be coated with a hydrophillic substance.

Cellulose based medium can be used for the separation of large molecules as they contain large pores. Protein binding in this medium is high and has low hydrophobic character. DEAE is an anion exchange matrix that is produced from a positive side group of diethylaminoethyl bound to cellulose or Sephadex.

Agarose gel based medium contain large pores as well but their substitution ability is lower in comparison to dextrans. The ability of the medium to swell in liquid is based on the cross-linking of these substances, the pH and the ion concentrations of the buffers used.

Incorporation of high temperature and pressure allows a significant increase in the efficiency of ion chromatography, along with a decrease in time. Temperature has an influence of selectivity due to its effects on retention properties. The retention factor $(k = (t_R^g - t_M^g)/(t_M^g - t_{ext}))$ increases with temperature for small ions, and the opposite trend is observed for larger ions.

Despite ion selectivity in different mediums, further research is being done to perform ion exchange chromatography through the range of 40-175 °C.

An appropriate solvent can be chosen based on observations of how column particles behave in a solvent. Using an optical microscope, one can easily distinguish a desirable dispersed state of slurry from aggregated particles.

Weak and Strong Ion Exchangers

A "strong" ion exchanger will not lose the charge on its matrix once the column is equilibrated and so a wide range of pH buffers can be used. "Weak" ion exchangers have a range of pH values in which they will maintain their charge. If the pH of the buffer used for a weak ion exchange column goes out of the capacity range of the matrix, the column will lose its charge distribution and the molecule of interest may be lost. Despite the smaller pH range of weak ion exchangers, they are often used over strong ion exchangers due to their having greater specificity. In some experiments, the retention times of weak ion exchangers are just long enough to obtain desired data at a high specificity.

Resins of ion exchange columns may include functional groups such as weak/strong acids and weak/strong bases. There are also special columns that have resins with amphoteric functional groups that can exchange both cations and anions. Examples of functional groups of strong ion exchange resins are quaternary ammonium (Q), which is an anion exchanger, and sulfonic acid (S), which is a cation exchanger. These types of exchangers can maintain their charge density over a pH range of 0-14. Examples of functional groups of Weak ion exchange resins include diethylaminoethyl (DEAE), which is an anion exchanger, and carboxymethyl (CM), which is a cation exchanger. These two types of exchangers can maintain the charge density of their columns over a pH range of 5-9.

In ion chromatography, the interaction of the solute ions and the stationary phase based on their charges determines which ions will bind and to what degree. When the stationary phase features positive groups which attracts anions, it is called an anion exchanger; when there are negative groups on the stationary phase, cations are attracted and it is a cation exchanger. The attraction between ions and stationary phase also depends on the resin, organic particles used as ion exchangers.

Each resin features relative selectivity which varies based on the solute ions present who will compete to bind to the resin group on the stationary phase. The selectivity coefficient, the equivalent to the equilibrium constant, is determined via a ratio of the concentrations between the resin and each ion, however, the general trend is that ion exchangers prefer binding to the ion with a higher charge, smaller hydrated radius, and higher polarizability, or the ability for the electron cloud of an ion to be disrupted by other charges. Despite this selectivity, excess amounts of an ion with a lower selectivity introduced to the column would cause the lesser ion to bind more to the stationary phase as the selectivity coefficient allows fluctuations in the binding reaction that takes place during ion exchange chromatography.

Typical Technique

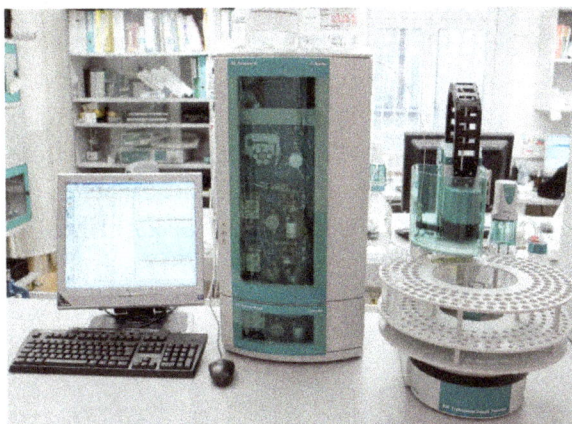

Metrohm 850 Ion chromatography system

A sample is introduced, either manually or with an autosampler, into a sample loop of known volume. A buffered aqueous solution known as the mobile phase carries the sample from the loop onto a column that contains some form of stationary phase material. This is typically a resin or gel matrix consisting of agarose or cellulose beads with covalently bonded charged functional groups. Equilibration of the stationary phase is needed in order to obtain the desired charge of the column. If the column is not properly equilibrated the desired molecule may not bind strongly to the column. The target analytes (anions or cations) are retained on the stationary phase but can be eluted by increasing the concentration of a similarly charged species that displaces the analyte ions from the stationary phase. For example, in cation exchange chromatography, the positively charged analyte can be displaced by adding positively charged sodium ions. The analytes of interest must then be detected by some means, typically by conductivity or UV/visible light absorbance.

Control an IC system usually requires a chromatography data system (CDS). In addition to IC systems, some of these CDSs can also control gas chromatography (GC) and HPLC.

Membrane Exchange Chromatography

A type of ion exchange chromatography, membrane exchange is a relatively new method of purification designed to overcome limitations of using columns packed with beads. Membrane Chromatographic devices are cheap to mass-produce and disposable unlike other chromatography devices that require maintenance and time to revalidate. There are three types of membrane absorbers that are typically used when separating substances. The three types are flat sheet, hollow fibre, and radial flow. The most common absorber and best suited for membrane chromatography is multiple flat sheets because it has more absorbent volume. It can be used to overcome mass transfer limitations and pressure drop, making it especially advantageous for isolating and purifying viruses, plasmid DNA, and other large macromolecules. The column is packed with microporous membranes with internal pores which contain adsorptive moieties that can bind the target protein. Adsorptive membranes are available in a variety of geometries and chemistry which allows them to be used for purification and also fractionation, concentration, and clarification in an efficiency that is 10 fold that of using beads. Membranes can be prepared through isolation of the membrane itself, where membranes are cut into squares and immobilized. A more recent method involved the use of live cells that are attached to a support membrane and are used for identification and clarification of signaling molecules.

Separating Proteins

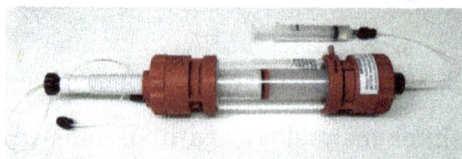

Preparative-scale ion exchange column used for protein purification.

In ion exchange chromatography can be used to separate proteins because proteins are composed of charged functional groups. The ions of interest (in this case charged proteins) are exchanged for another ion, (which is usually salt) on a charged solid support. Usually the solutes are in a liquid phase, which tends to be water. Take for example proteins in water, which would be a liquid phase that is passed through a column. The column is commonly known as the solid phase since it is filled with porous synthetic particles that are of a particular charge. These porous particles are also referred to as beads, may be aminated (containing amino groups) or have metal ions in order to have a charge. The column can be prepared using porous polymers, for macromolecules over 105 the optimum size of the porous particle is about 1μm2. This is because slow diffusion of the solutes within the pores does not restrict the separation quality. . The beads containing positively charged groups are attracted to the negatively charged protein, commonly referred to as anion exchange resins. Proteins that have negatively at pH 7 (pH of water) charged side chains are glutamate and aspartate. Beads that are negatively charged will be cation exchange resins, as positively charged protein will attracted. The proteins that have positively charged side chains at pH 7 are lysine, histidine and asparagine.

The isoelectric point is the pH at which a compound in this case a protein has no net charge. A protein's isoelectric point or PI can be determines using the pka of the side chains, if the amino (positive chain) is able to cancel out the carbonyl (negative) chain, the protein would be at its PI. Using buffers instead of water for proteins that do not have a charge at pH 7, is a good idea as it enables the manipulation of pH to alter ionic interactions between the proteins and the beads. Weakly acid or basic proteins (such as leucine, proline, alanine, valine, glycine...etc.) are able to have a charge of the pH is high enough to deprotonate the amino group. Separation can be achieved based on the natural isoelectric point of the protein. Alternatively a peptide tag can be genetically added to the protein to give the protein an isoelectric point away from most natural proteins (e.g., 6 arginines for binding to a cation-exchange resin or 6 glutamates for binding to an anion-exchange resin such as DEAE-Sepharose).

Svec, F., & Frechet, J. M. (1992). Continuous rods of macroporous polymer as high-performance liquid chromatography separation media. Analytical Chemistry Anal. Chem., 64(7), 820-822. doi:10.1021/ac00031a022 Garrett, Reginald H.; Grisham, Charles M. (2009). Biochemistry (4th ed. ed.). Pacific Grove, Calif.: Brooks/Cole. p. 71-75. ISBN 978-0-495-11464-2. Dasgupta, P. K. (1992). Ion Chromatography The State of the Art. Analytical Chemistry Anal. Chem., 64(15). doi:10.1021/ac00039a722 Elution by increasing ionic strength of the mobile phase is more subtle. It works because ions from the mobile phase interact with the immobilized ions on the stationary phase, thus "shielding" the stationary phase from the protein, and letting the protein elute.

Elution from ion-exchange columns can be sensitive to changes of a single charge-chromatofocusing. Ion-exchange chromatography is also useful in the isolation of specific multimeric protein assemblies, allowing purification of specific complexes according to both the number and the position of charged peptide tags. Svec, F., & Frechet, J. M. (1992). Continuous rods of macroporous polymer as high-performance liquid chromatography separation media. Analytical Chemistry Anal. Chem., 64(7), 820-822. doi:10.1021/ac00031a022 Garrett, Reginald H.; Grisham, Charles M. (2009). Biochemistry (4th ed. ed.). Pacific Grove, Calif.: Brooks/Cole. p. 71-75. ISBN 978-0-495-11464-2.

Gibbs-Donnan Effect

In ion exchange chromatography, the Gibbs–Donnan effect is observed when the pH of the applied buffer and the ion exchanger differ, even up to one pH unit. For example, in anion-exchange columns, the ion exchangers repeal protons so the pH of the buffer near the column differs is higher than the rest of the solvent. As a result, an experimenter has to be careful that the protein(s) of interest is stable and properly charged in the "actual" pH.

This effect comes as a result of two similarly charged particles, one from the resin and one from the solution, failing to distribute properly between the two sides; there is a selective uptake of one ion over another. For example, in a sulphonated polystyrene resin, a cation exchange resin, the chlorine ion of a hydrochloric acid buffer should equilibrate into the resin. However, since the concentration of the sulphonic acid in the resin is high, the hydrogen of HCl has no tendency to enter the column. This, combined with the need of electroneutrality, leads to a minimum amount of hydrogen and chlorine entering the resin.

Uses

Clinical Utility

A use of ion chromatography can be seen in the argentation ion chromatography.Usually silver and compounds containing acetylenic and ethylenic bonds have very weak interactions. This phenomenon has been widely tested on olefin compounds. The ion complexes the olefins make with silver ions are weak and made based on the overlapping of pi, sigma, and d orbitals and available electrons therefore cause no real changes in the double bond. This behavior was manipulated to separate lipids, mainly fatty acids from mixtures in to fractions with differing number of double bonds using silver ions. The ion resins were impregnated with silver ions, which were then exposed to various acids (silicic acid) to elute fatty acids of different characteristics.

Detection limits as low as 1 μM can be obtained for alkali metal ions. It may be used for measurement of HbA1c, porphyrin and with water purification. Ion Exchange Resins(IER) have been widely used especially in medicines due to its high capacity and the uncomplicated system of the separation process. One of the synthetic uses is to use Ion Exchange Resins for kidney dialysis. This method is used to separate the blood elements by using the cellulose membraned artificial kidney.

Another clinical application of ion chromatography is in the rapid anion exchange chromatography technique used to separate creatine kinase (CK) isoenzymes from human serum and tissue sourced in autopsy material (mostly CK rich tissues were used such as cardiac muscle and brain).These isoenzymes include MM, MB, and BB, which all carry out the same function given different amino acid sequences. The functions of these isoenzymes are to convert creatine, using ATP, into phosphocreatine expelling ADP. Mini columns were filled with DEAE-Sephadex A-50 and further eluted with tris- buffer sodium chloride at various concentrations (each concentration was chosen advantageously to manipulate elution). Human tissue extract was inserted in columns for separation. All fractions were analyzed to see total CK activity and it was found that each source of CK isoenzymes had characteristic isoenzymes found within. Firstly, CK-MM was eluted, then CK-MB, followed by CK-BB. Therefore, the isoenzymes found in each sample could be used to identify the source, as they were tissue specific.

Using the information from results, correlation could be made about the diagnosis of patients and the kind of CK isoenzymes found in most abundant activity. From the finding, about 35 out of 71 patients studied suffered from heart attack (myocardial infarction) also contained an abundant amount of the CK-MM and CK-MB isoenzymes. Findings further show that many other diagnosis including renal failure, cerebrovascular disease, and pulmonary disease were only found to have the CK-MM isoenzyme and no other isoenzyme. The results from this study indicate correlations between various diseases and the CK isoenzymes found which confirms previous test results using various techniques. Studies about CK-MB found in heart attack victims have expanded since this study and application of ion chromatography.

Industrial Applications

Since 1975 ion chromatography has been widely used in many branches of industry. The main beneficial advantages are reliability, very good accuracy and precision, high selectivity, high speed, high separation efficiency, and low cost of consumables. The most significant development related to ion chromatography are new sample preparation methods; improving the speed and selectivity of analytes separation; lowering of limits of detection and limits of quantification; extending the scope of applications; development of new standard methods; miniaturization and extending the scope of the analysis of a new group of substances. Allows for quantitative testing of electrolyte and proprietary additives of electroplating baths. It is an advancement of qualitative hull cell testing or less accurate UV testing. Ions, catalysts, brighteners and accelerators can be measured. Ion exchange chromatography has gradually become a widely known, universal technique for the detection of both anionic and cationic species. Applications for such purposes have been developed, or are under development, for a variety of fields of interest, and in particular, the pharmaceutical industry. The usage of ion exchange chromatography in pharmaceuticals has increased in recent years, and in 2006, a chapter on ion exchange chromatography was officially added to the United States Pharmacopia-National Formulary (USP-NF). Furthermore, in 2009 release of the USP-NF, the United States Pharmacopia made several analyses of ion chromatography available using two techniques: conductivity detection, as well as pulse amperometric detection. Majority of these applications are primarily used for measuring and analyzing residual limits in pharmaceuticals, including detecting the limits of oxalate, iodide, sulfate, sulfamate, phosphate, as well as various electrolytes including potassium, and sodium. In total, the 2009 edition of the USP-NF officially released twenty eight methods of detection for the analysis of active compounds, or components of active compounds, using either conductivity detection or pulse amperometric detection.

Drug Development

There has been a growing interest in the application of IC in the analysis of pharmaceutical drugs. IC is used in different aspects of product development and quality control testing. For example, IC is used to improve stabilities and solubility properties of pharmaceutical active drugs molecules as well as used to detect systems that have higher tolerance for organic solvents. IC has been used for the determination of analytes as a part of a dissolution test. For instance, calcium dissolution tests have shown that other ions present in the medium can be well resolved among themselves and also from the calcium ion. Therefore, IC has been employed in drugs in the form of tablets and capsules in order to determine the amount of drug dissolve with time. IC is also widely used for detection and quantification of excipients or inactive ingredients used in pharmaceutical formulations. Detection of sugar and sugar alcohol in such formulations through IC has been done due to these polar groups getting resolved in ion column. IC methodology also established in analysis of impurities in drug substances and products. Impurities or any components that are not part of the drug chemical entity are evaluated and they give insights about the maximum and minimum amounts of drug that should be administered in a patient per day.

Anion-exchange Chromatography

Anion-exchange chromatography is a process that separates substances based on their charges using an ion-exchange resin containing positively charged groups, such as diethyl-aminoethyl groups (DEAE). In solution, the resin is coated with negatively charged counter-ions (anions). Anion exchange resins will bind to negatively charged molecules, displacing the counter-ion. Anion exchange chromatography is commonly used to purify proteins, amino acids, sugars/carbohydrates and other acidic substances with a negative charge at higher pH levels. The tightness of the binding between the substance and the resin is based on the strength of the negative charge of the substance.

General Technique for Protein Purification

A slurry of resin, such as DEAE-Sephadex is poured into the column. After it settles, the column is pre-equilibrated in buffer before the protein mixture is applied. Unbound proteins are collected in the flow-through and/or in subsequent buffer washes. Proteins that bind to the resin are retained and can be eluted in one of two ways. First, the salt concentration in the elution buffer is gradually increased. The negative ions in the salt solution (e.g. Cl-) compete with protein in binding to the resin. Second, the pH of the solution can be gradually decreased which results in a more positive charge on the protein, releasing it from the resin. As buffer elutes from the column, the samples are collected using a fraction collector.

Mixed-mode Chromatography

Mixed-mode chromatography (MMC), or multimodal chromatography, refers to chromatographic methods that utilize more than one form of interaction between the stationary phase and analytes in order to achieve their separation. What is distinct from conventional single-mode chromatography is that the secondary interactions in MMC cannot be too weak, and thus they also contribute to the retention of the solutes.

History

Before MMC was considered as a chromatographic approach, secondary interactions were generally believed to be the main cause of peak tailing. However, it was discovered afterwards that secondary interactions can be applied for improving separation power. In 1986, Regnier's group firstly synthesized a stationary phase that had characteristics of anion exchange chromatography (AEX) and hydrophobic interaction chromatography (HIC) on protein separation. In 1998, a new form of MMC, hydrophobic charge induction chromatography (HCIC), was proposed by Burton and Harding. In the same year, conjoint liquid chromatography (CLC), which combines different types of monolithic convective interaction media (CIM) disks in the same housing, was introduced by Štrancar et al. In 1999, Yates' group loaded strong-cation exchange (SCX) and reversed phase liquid chromatography (RPLC) stationary phases sequentially into a capillary column coupled with tandem mass spectrometry (MS/MS) in the analysis of peptides, which became one of the most efficient technique in proteomics afterwards. In 2009, Geng's group first achieved online two-dimensional (2D) separation of intact proteins using a single column possessing separation features of weak-cation exchange chromatography (WCX) and HIC (termed as two-dimensional liquid chromatography using a single column, (2D-LC-1C).

Advantages

Higher selectivity: for example, positive, negative and neutral substances could be separated by a reversed phase (RP)/anion-cation exchange (ACE) column in a single run. Higher loading capacity, for example, loading capacity of ACE/ hydrophilic interaction chromatography (HILIC) increased 10-100 times when compared with RPLC, which offered a new selection and idea for developing semi-preparative and preparative chromatography.

One mixed-mode column can replace two or even more single mode columns, which is economic and eco-friendly for employing the stationary phase more sufficiently and reducing the consuming and 'waste' of raw materials.

Single mixed-mode column can be applied for on-line two-dimensional (2D) analysis in a sealed system via establishing corresponding chromatographic system or off-line 2D analysis as two columns.

Classification of MMC

MMC can be classified into physical MMC and chemical MMC. In the former method, the stationary phase is constructed of two or more types of packing materials. In the chemical method, just one type of packing material containing two or more functionalities is used.

Physical Methods

The simplest approach is to connect two commercial columns in series, which is termed a "tandem column". Another approach is "biphasic column", by packing two stationary phases separately in two ends of the same column. The third approach is to homogenize two or more different types of stationary phases in a single column, which is termed a "hybrid column" or "mixed-bed column".

Chemical Methods

IEC/HIC

Since IEC and HIC conditions are the closest ones to physiological conditions which are fit for maintaining biological activity, the combinations of them are widely used in the separation of biological products. IEC/HIC MMC has improved separation power and selectivity on the grounds that it applies both electrostatic and hydrophobic interactions. One such example is Nuvia(tm) cPrime(tm), from Bio-Rad Laboratories.

IEC/RPLC

IEC/RP MMC combines the advantages of RPLC and IEC. For example, WAX/RP has increased separation power and degree of freedom in adjusting the separation selectivity when compared with single WAX or RPLC.

HILIC/RPLC

Liu et al. synthesized a HILIC/RP stationary phase which could show RPLC or HILIC retention by adjusting the organic phase in mobile phase.

HILIC/IEC

Mant et al. reported that HILIC/CEX offered unique selectivity, stronger separation power and wider range of applications compared to RPLC for peptide separations.

SEC/IEC

Hydrophobic interactions in protein SEC are relatively weak at low ionic strength, electrostatic effects may contribute significantly to retention, and this allows us to use an SEC column as a weak ion exchanger.

Chromatography in Blood Processing

Chromatographic techniques have been used in blood processing and purification since the 1980s. It has emerged as an effective method of purifying blood components for therapeutic use.

Human Blood Plasma

Blood plasma is the liquid component of blood, which contains dissolved proteins, nutrients, ions, and other soluble components. In whole blood, red blood cells, leukocytes, and platelets are suspended within the plasma. The goal of plasma purification and processing is to extract specific materials that are present in blood, and use them for restoration and repair. There are several components that make up blood plasma, one of which is the protein albumin. Albumin is a highly water-soluble protein with considerable structural stability. It serves as a transportation device for materials such as hormones, enzymes, fatty acids, metal ions, and medicinal products. It is also used for therapeutic purposes, being essential in restoration and maintenance of circulating blood volume in imperative situations such as severe trauma or surgery. With little room for error, extremely pure samples that are lacking impurities needs to be at hand in good amount.

Development of Chromatography

Traditionally, the Cohn process incorporating cold ethanol fractionation has been used for albumin purification. However, chromatographic methods for separation started being adopted in the early 1980s. Developments were ongoing in the time period between when Cohn fractionation started being used, in 1946, and when chromatography started being used, in 1983. In 1962, the Kistler & Nistchmann process was created which was a spinoff of the Cohn process. Chromatographic processes began to take shape in 1983. In the 1990s, the Zenalb and the CSL Albumex processes were created which incorporated chromatography with a few variations.

The general approach to using chromatography for plasma fractionation for albumin is: recovery of supernatant I, delipidation, anion exchange chromatography, cation exchange chromatography, and gel filtration chromatography.

The recovered purified material is formulated with combinations of sodium octanoate and sodium N-acetyl tryptophanate and then subjected to viral inactivation procedures, including pasteurisation at 60 °C.

This is a more efficient alternative than the Cohn process for four main reasons: 1) smooth automation and a relatively inexpensive plant was needed, 2) easier to sterilize equipment and maintain a good manufacturing environment, 3) chromatographic processes are less damaging to the albumin protein, and 4) a more successful albumin end result can be achieved.

Compared with the Cohn process, the albumin purity went up from about 95% to 98% using chromatography, and the yield increased from about 65% to 85%. Small percentage increases make a difference in regard to sensitive measurements like purity. There is one big drawback in using chromatography, which has to do with the economics of the process. Although the method was efficient from the processing aspect, acquiring the necessary equipment is a big task. Large machinery is necessary, and for a long time the lack of equipment availability was not conducive to its widespread use. The components are more readily available now but it is still a work in progress.

Bridging Methods

Integrating traditional and modern methods is a useful way to process albumin.

There are three main steps that combine Cohn fractionation with chromatography: 1) factors I, II, and III are removed via cold ethanol fractionation, 2) Sepharose fast flow ion exchange and sepharose fast flow chromatography procedures are run, and 3) gel filtration is run. The result is albumin with 9% lower aluminum levels with a processing time that is almost twice as fast.

Although it was hard to make chromatographic processing methods widely adopted, global expansion is a work in progress. Various blood components must be readily available at various medical treatment centers around the world. The Institute of Transfusion Medicine in Skopje, Macedonia is a plasma fractionation center in the Balkans. Their modernized albumin purification process consists of five steps:

1. starting material is plasma that has been pretreated by centrifugation,

2. a round of gel filtration is run,

3. ion exchange on DEAE Sepharose is run to bind the albumin to the column,

4. albumin is eluted with a sodium acetate buffer, and

5. final polishing with gel filtration.

The end result is a highly pure and safe batch of albumin that is 100% non-pyrogenic, sterile, and free of active HIV virus. The product purity is greater than 98% and the protein content is about 50 g/L.

Non-chromatographic Processing Methods

Other plasma processing methods exist, but generally do not provide the resolution or purity of chromatographic methods. Two-phase liquid extraction may be performed using polyethylene glycol (PEG)-phosphate Aqueous two-phase systems, with a PEG-rich top layer and a phosphate-rich bottom layer. Although this method is somewhat useful for protein recovery, it does not work as well for the recovery of other blood components. Membrane fractionation has the advantage of minimal protein loss yet high removal of pathological plasma components. This method incorporates processes such as thermofiltration and applying pulsate flow. The latest two-stage membrane system utilizes a high flow recirculation circuit that is effective for removal of LDL cholesterol. It may prove useful for patients that have clogged arteries and other cardiovascular problems involving cholesterol. Batch adsorption, e.g. onto ion exchange media, is only useful when dealing with smaller samples of plasma, typically 200 mL or less. Batch adsorption recovers the product in a larger volume of elution buffer than does column chromatography or frontal chromatography, and the resulting more dilute product requires concentration, typically on a membrane system, which can lead to loss of product by irreversible adsorption to the membrane.

Reversed-phase Chromatography

Reversed-phase chromatography (also called RPC, reverse-phase chromatography, or hydrophobic chromatography) includes any chromatographic method that uses a hydrophobic stationary phase. RPC refers to liquid (rather than gas) chromatography.

The term "reversed-phase" has a historical background. In the 1970s, most liquid chromatography was performed using a solid support stationary phase (also called a "column") containing unmodified silica or alumina resins. This method is now called "normal phase chromatography". In normal phase chromatography, the stationary phase is hydrophilic and therefore has a strong affinity for hydrophilic molecules in the mobile phase. Thus, the hydrophilic molecules in the mobile phase tend to bind (or "adsorb") to the column, while the hydrophobic molecules pass through the column and are eluted first. In normal phase chromatography, hydrophilic molecules can be eluted from the column by increasing the polarity of the solution in the mobile phase.

The introduction of a technique using alkyl chains covalently bonded to the solid support created a hydrophobic stationary phase, which has a stronger affinity for hydrophobic compounds. The use of a hydro*phobic* stationary phase can be considered the opposite, or "reverse", of normal phase chromatography - hence the term "reversed-phase chromatography". Reversed-phase chromatography employs a polar (aqueous) mobile phase. As a result, hydrophobic molecules in the polar mobile phase tend to adsorb to the hydrophobic stationary phase, and hydro*philic* molecules in the mobile phase will pass through the column and are eluted first. Hydrophobic molecules can be eluted from the column by decreasing the polarity of the mobile phase using an organic (non-polar) solvent, which reduces hydrophobic interactions. The more hydrophobic the molecule, the more strongly it will bind to the stationary phase, and the higher the concentration of organic solvent that will be required to elute the molecule.

Many of the mathematical and experimental considerations used in other chromatographic methods also apply to RPC (for example, the separation resolution is dependent on the length of the column). It can be used for the separation of a wide variety of molecules. It is not typically used for separation of proteins, because the organic solvents used in RPC can denature many proteins. For this reason, normal phase chromatography is more commonly used for separation of proteins.

Today, RPC is a frequently used analytical technique. There are a variety of stationary phases available for use in RPC, allowing great flexibility in the development of separation methods.

Stationary Phases

Silica-based Stationary Phases

Any inert non-polar substance that achieves sufficient packing can be used for reversed-phase chromatography. The most popular column is an octadecyl carbon chain (C18)-bonded silica (USP classification L1) with 297 columns commercially available. This is followed by C8-bonded silica (L7 - 166 columns), pure silica (L3 - 88 columns), cyano-bonded silica (L10 - 73 columns) and phenyl-bonded silica (L11 - 72 columns). Note that C18, C8 and phenyl are dedicated reversed-phase resins, while cyano columns can be used in a reversed-phase mode depending on analyte and mobile phase conditions. It should be noted at this point that not all C18 columns have identical retention properties. Surface functionalization of silica can be performed in a monomeric or a polymeric reaction with different short-chain organosilanes used in a second step to cover remaining silanol groups (end-capping). While the overall retention mechanism remains the same, subtle differences in the surface chemistries of different stationary phases will lead to changes in selectivity.

Modern columns have different polarity. PFP is pentafluorphenyl. CN is cyano. NH2 is amino. ODS is octadecyl or C18. ODCN is a mixed mode column consisting of C18 and nitrile. SCX is strong cationic exchange (used for separation of organic amines). SAX is strong anionic exchange (used for separation of carboxylic acid compounds).

Mobile Phases

Mixtures of water or aqueous buffers and organic solvents are used to elute analytes from a reversed-phase column. The solvents must be miscible with water, and the most common organic solvents used are acetonitrile, methanol, and tetrahydrofuran (THF). Other solvents can be used such as ethanol or 2-propanol (isopropyl alcohol). Elution can be performed isocratically (the water-solvent composition does not change during the separation process) or by using a solution gradient (the water-solvent composition changes during the separation process, usually by decreasing the polarity). The pH of the mobile phase can have an important role on the retention of an analyte and can change the selectivity of certain analytes.

Charged analytes can be separated on a reversed-phase column by the use of ion-pairing (also called ion-interaction). This technique is known as reversed-phase ion-pairing chromatography.

Aqueous Normal-phase Chromatography

Aqueous normal-phase chromatography (ANP) is a chromatographic technique which involves the mobile phase region between reversed-phase chromatography (RP) and organic normal-phase chromatography (ONP).

Principle

In normal-phase chromatography, the stationary phase is polar and the mobile phase is nonpolar. In reversed phase we have just the opposite; the stationary phase is nonpolar and the mobile phase is polar. Typical stationary phases for normal-phase chromatography are silica or organic moieties with cyano and amino functional groups. For reversed phase, alkyl hydrocarbons are the preferred stationary phase; octadecyl (C18) is the most common stationary phase, but octyl (C8) and butyl (C4) are also used in some applications. The designations for the reversed phase materials refer to the length of the hydrocarbon chain.

In normal-phase chromatography, the least polar compounds elute first and the most polar compounds elute last. The mobile phase consists of a nonpolar solvent such as hexane or heptane mixed with a slightly more polar solvent such as isopropanol, ethyl acetate or chloroform. Retention decreases as the amount of polar solvent in the mobile phase increases. In reversed phase chromatography, the most polar compounds elute first with the most nonpolar compounds eluting last. The mobile phase is generally a binary mixture of water and a miscible polar organic solvent like methanol, acetonitrile or THF. Retention increases as the amount of the polar solvent (water) in the mobile phase increases. Normal phase chromatography, an adsorptive mechanism, is used for the analysis of solutes readily soluble in organic solvents, based on their polar differences such as amines, acids, metal complexes, etc.. Reversed-phase chromatography, a partition mechanism, is typically used for separations by non-polar differences.

The "hydride surface" distinguishes the support material from other silica materials; most silica materials used for chromatography have a surface composed primarily of silanols (-Si-OH). In a "hydride surface" the terminal groups are primarily -Si-H. The hydride surface can also be functionalized with carboxylic acids and long-chain alkyl groups. Mobile phases for ANPC are based on an organic solvent (such as methanol or acetonitrile) with a small amount of water; thus, the mobile phase is both "aqueous" (water is present) and "normal" (less polar than the stationary phase). Thus, polar solutes (such as acids and amines) are most strongly retained, with retention decreasing as the amount of water in the mobile phase increases.

Typically the amount of the nonpolar component in the mobile phase must be 60% or greater with the exact point of increased retention depending on the solute and the organic component of the mobile phase. A true ANP stationary phase will be able to function in both the reversed phase and normal phase modes with only the amount of water in the eluent varying. Thus a continuum of solvents can be used from 100% aqueous to pure organic. ANP retention has been demonstrated for a variety of polar compounds on the hydride based stationary phases. Recent investigations have demonstrated that silica hydride materials have a very thin water layer (about 0.5 monolayer)in comparison to HILIC phases that can have from 6-8 monolayers . In addition the substantial negative charge on the surface of hydride phases is the result of hydroxide ion adsorption from the solvent rather than silanols .

Features

An interesting feature of these phases is that both polar and nonpolar compounds can be retained over some range of mobile phase composition (organic/aqueous). The retention mechanism of polar compounds has recently been shown to be the result of the formation of a hydroxide layer on the surface of the silica hydride . Thus positively charged analytes are attracted to the negatively charged surface and other polar analytes are likely to be retained through displacement of hydroxide or other charged species on the surface. This property distinguishes it from a pure HILIC (hydrophilic interaction chromatography) columns where separation by polar differences is obtained through partitioning into a water-rich layer on the surface, or a pure RP stationary phase on which separation by nonpolar differences in solutes is obtained with very limited secondary mechanisms operating.

Another important feature of the hydride-based phases is that for many analyses it is usually not necessary to use a high pH mobile phase to analyze polar compounds such as bases. The aqueous component of the mobile phase usually contains from 0.1 to 0.5% formic or acetic acid, which is compatible with detector techniques that include mass spectral analysis.

Centrifugal Partition Chromatography

Centrifugal partition chromatography (CPC) is a special chromatographic technique where both stationary and mobile phase are liquid, and the stationary phase is immobilized by a strong centrifugal force. CPC consists of a series-connected network of extraction cells, which operates as elemental extractors, and the efficiency is guaranteed by the cascade.

Operation

The extracion cells consists of hollow bodies with inlets and oulets of liquid connection. The cells are first filled with the liquid chosen to be the stationary phase. Under rotation, the pumping of the mobile phase is started, which enters the cells from the inlet. When entering the flow of mobiles phase forms small droplets according to the Stokes' law, which is called atomization. These droplets fall through the stationary phase, creating a high interfacial area, which is called the extraction. At the end of the cells, these droplets unite due to the surface tension, which is called settling.

When a sample mixture is injected as a plug into the flow of mobile phase the compounds of the mixtures elute according to their partition coefficients:

$$V_{elution} = V_{dead-volume} + K_{upper/lower} * V_{stationary-phase}$$

As CPC requies only a biphasic mixture of solvents, by varying the constitution of the solvent system it is possible to tune the partition coefficients of different compounds so, that separation is guaranteed by the high selectivity.

Comparison with Countercurrent Chromatography

Countercurrent chromatography (CCC) and centrifugal partition chromatography are two different instrumental realization of the same liquid–liquid chomatographic theory. CCC usually uses a planetary gear motion without rotary seals, while CPC uses circular rotation with rotary seals for liquid connection. CCC has interchagning mixing and settling zones in the coil tube, so atomization, extraction and settling are time and zone separated. Inside CPC all three steps happen continuously in one time, inside the cells.

Adventages of CPC:

- Higher flow rate for same volume size Laboratory scale example: 250 mL CPC has optimal flow rate of 5–15 mL/min, 250 mL CCC has optimal flow rate of 1–3 mL/min. Process scale example: 25 L CCC has optimal flow rate of 100–300 ml/min, 25 L CPC has optimal flow rate of 1000–3000 ml/min.

- Higher productivity (due to higher flow rate and faster separation time)

- Scalable up to tonnes per month

- Better stationary phase retention for most phases

Disadvantages of CPC:

- Higher pressure than CCC (typical operation pressures of 40160 bar vs 5–25 bar)

- Rotary seal wear over time

Laboratory Scale

CPC has been extensively used for isolation and purification of natural products for 40 years. Due to the ability to get very high selectivity, and the ability to tolerate samples containing particulated matter, it is possible to work with direct extracts of biomass, opposed to traditional liquid chromatography, where impurities degrade the solid stationary phase so that separation become impossible.

There are numerous laboratory scale CPC manufacturers around the world, like Armen Instrument (Gilson), RotaChrom, Kromaton (Rousselet Robatel), and AECS-QUIKPREP. These instruments operate at flow rates of 1–500 mL/min. with stationary phase retentions of 40–80% and column volumes of 25 mL to 25 L.

Production Scale

As CPC does not uses any solid stationary phase it guarantees a cost-effective separation for the highest industiral levels. Opposed to CCC it is possible to get very high flow rates (for example 10 liters / min) with active stationary phase ratio of >80%, which guarantees good separation and high productivity. As in CPC material is dissolved, and loaded the column in mass / volume units, loading capability can be much higher than standard solid-liquid chromatographic techniques, where material is loaded to the active surface area of the stationary phase, which takes up less than 10% of the column.

Industrial instrument differ from laboraotry scale instruments by the applicable flow rate with staisfactory stationary phase retention (70-90%). Industrial instruments have flow rates of multiple liter / minutes, while able to purify materials from 10 kg to tonnes per month.

Two-dimensional Chromatography

Two-dimensional chromatography is a type of chromatographic technique in which the injected sample is separated by passing through two different separation stages. This is done by injecting the eluent from the first column onto a second column. Typically the second column has a different separation mechanism, so that bands that are poorly resolved from the first column may be completely separated in the second column. (For instance, a C18 reversed-phase chromatography column may be followed by a phenyl column.) Alternately, the two columns might run at different temperatures. The second stage of the separation must be run much faster than the first, since there is still only a single detector. The plane surface is amenable to sequential development in two directions using two different solvents.

Examples

Two-dimensional separations can be carried out in gas chromatography or liquid chromatography. Various different coupling strategies have been developed to "resample" from the first column into the second.

The chief advantage of two-dimensional techniques is that they offer a large increase in peak capacity, without requiring extremely efficient separations in either column. (For instance, if the first column offers a peak capacity (k_1)of 100 for a 10-minute separation, and the second column offers a peak capacity of 5 (k_2) in a 5-second separation, then the combined peak capacity may approach $k_1 \times k_2 = 500$, with the total separation time still ~ 10 minutes). 2D separations have been applied to the analysis of gasoline and other petroleum mixtures, and more recently to protein mixtures.

Paper Chromatography

Paper chromatography is an analytical method used to separate colored chemicals or substances. It is primarily used as a teaching tool, having been replaced by other chromatography methods, such as thin-layer chromatography. A paper chromatography variant, two-dimensional chromatography involves using two solvents and rotating the paper 90° in between. This is useful for separating complex mixtures of compounds having similar polarity, for example, amino acids. The setup has three components. The mobile phase is a solution that travels up the stationary phase, due to capillary action. The mobile phase is generally an alcohol solvent mixture, while the stationary phase is a strip of chromatography paper, also called a chromatogram.

R_f Value, Solutes, and Solvents

The retention factor (R_f) may be defined as the ratio of the distance traveled by the substance to the distance traveled by the solvent. R_f values are usually expressed as a fraction of two decimal places. If R_f value of a solution is zero, the solute remains in the stationary phase and thus it is immobile. If R_f value = 1 then the solute has no affinity for the stationary phase and travels with the solvent front. To calculate the R_f value, take the distance traveled by the substance divided by the distance traveled by the solvent (as mentioned earlier in terms of ratios). For example, if a compound travels

9.9 cm and the solvent front travels 12.7 cm, (9.9/12.7) the R_f value = 0.779 or 0.78. R_f value depends on temperature and the solvent used in experiment, so several solvents offer several R_f values for the same mixture of compound. A solvent in chromatography is the liquid the paper is placed in, and the solute is the ink which is being separated.

Pigments and Polarity

Paper chromatography is one method for testing the purity of compounds and identifying substances. Paper chromatography is a useful technique because it is relatively quick and requires only small quantities of material. Separations in paper chromatography involve the same principles as those in thin layer chromatography, as it is a type of thin layer chromatography. In paper chromatography, substances are distributed between a stationary phase and a mobile phase. The stationary phase is the water trapped between the cellulose fibers of the paper. The mobile phase is a developing solution that travels up the stationary phase, carrying the samples with it. Components of the sample will separate readily according to how strongly they adsorb onto the stationary phase versus how readily they dissolve in the mobile phase.

When a colored chemical sample is placed on a filter paper, the colors separate from the sample by placing one end of the paper in a solvent. The solvent diffuses up the paper, dissolving the various molecules in the sample according to the polarities of the molecules and the solvent. If the sample contains more than one color, that means it must have more than one kind of molecule. Because of the different chemical structures of each kind of molecule, the chances are very high that each molecule will have at least a slightly different polarity, giving each molecule a different solubility in the solvent. The unequal solubility causes the various color molecules to leave solution at different places as the solvent continues to move up the paper. The more soluble a molecule is, the higher it will migrate up the paper. If a chemical is very non-polar it will not dissolve at all in a very polar solvent. This is the same for a very polar chemical and a very non-polar solvent.

It is very important to note that when using water (a very polar substance) as a solvent, the more polar the color, the higher it will rise on the paper.

Types of Paper Chromatography

1. Descending Paper Chromatography-In this type, development of the chromatogram is done by allowing the solvent to travel down the paper.

2. Ascending Paper Chromatography-Here the solvent travels up the chromatographic paper. Both Descending and Ascending Paper Chromatography are used for the separation of organic and inorganic substances.

3. Ascending-Descending Paper Chromatography-It is the hybrid of both of the above techniques. The upper part of Ascending Chromatography can be folded over a rod in order to allow the paper to become Descending after crossing the rod.

4. Radial Paper Chromatography-It is also called Circular Chromatography. Here a circular filter paper is taken and the sample is deposited at the center of the paper. After drying the spot, the filter paper is tied horizontally on a Petri dish containing solvent, so that the wick of the paper is dipped in the solvent. The solvent rises through the wick and the components are separated into concentric circles.

5. Two-Dimensional Paper Chromatography-In this technique a square or rectangular paper is used. Here the sample is applied to one of the corners and development is performed at right angle to the direction of the first run.

History

The discovery of paper chromatography in 1943 by Martin and Synge provided, for the first time, the means of surveying constituents of plants and for their separation and identification. There was an explosion of activity in this field after 1945.

Thin-layer Chromatography

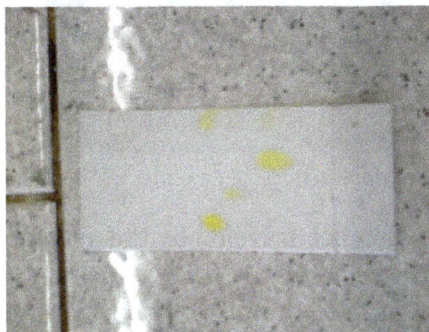

TLC of three standards (ortho-, meta- and para-isomers) and a sample

Thin-layer chromatography (TLC) is a chromatography technique used to separate non-volatile mixtures. Thin-layer chromatography is performed on a sheet of glass, plastic, or aluminium foil, which is coated with a thin layer of adsorbent material, usually silica gel, aluminium oxide (alumina), or cellulose. This layer of adsorbent is known as the stationary phase.

After the sample has been applied on the plate, a solvent or solvent mixture (known as the mobile phase) is drawn up the plate via capillary action. Because different analytes ascend the TLC plate at different rates, separation is achieved. The mobile phase has different properties from the stationary phase. For example, with silica gel, a very polar substance, non-polar mobile phases such as heptane are used. The mobile phase may be a mixture, allowing chemists to fine-tune the bulk properties of the mobile phase.

After the experiment, the spots are visualized. Often this can be done simply by projecting ultraviolet light onto the sheet; the sheets are treated with a phosphor, and dark spots appear on the sheet where compounds absorb the light impinging on a certain area. Chemical processes can also be used to visualize spots; anisaldehyde, for example, forms colored adducts with many compounds, and sulfuric acid will char most organic compounds, leaving a dark spot on the sheet.

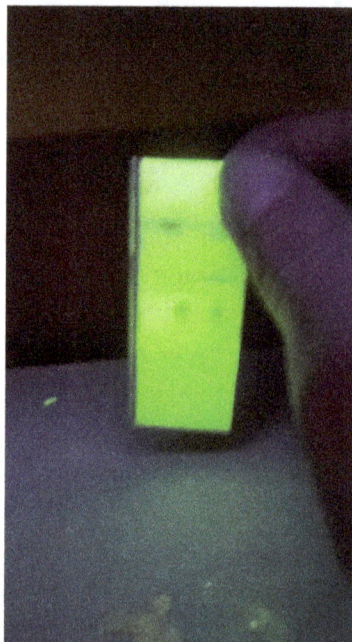

fluorescent tlc plate under UV light

To quantify the results, the distance traveled by the substance being considered is divided by the total distance traveled by the mobile phase. (The mobile phase must not be allowed to reach the end of the stationary phase.) This ratio is called the retention factor or R_f. In general, a substance whose structure resembles the stationary phase will have low R_f, while one that has a similar structure to the mobile phase will have high retention factor. Retention factors are characteristic, but will change depending on the exact condition of the mobile and stationary phase. For this reason, chemists usually apply a sample of a known compound to the sheet before running the experiment.

Thin-layer chromatography can be used to monitor the progress of a reaction, identify compounds present in a given mixture, and determine the purity of a substance. Specific examples of these applications include: analyzing ceramides and fatty acids, detection of pesticides or insecticides in food and water, analyzing the dye composition of fibers in forensics, assaying the radiochemical purity of radiopharmaceuticals, or identification of medicinal plants and their constituents

A number of enhancements can be made to the original method to automate the different steps, to increase the resolution achieved with TLC and to allow more accurate quantitative analysis. This method is referred to as HPTLC, or "high-performance TLC".

HPTLC typically uses thinner layers of stationary phase and smaller sample volumes, thus reducing the loss of resolution due to diffusion.

Plate Preparation

TLC plates are usually commercially available, with standard particle size ranges to improve reproducibility. They are prepared by mixing the adsorbent, such as silica gel, with a small amount of inert binder like calcium sulfate (gypsum) and water. This mixture is spread as a thick slurry on an unreactive carrier sheet, usually glass, thick aluminum foil, or plastic. The resultant plate is dried and *activated* by heating in an oven for thirty minutes at 110 °C. The thickness of the absorbent layer is typically around 0.1 – 0.25 mm for analytical purposes and around 0.5 – 2.0 mm for preparative TLC.

Technique

The process is similar to paper chromatography with the advantage of faster runs, better separations, and the choice between different stationary phases. Because of its simplicity and speed TLC is often used for monitoring chemical reactions and for the qualitative analysis of reaction products. Plates can be labeled before or after the chromatography process using a pencil or other implement that will not interfere or react with the process.

To run a thin layer chromatography plate, the following procedure is carried out:

- Using a capillary, a small spot of solution containing the sample is applied to a plate, about 1.5 centimeters from the bottom edge. The solvent is allowed to completely evaporate off to prevent it from interfering with sample's interactions with the mobile phase in the next step. If a non-volatile solvent was used to apply the sample, the plate needs to be dried in a vacuum chamber. This step is often repeated to ensure there is enough analyte at the starting spot on the plate to obtain a visible result. Different samples can be placed in a row of spots the same distance from the bottom edge, each of which will move in its own adjacent lane from its own starting point.

- A small amount of an appropriate solvent (eluent) is poured into a glass beaker or any other suitable transparent container (separation chamber) to a depth of less than 1 centimeter. A strip of filter paper (aka "wick") is put into the chamber so that its bottom touches the solvent and the paper lies on the chamber wall and reaches almost to the top of the container. The container is closed with a cover glass or any other lid and is left for a few minutes to let the solvent vapors ascend the filter paper and saturate the air in the chamber. (Failure to saturate the chamber will result in poor separation and non-reproducible results).

- The TLC plate is then placed in the chamber so that the spot(s) of the sample do not touch the surface of the eluent in the chamber, and the lid is closed. The solvent moves up the plate by capillary action, meets the sample mixture and carries it up the plate (elutes the sample). The plate should be removed from the chamber before the solvent front reaches the top of the stationary phase (continuation of the elution will give a misleading result) and dried.

- Without delay, the *solvent front*, the furthest extent of solvent up the plate, is marked.

- The plate is visualized. As some plates are pre-coated with a phosphor such as zinc sulfide, allowing many compounds to be visualized by using ultraviolet light; dark spots appear where the compounds block the UV light from striking the plate. Alternatively, plates can be sprayed or immersed in chemicals after elution. Various visualising agents react with the spots to produce visible results.

Separation Process and Principle

Different compounds in the sample mixture travel at different rates due to the differences in their attraction to the stationary phase and because of differences in solubility in the solvent. By changing the solvent, or perhaps using a mixture, the separation of components (measured by the R_f value) can be adjusted. Also, the separation achieved with a TLC plate can be used to estimate the separation of a flash chromatography column. (A compound elutes from a column when the amount of solvent collected is equal to $1/R_f$.) Chemists often use TLC to develop a protocol for separation by chromatography and they use TLC to determine which fractions contain the desired compounds.

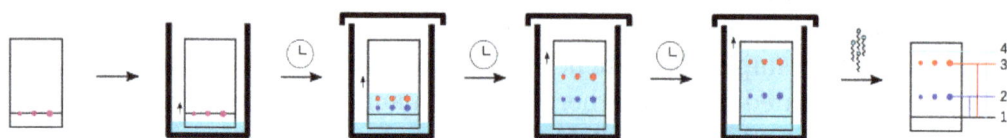

Development of a TLC plate, a purple spot separates into a red and blue spot

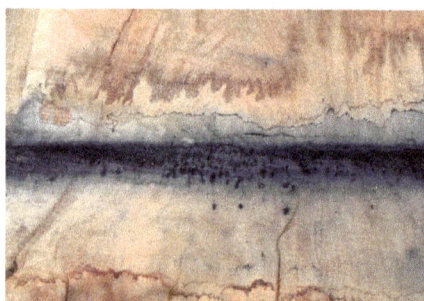

Surface of a freshly cut plank of Eucalyptus camaldulensis displaying thin-layer chromatography. The horizontal blue strip is from a reaction between the iron bandsaw supports and the acidic timber

Separation of compounds is based on the competition of the solute and the mobile phase for binding places on the stationary phase. For instance, if normal-phase silica gel is used as the stationary phase, it can be considered polar. Given two compounds that differ in polarity, the more polar compound has a stronger interaction with the silica and is, therefore, more capable to dispel the mobile phase from the binding places. As a consequence, the less polar compound moves higher up the plate (resulting in a higher R_f value). If the mobile phase is changed to a more polar solvent or mixture of solvents, it is more capable of dispelling solutes from the silica binding places, and all compounds on the TLC plate will move higher up the plate. It is commonly said that "strong" solvents (eluents) push the analyzed compounds up the plate, whereas "weak" eluents barely move them. The order of strength/weakness depends on the coating (stationary phase) of the TLC plate. For silica gel-coated TLC plates, the eluent strength increases in the following order: perfluoroalkane (weakest), hexane, pentane, carbon tetrachloride, benzene/toluene, dichloromethane, diethyl ether, ethyl acetate, acetonitrile, acetone, 2-propanol/*n*-butanol, water, methanol, triethylamine, acetic acid, formic acid (strongest). For C18-coated plates the order is reverse. In other words, when the stationary phase is polar and the mobile phase is polar, the method is *normal-phase* as opposed to *reverse-phase*. This means that if a mixture of ethyl acetate and hexane as the mobile phase is used, adding more ethyl acetate results in higher R_f values for all compounds on the TLC plate. Changing the polarity of the mobile phase will normally not result in reversed order of running of the compounds on the TLC plate. An eluotropic series can be used as a guide in selecting a mobile phase. If a reversed order of running of the compounds is desired, an apolar stationary phase should be used, such as C18-functionalized silica.

Analysis

As the chemicals being separated may be colorless, several methods exist to visualize the spots:

- fluorescent analytes like quinine may be detected under blacklight (366 nm)

- Often a small amount of a fluorescent compound, usually manganese-activated zinc silicate, is added to the adsorbent that allows the visualization of spots under UV-C light (254 nm). The adsorbent layer will thus fluoresce light-green by itself, but spots of analyte quench this fluorescence.

- Iodine vapors are a general unspecific color reagent

- Specific color reagents into which the TLC plate is dipped or which are sprayed onto the plate exist.

    o Potassium permanganate - oxidation

    o Bromine

- In the case of lipids, the chromatogram may be transferred to a PVDF membrane and then subjected to further analysis, for example mass spectrometry, a technique known as Far-Eastern blotting.

Once visible, the R_f value, or retardation factor, of each spot can be determined by dividing the distance the product traveled by the distance the solvent front traveled using the initial spotting site as reference. These values depend on the solvent used and the type of TLC plate and are not physical constants.

Applications

Characterization

In organic chemistry, reactions are qualitatively monitored with TLC. Spots sampled with a capillary tube are placed on the plate: a spot of starting material, a spot from the reaction mixture, and a cross-spot with both. A small (3 by 7 cm) TLC plate takes a couple of minutes to run. The analysis is qualitative, and it will show if the starting material has disappeared, i.e. the reaction is complete, if any product has appeared, and how many products are generated (although this might be underestimated due to co-elution). Unfortunately, TLCs from low-temperature reactions may give misleading results, because the sample is warmed to room temperature in the capillary, which can alter the reaction—the warmed sample analyzed by TLC is not the same as what is in the low-temperature flask. One such reaction is the DIBALH reduction of ester to aldehyde.

In one study TLC has been applied in the screening of organic reactions, for example in the fine-tuning of BINAP synthesis from 2-naphthol. In this method, the alcohol and catalyst solution (for instance iron(III) chloride) are placed separately on the base line, then reacted, and then instantly analyzed.

A special application of TLC is in the characterization of radiolabeled compounds, where it is used to determine radiochemical purity. The TLC sheet is visualized using a sheet of photographic film or an instrument capable of measuring radioactivity. It may be visualized using other means as well. This method is much more sensitive than the others and can be used to detect an extremely small amount of a compound, provided that it carries a radioactive atom.

Isolation

Since different compounds will travel a different distance in the stationary phase, chromatography can in effect be used as an isolation technique. The separated compounds each occupying a specific area on the plate, they can be scraped away, put in another solvent to separate them from the stationary phase and used for further analysis. As an example, in the chromatography of an extract of green leaves (for example spinach) in 7 stages of development, Carotene elutes quickly and is only visible until step 2. Chlorophyll A and B are halfway in the final step and lutein the first compound staining yellow. Once the chromatography is over, the carotene can be removed from the plate, put back into a solvent and ran into a spectrophotometer to characterize its wavelength absorption.

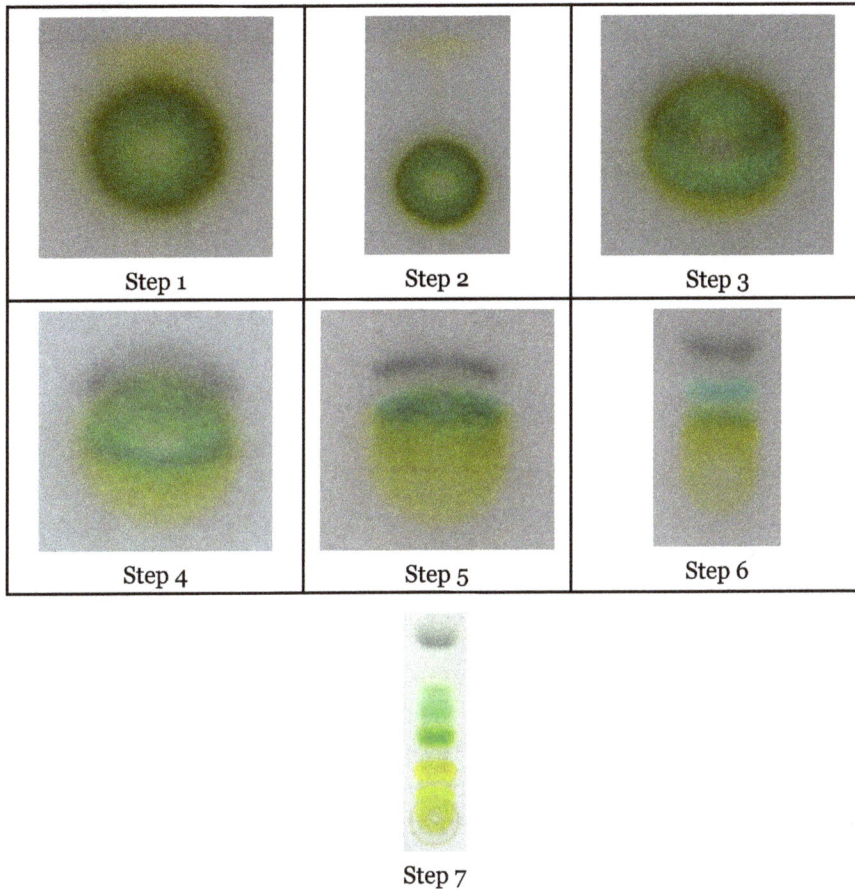

Step 1   Step 2   Step 3

Step 4   Step 5   Step 6

Step 7

Displacement Chromatography

Displacement chromatography is a chromatography technique in which a sample is placed onto the head of the column and is then displaced by a solute that is more strongly sorbed than the components of the original mixture. The result is that the components are resolved into consecutive "rectangular" zones of highly concentrated pure substances rather than solvent-separated "peaks". It is primarily a preparative technique; higher product concentration, higher purity, and increased throughput may be obtained compared to other modes of chromatography.

Discovery

The advent of displacement chromatography can be attributed to Arne Tiselius, who in 1943 first classified the modes of chromatography as frontal, elution, and displacement. Displacement chromatography found a variety of applications including isolation of transuranic elements and biochemical entities. The technique was redeveloped by Csaba Horváth, who employed modern high-pressure columns and equipment. It has since found many applications, particularly in the realm of biological macromolecule purification.

Principle

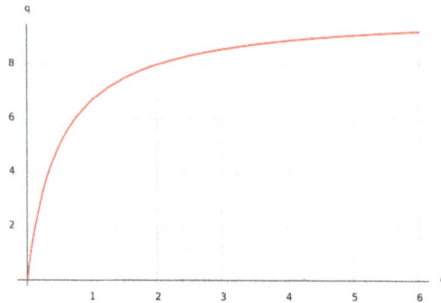

Example of a Langmuir isotherm, for a population of binding sites having uniform affinity. In this case the vertical axis represents the amount bound per unit of stationary phase, the horizontal axis the concentration in the mobile phase. In this case, the dissociation constant is 0.5 and the capacity 10; units are arbitrary.

The basic principle of displacement chromatography is: there are only a finite number of binding sites for solutes on the matrix (the stationary phase), and if a site is occupied by one molecule, it is unavailable to others. As in any chromatography, equilibrium is established between molecules of a given kind bound to the matrix and those of the same kind free in solution. Because the number of binding sites is finite, when the concentration of molecules free in solution is large relative to the dissociation constant for the sites, those sites will mostly be filled. This results in a downward-curvature in the plot of bound *vs* free solute, in the simplest case giving a Langmuir isotherm. A molecule with a high affinity for the matrix (the displacer) will compete more effectively for binding sites, leaving the mobile phase enriched in the lower-affinity solute. Flow of mobile phase through the column preferentially carries off the lower-affinity solute and thus at high concentration the higher-affinity solute will eventually displace all molecules with lesser affinities.

Mode of Operation

Loading

At the beginning of the run, a mixture of solutes to be separated is applied to the column, under conditions selected to promote high retention. The higher-affinity solutes are preferentially retained near the head of the column, with the lower-affinity solutes moving farther downstream. The fastest moving component begins to form a pure zone downstream. The other components also begin to form zones, but the continued supply of the mixed feed at head of the column prevents full resolution.

Displacement

After the entire sample is loaded, the feed is switched to the displacer, chosen to have higher affinity than any sample component. The displacer forms a sharp-edged zone at the head of the column, pushing the other components downstream. Each sample component now acts as a displacer for the lower-affinity solutes, and the solutes sort themselves out into a series of contiguous bands (a "displacement train"), all moving downstream at the rate set by the displacer. The size and loading of the column are chosen to let this sorting process reach completion before the components reach the bottom of the column. The solutes appear at the bottom of the column as a series of contiguous zones, each consisting of one purified component, with the concentration within each individual zone effectively uniform.

Regeneration

After the last solute has been eluted, it is necessary to strip the displacer from the column. Since the displacer was chosen for high affinity, this can pose a challenge. On reverse-phase materials, a wash with a high percentage of organic solvent may suffice. Large pH shifts are also often employed. One effective strategy is to remove the displacer by chemical reaction; for instance if hydrogen ion was used as displacer it can be removed by reaction with hydroxide, or a polyvalent metal ion can be removed by reaction with a chelating agent. For some matrices, reactive groups on the stationary phase can be titrated to temporarily eliminate the binding sites, for instance weak-acid ion exchangers or chelating resins can be converted to the protonated form. For gel-type ion exchangers, selectivity reversal at very high ionic strength can also provide a solution. Sometimes the displacer is specifically designed with a titratable functional group to shift its affinity. After the displacer is washed out, the column is washed as needed to restore it to its initial state for the next run.

Comparison with Elution Chromatography

Common Fundamentals

In any form of chromatography, the rate at which the solute moves down the column is a direct reflection of the percentage of time the solute spends in the mobile phase. To achieve separation in either elution or displacement chromatography, there must be appreciable differences in the affinity of the respective solutes for the stationary phase. Both methods rely on movement down the column to amplify the effect of small differences in distribution between the two phases. Distribution between the mobile and stationary phases is described by the binding isotherm, a plot of solute bound to (or partitioned into) the stationary phase as a function of concentration in the mobile phase. The isotherm is often linear, or approximately so, at low concentrations, but commonly curves (concave-downward) at higher concentrations as the stationary phase becomes saturated.

Characteristics of Elution Mode

In elution mode, solutes are applied to the column as narrow bands and, at low concentration, move down the column as approximately Gaussian peaks. These peaks continue to broaden as they travel, in proportion to the square root of the distance traveled. For two substances to be resolved, they must migrate down the column at sufficiently different rates to overcome the effects of band spreading. Operating at high concentration, where the isotherm is curved, is disadvantageous in elution chromatography because the rate of travel then depends on concentration, causing the peaks to spread and distort.

Retention in elution chromatography is usually controlled by adjusting the composition of the mobile phase (in terms of solvent composition, pH, ionic strength, and so forth) according to the type of stationary phase employed and the particular solutes to be separated. The mobile phase components generally have lower affinity for the stationary phase than do the solutes being separated, but are present at higher concentration and achieve their effects due to mass action. Resolution in elution chromatography is generally better when peaks are strongly retained, but conditions that give good resolution of early peaks lead to long run-times and excessive broadening of later peaks unless gradient elution is employed. Gradient equipment adds complexity and expense, particularly at large scale.

Advantages and Disadvantages of Displacement Mode

In contrast to elution chromatography, solutes separated in displacement mode form sharp-edged zones rather than spreading peaks. Zone boundaries in displacement chromatography are self-sharpening: if a molecule for some reason gets ahead of its band, it enters a zone in which it is more strongly retained, and will then run more slowly until its zone catches up. Furthermore, because displacement chromatography takes advantage of the non-linearity of the isotherms, loadings are deliberately high; more material can be separated on a given column, in a given time, with the purified components recovered at significantly higher concentrations. Retention conditions can still be adjusted, but the displacer controls the migration rate of the solutes. The displacer is selected to have higher affinity for the stationary phase than does any of the solutes being separated, and its concentration is set to approach saturation of the stationary phase and to give the desired migration rate of the concentration wave. High-retention conditions can be employed without gradient operation, because the displacer ensures removal of all solutes of interest in the designed run time.

Because of the concentrating effect of loading the column under high-retention conditions, displacement chromatography is well suited to purify components from dilute feed streams. However, it is also possible to concentrate material from a dilute stream at the head of a chromatographic column and then switch conditions to elute the adsorbed material in conventional isocratic or gradient modes. Therefore this approach is not unique to displacement chromatography, although the higher loading capacity and less dilution allow greater concentration in displacement mode.

A disadvantage of displacement chromatography is that non-idealities always give rise to an overlap zone between each pair of components; this mixed zone must be collected separately for recycle or discard to preserve the purity of the separated materials. The strategy of adding spacer molecules to form zones between the components (sometimes termed "carrier displacement chromatography") has been investigated and can be useful when suitable, readily removable spacers are found. Another disadvantage is that the raw chromatogram, for instance a plot of absorbance or refractive index *vs* elution volume, can be difficult to interpret for contiguous zones, especially if the displacement train is not fully developed. Documentation and troubleshooting may require additional chemical analysis to establish the distribution of a given component. Another disadvantage is that the time required for regeneration limits throughput.

According to John C. Ford's article in the *Encyclopedia of Chromatography*, theoretical studies indicate that at least for some systems, optimized overloaded elution chromatography offers higher throughput than displacement chromatography, though limited experimental tests suggest that displacement chromatography is superior (at least before consideration of regeneration time).

Applications

Historically, displacement chromatography was applied to preparative separations of amino acids and rare earth elements and has also been investigated for isotope separation.

Proteins

The chromatographic purification of proteins from complex mixtures can be quite challenging, particularly when the mixtures contain similarly retained proteins or when it is desired to enrich trace components in the feed. Further, column loading is often limited when high resolutions are required using traditional modes of chromatography (e.g. linear gradient, isocratic chromatography). In these cases, displacement chromatography is an efficient technique for the purification of proteins from complex mixtures at high column loadings in a variety of applications.

An important advance in the state of the art of displacement chromatography was the development of low molecular mass displacers for protein purification in ion exchange systems. This research was significant in that it represented a major departure from the conventional wisdom that large polyelectrolyte polymers are required to displace proteins in ion exchange systems.

Low molecular mass displacers have significant operational advantages as compared to large polyelectrolyte displacers. For example, if there is any overlap between the displacer and the protein of interest, these low molecular mass materials can be readily separated from the purified protein during post-displacement processing using standard size-based purification methods (e.g. size exclusion chromatography, ultrafiltration). In addition, the salt-dependent adsorption behavior of these low MW displacers greatly facilitates column regeneration. These displacers have been employed for a wide variety of high resolution separations in ion exchange systems. In addition, the utility of displacement chromatography for the purification of recombinant growth factors, antigenic vaccine proteins and antisense oligonucleotides has also been demonstrated. There are several examples in which displacement chromatography has been applied to the purification of proteins using ion exchange, hydrophobic interaction, as well as reversed-phase chromatography.

Displacement chromatography is well suited for obtaining mg quantities of purified proteins from complex mixtures using standard analytical chromatography columns at the bench scale. It is also particularly well suited for enriching trace components in the feed. Displacement chromatography can be readily carried out using a variety of resin systems including, ion exchange, HIC and RPLC

Two-dimensional Chromatography

Two-dimensional chromatography represents the most thorough and rigorous approach to evaluation of the proteome. While previously accepted approaches have utilized elution mode chromatographic approaches such as cation exchange to reversed phase HPLC, yields are typically very low requiring analytical sensitivities in the picomolar to femtomolar range. As displacement chromatography offers the advantage of concentration of trace components, two dimensional chromatography utilizing displacement rather than elution mode in the upstream chromatography step represents a potentially powerful tool for analysis of trace components, modifications, and identification of minor expressed components of the proteome.

References

- Bartle, K. D. Capillary ElectroChromatography Published by The Royal Society of Chemistry, Cambridge. ISBN 0-85404-530-9

- Garrett, Reginald H.; Grisham, Charles M. (2013). Biochemistry (5th ed.). Belmont, CA: Brooks/ Cole, Cengage Learning. p. 108. ISBN 9781133106296.

- Ballou, David P.; Benore, Marilee; Ninfa, Alexander J. (2008). Fundamental laboratory approaches for biochemistry and biotechnology. (2nd ed.). Hoboken, N.J.: Wiley. p. 129. ISBN 9780470087664.

- Rouessac, Annick; Rouessac, Francis (2000). Chemical analysis : modern instrumental methods and techniques (Engl. ed.). Chichester [u.a.]: Wiley. pp. 101–103. ISBN 0471972614.

- Ninfa, Alexander; Ballou, David; Benore, Marilee (May 26, 2009). Fundamental Laboratory Approaches for Biochemistry and Biotechnology. Wiley. pp. 143–145. ISBN 0470087668.

- Appling, Dean; Anthony-Cahill, Spencer; Mathews, Christopher (2016). Biochemistry: Concepts and Connections. New Jersey: Pearson. p. 134. ISBN 9780321839923.

- Harris, Daniel C. (2010). Quantitative Chemical Analysis. New York: W. H. Freeman and Company. p. 637. ISBN 1429218150.

- Heftmann, Erich (2004). Chromatography. Fundamentals and Applications of Chromatography and Related Differential Migration Methods: Applications. Amsterdam: Elsevier. p. 716. ISBN 9780080472256.

- Bhattacharyya, Lokesh; Rohrer, Jeffrey (2012). Applications of Ion Chromatography in the Analysis of Pharmaceutical and Biological Products. Wiley. p. 247. ISBN 0470467096.

4

Liquid Chromatography: A Comprehensive Study

Liquid chromatography is a technique that is used in combining the capabilities of liquid chromatography with the capabilities of mass spectrometry. It's a very powerful technique that is found to be useful in a number of applications. The chapter concentrates on topics such as ion suppression in liquid chromatography–mass spectrometry, fast protein liquid chromatography, countercurrent chromatography, column chromatography etc. The section on liquid chromatography offers an insightful focus, keeping in mind the complex topic.

Liquid Chromatography–mass Spectrometry

Liquid chromatography–mass spectrometry (LC-MS, or alternatively HPLC-MS) is an analytical chemistry technique that combines the physical separation capabilities of liquid chromatography (or HPLC) with the mass analysis capabilities of mass spectrometry (MS). LC-MS is a powerful technique that has very high sensitivity, making it useful in many applications. Its application is oriented towards the separation, general detection and potential identification of chemicals of particular masses in the presence of other chemicals (i.e., in complex mixtures), e.g., natural products from natural-products extracts, and pure substances from mixtures of chemical intermediates. Preparative LC-MS systems can be used for rapid mass-directed purification of specific substances from such mixtures that are important in basic research, and pharmaceutical, agrochemical, food, and other industries. Like gas chromatography–mass spectrometry, it allows analysis and detection even of tiny amounts of a substance.

Liquid Chromatography

Present day liquid chromatography generally utilizes very small particles packed and operating at relatively high pressure, and is referred to as high performance liquid chromatography (HPLC); modern LC-MS methods use HPLC instrumentation, essentially exclusively, for sample introduction. In HPLC, the sample is forced by a liquid at high pressure (the mobile phase) through a column that is packed with a stationary phase generally composed of irregularly or spherically shaped particles chosen or derivatized to accomplish particular types of separations. HPLC methods are historically divided into two different sub-classes based on stationary phases and the corresponding required polarity of the mobile phase. Use of octadecylsilyl (C18) and related organic-modified particles as stationary phase with pure or pH-adjusted water-organic mixtures such as water-acetonitrile and water-methanol are used in techniques termed reversed phase liquid chromatography (RP-LC). Use of materials such as silica gel as stationary phase with neat or mixed organic mixtures are used in techniques termed normal phase liquid chromatography (NP-LC). RP-LC is most often used as the means to introduce samples into the MS, in LC-MS instrumentation.

Flow Splitting

When standard bore (4.6 mm) columns are used the flow is often split ~10:1. This can be beneficial by allowing the use of other techniques in tandem such as MS and UV detection. However splitting the flow to UV will decrease the sensitivity of spectrophotometric detectors. The mass spectrometry on the other hand will give improved sensitivity at flow rates of 200 μL/min or less.

Mass Spectrometry

Mass spectrometry (MS) is an analytical technique that measures the mass-to-charge ratio of charged particles. It is used for determining masses of particles, for determining the elemental composition of a sample or molecule, and for elucidating the chemical structures of molecules, such as peptides and other chemical compounds. MS works by ionizing chemical compounds to generate charged molecules or molecule fragments and measuring their mass-to-charge ratios. In a typical MS procedure, a sample is loaded onto the MS instrument and undergoes vaporization. The components of the sample are ionized by one of a variety of methods (e.g., by impacting them with an electron beam, using uv lights as a photon beam, laser beam or corona discharge etc.), which results in the formation of charged particles (ions). The ions are separated according to their mass-to-charge ratio in an analyzer by electromagnetic fields. The ions are detected, usually by a quantitative method. The ion signal is processed into mass spectra.

Additionally, MS instruments consist of three modules. An ion source, which can convert gas phase sample molecules into ions (or, in the case of electrospray ionization, move ions that exist in solution into the gas phase). A mass analyzer, which sorts the ions by their masses by applying electromagnetic fields. A detector, which measures the value of an indicator quantity and thus provides data for calculating the abundances of each ion present

The technique has both qualitative and quantitative uses. These include identifying unknown compounds, determining the isotopic composition of elements in a molecule, and determining the structure of a compound by observing its fragmentation. Other uses include quantifying the amount of a compound in a sample or studying the fundamentals of gas phase ion chemistry (the chemistry of ions and neutrals in a vacuum).

MS is now in very common use in analytical laboratories that study physical, chemical, or biological properties of a great variety of compounds.

Mass Analyzer

There are many different mass analyzers that can be used in LC/MS. Single quadrupole, triple quadrupole, ion trap, time of flight (TOF) and quadrupole-time of flight (Q-TOF).

Interface

Understandably the interface between a liquid phase technique which continuously flows liquid, and a gas phase technique carried out in a vacuum was difficult for a long time. The advent of electrospray ionization changed this. The interface is most often an electrospray ion source or variant such as a nanospray source; however atmospheric pressure chemical ionization interface is also used. Various deposition and drying techniques have also been used such as using moving belts; however the most common of these is off-line MALDI deposition. A new approach still under development called Direct-EI LC-MS interface, couples a nano HPLC system and an electron ionization equipped mass spectrometer.

Applications

Pharmacokinetics

LC-MS is very commonly used in pharmacokinetic studies of pharmaceuticals and is thus the most frequently used technique in the field of bioanalysis. These studies give information about how quickly a drug will be cleared from the hepatic blood flow, and organs of the body. MS is used for this due to high sensitivity and exceptional specificity compared to UV (as long as the analyte can be suitably ionised), and short analysis time.

The major advantage MS has is the use of tandem MS-MS. The detector may be programmed to select certain ions to fragment. The process is essentially a selection technique, but is in fact more complex. The measured quantity is the sum of molecule fragments chosen by the operator. As long as there are no interferences or ion suppression, the LC separation can be quite quick.

Proteomics/Metabolomics

LC-MS is also used in proteomics where again components of a complex mixture must be detected and identified in some manner. The bottom-up proteomics LC-MS approach to proteomics generally involves protease digestion and denaturation (usually trypsin as a protease, urea to denature tertiary structure and iodoacetamide to cap cysteine residues) followed by LC-MS with peptide mass fingerprinting or LC-MS/MS (tandem MS) to derive sequence of individual peptides. LC-MS/MS is most commonly used for proteomic analysis of complex samples where peptide masses may overlap even with a high-resolution mass spectrometer. Samples of complex biological fluids like human serum may be run in a modern LC-MS/MS system and result in over 1000 proteins being identified, provided that the sample was first separated on an SDS-PAGE gel or HPLC-SCX.

Profiling of secondary metabolites in plants or food like phenolics can be achieved with liquid chromatography–mass spectrometry.

Drug Development

LC-MS is frequently used in drug development at many different stages including peptide mapping, glycoprotein mapping, natural products dereplication, bioaffinity screening, *in vivo* drug screening, metabolic stability screening, metabolite identification, impurity identification, quantitative bioanalysis, and quality control.

Ion Suppression in Liquid Chromatography–mass Spectrometry

Ion suppression in LC-MS and LC-MS/MS refers to reduced detector response, or signal:noise as a manifested effect of competition for ionisation efficiency in the ionisation source, between the analyte(s) of interest and other endogenous or exogenous (e.g. plasticisers extracted from plastic tubes, mobile phase additives) species which have not been removed from the sample matrix during sample preparation. Ion suppression is not strictly a problem unless interfering compounds elute at the same time as the analyte of interest. In cases where ion suppressing species do co-elute with an analyte, the effects on the important analytical parameters including precision, accuracy and limit of detection (analytical sensitivity) can be extensive, severely limiting the validity of an assay's results.

History

At its inception as a tool of analytical chemistry, LC-MS/MS spread rapidly and indeed continues to do so in (amongst others) bioanalytical fields, owing to its selectivity for analytes of interest. Indeed, in many cases this selectivity can lead to a misconception that it is always possible to simplify or (on occasion) almost completely remove the necessity for extensive sample preparation. Consequently, LC-MS/MS has become the analytical tool of choice for bioanalysis owing to its impressive sensitivity and selectivity over other, more conventional chromatographic approaches. However, during and after uptake by bioanalytical laboratories the world wide, it became apparent that there were inherent problems with detection of relatively small analyte concentrations in the complex sample matrices associated with biological fluids (e.g. blood and urine).

Proposed Mechanisms of Ion Suppression

Put simply, ion suppression describes the adverse effect on detector response due to reduced ionisation efficiency for analyte(s) of interest, resulting from the presence of species in the sample matrix which compete for ionisation, or inhibit efficient ionisation in other ways. Use of MS/MS as a means of detection may give the impression that there are no interfering species present, since no chromatographic impurities are detected. However, species which are not isobaric may still have an adverse effect on the sensitivity, accuracy and precision of the assay owing to suppression of the ionisation of the analyte of interest.

Although the precise chemical and physical factors involved in ion suppression are not fully understood, it has been proposed that basicity, high concentration, mass and more intuitively, co-elution with the analyte of interest are factors which should not be ignored.

The most common atmospheric pressure ionisation techniques used in LC-MS/MS are electrospray ionization (ESI) and atmospheric pressure chemical ionization (APCI). APCI is less prone to pronounced ion suppression than ESI, an inherent property of the respective ionisation mechanisms.

In APCI, the sole source of ion suppression can be attributed to the loss of efficiency during charge transfer from the corona discharge needle to the analyte of interest in the presence of other sample components.

ESI has a more complex ionisation mechanism, relying heavily on droplet charge excess and as such there are many more factors to consider when exploring the cause of ion suppression. It has been widely observed that for many analytes, at high concentrations, ESI exhibits a loss of detector response linearity, perhaps due to reduced charge excess caused by analyte saturation at the droplet surface, inhibiting subsequent ejection of gas phase ions from further inside the droplet. Thus competition for space and/or charge may be considered as a source of ion suppression in ESI. Both physical and chemical properties of analytes (e.g. basicity and surface activity) determine their inherent ionisation efficiency. Biological sample matrices naturally tend to contain many endogenous species with high basicity and surface activity, hence the total concentration of these species in the sample will quickly reach levels at which ion suppression should be expected.

Another explanation of ion suppression in ESI considers the physical properties of the droplet itself rather than the species present. High concentrations of interfering components give rise to an increased surface tension and viscosity, giving a reduction in desolvation (solvent evaporation), which is known to have a marked effect of ionisation efficiency.

The third proposed theory for ion suppression in ESI relates to the presence of non-volatile species which can either cause co-precipitation of analyte in the droplet (thus preventing ionisation) or prevent the contraction of droplet size to the critical radius required for the ion evaporation and/or charge residue mechanisms to form gas phase ions efficiently.

It is worthwhile to consider that the degree of ion suppression may be dependent on the concentration of the analyte being monitored. A higher analyte/matrix ratio can give a reduced effect of ion suppression.

Assessment of Ion Suppression during Method Validation

Since it is accepted that ion suppression has the potential to affect the other analytical parameters of any assay, a prudent approach to any LC-MS method development should include an evaluation of ion-suppression. There are two accepted protocols by which this may be achieved, described as follows.

Monitoring of Detector Response under Constant Infusion

The more comprehensive approach to assessment of ion suppression is to constantly infuse an appropriate concentration into the mobile phase flow, downstream from the analytical column, using a syringe pump and a 'tee union'. A typical sample should then be injected through the HPLC inlet as per the usual analytical parameters.

Monitoring of detector response during this experiment should yield a constant signal appropriate to the concentration of infused species. Once the sample has been injected, a drop in signal intensity (or a negative response) should be observed any time a species is ionised in the ion source. This should allow the retention time of any such species under the analytical parameters of the assay to be determined. Any species causing a negative response may be considered to be contributing to ion suppression, but only if such species co-elute with the analyte of interest.

It is also important to consider that species contributing to ion suppression may be retained by the column to a much greater extent than the analyte of interest. To this end, the detector response should be monitored for several times the usual chromatographic run time to ensure that ion suppression will not affect subsequent injections.

Preparation of Spiked Plasma Samples

Another approach to evaluation of ion suppression is to make a comparison between:

- Detector response of calibration standard (either aqueous or in another suitable solvent) - This gives the best case scenario for detector response, i.e. under conditions of zero ion suppression

- Pre-prepared sample matrix spiked with an identical concentration of analyte - This demonstrates the effect of ion suppression

- Detector response of sample matrix spiked with an identical concentration of analyte and subjected to the usual sample preparation procedure - This can demonstrate the difference between any signal loss due to under-recovery during the sample preparation process and true ion suppression

Approaches to Negating Ion Suppression

There are several strategies for removal and/or negation of ion suppression. These approaches may require in-depth understanding of the ionisation mechanisms involved in different ionisation sources or may be completely independent of the physical factors involved.

Chromatographic Separation

If the chromatographic separation can be modified to prevent coelution of suppressing species then other approaches need not be considered. The effect of chromatographic modification may be evaluated using the detector response monitoring under constant infusion approach described previously.

Sample Preparation

An effective sample preparation protocol, usually involving either liquid-liquid extraction (LLE) or solid phase extraction (SPE) and frequently derivatisation can remove ion suppressing species from the sample matrix prior to analysis. These common approaches may also remove other interferences, such as isobaric species.

Protein precipitation is another method that can be employed for small molecule analysis. Removal of all protein species from the sample matrix may be effective in some cases, although for many analytes, ion suppressing species are not of protein origin and so this technique is often used in conjunction with extraction and derivatisation.

Sample Concentration and Mobile Phase Flow Rate

Dilution of sample or reducing the volume of sample injected may give a reduction of ion suppression by reducing the quantity of interfering species present, although the quantity of analyte of interest will also be reduced, making this an undesirable approach for trace analysis.

Similar is the effect of reducing the mobile phase flow rate to the nanolitre-per-minute range since, in addition to resulting in improved desolvation, the smaller droplets formed are more tolerant to the presence of non-volatile species in the sample matrix.

Calibration Techniques to Compensate for Ion Suppression

It is not always possible to eliminate ion suppression by sample preparation and/or chromatographic resolution. In such cases it may be possible to compensate for the effects of ion suppression on accuracy and precision (although not for analytical sensitivity) by adopting complex calibration strategies.

Matrix Matched Calibration Standards

Using matrix matched calibration standards can compensate for ion suppression. Using this technique, calibration standards are prepared in identical sample matrix to that used for analysis (e.g. plasma) by spiking a normal sample with known concentrations of analyte. This is not always possible for biological samples, since the analyte of interest is often endogenously present in a clinically significant, albeit normal, quantity. For matrix matched calibration standards to be effective in compensating for ion suppression, the sample matrix must be free of the analyte of interest. Additionally, it is important that there is little variation in test sample composition since both the test sample and the prepared calibration sample must be affected in the same way by ion suppression. Again, in complex biological samples from different individuals, or even the same individual at a different time, there may be large fluctuations in the concentrations of ion suppressing species.

Standard Addition

The standard addition approach involves spiking the same sample extract with several known concentrations of analyte. This technique is more robust and effective than using matrix matched standards but is labor-intensive since each sample must be prepared several times to achieve a reliable calibration.

Internal Standard

In this approach, the sample is spiked with a species (internal standard) which is used to normalise the response of analyte, compensating for variables at any stage of the sample preparation and analysis, including ion suppression.

It is important that the internal standard displays very similar (ideally identical) properties, with respect to detector response (i.e. ionisation), as the analyte of interest. To simplify the selection of internal standard, most laboratories use an analogous stable isotope in an isotope dilution type analysis. The stable isotope is almost guaranteed to be chemically and physically as close as possible to the analyte of interest, hence producing an almost identical detector response in addition to behaving identically during sample preparation and chromatographic resolution. To this end, the ion suppression experienced by both the analyte and the internal standard should be identical. It is important to note that an excessively high concentration of stable isotope internal standard may cause ion suppression itself, since it will co-elute with the analyte of interest. Hence, the internal standard should be added at an appropriate concentration.

Choice of Ionization Source

APCI generally suffers less ion suppression than ESI, as discussed previously. Where possible, if ion suppression is unavoidable it may be advisable to switch from ESI to

APCI. If this is not possible, it may be useful to switch the ESI ionisation mode from positive to negative. Since less compounds are ionisable in negative ionisation mode, it is entirely possible that the ion suppressing species may be removed from the analysis. However, it should also be considered that the analyte of interest may not be ionised effectively in negative mode either, rendering this approach useless.

Fast Protein Liquid Chromatography

Fast protein liquid chromatography (FPLC), is a form of liquid chromatography that is often used to analyze or purify mixtures of proteins. As in other forms of chromatography, separation is possible because the different components of a mixture have different affinities for two materials, a moving fluid (the "mobile phase") and a porous solid (the stationary phase). In FPLC the mobile phase is an aqueous solution, or "buffer". The buffer flow rate is controlled by a positive-displacement pump and is normally kept constant, while the composition of the buffer can be varied by drawing fluids in different proportions from two or more external reservoirs. The stationary phase is a resin composed of beads, usually of cross-linked agarose, packed into a cylindrical glass or plastic column. FPLC resins are available in a wide range of bead sizes and surface ligands depending on the application.

In the most common FPLC strategy, ion exchange, a resin is chosen that the protein of interest will bind to the resin by a charge interaction while in buffer A (the running buffer) but become dissociated and return to solution in buffer B (the elution buffer). A mixture containing one or more proteins of interest is dissolved in 100% buffer A and pumped into the column. The proteins of interest bind to the resin while other components are carried out in the buffer. The total flow rate of the buffer is kept constant; however, the proportion of Buffer B (the "elution" buffer) is gradually increased from 0% to 100% according to a programmed change in concentration (the "gradient"). At some point during this process each of the bound proteins dissociates and appears in the effluent. The effluent passes through two detectors which measure salt concentration (by conductivity) and protein concentration (by absorption of ultraviolet light at a wavelength of 280nm). As each protein is eluted it appears in the effluent as a "peak" in protein concentration and can be collected for further use.

FPLC was developed and marketed in Sweden by Pharmacia in 1982 and was originally called fast performance liquid chromatography to contrast it with HPLC or high-performance liquid chromatography. FPLC is generally applied only to proteins; however, because of the wide choice of resins and buffers it has broad applications. In contrast to HPLC the buffer pressure used is relatively low, typically less than 5 bar, but the flow rate is relatively high, typically 1-5 ml/min. FPLC can be readily scaled from analysis of milligrams of mixtures in columns with a total volume of 5ml or less to industrial production of kilograms of purified protein in columns with volumes of many liters.

When used for analysis of mixtures the effluent is usually collected in fractions of 1-5 ml which can be further analyzed, e.g. by MALDI mass spectrometry. When used for protein purification there may be only two collection containers, one for the purified product and one for waste.

FPLC System Components

A typical laboratory FPLC consist of one or two high-precision pumps, a control unit, a column, a detection system and a fraction collector. Although it is possible to operate the system manually, the components are normally linked to a personal computer or, in older units, a microcontroller.

1. Pumps: Majority of systems utilize two two-cylinder piston pumps, one for each buffer, combining the output of both in a mixing chamber. Some simpler systems use a single peristaltic pump which draws both buffers from separate reservoirs through a proportioning valve and mixing chamber. In either case the system allows the fraction of each buffer entering the column to be continuously varied. The flow rate can go from a few milliliters per minute in bench-top systems to liters per minute for industrial scale purifications. The wide flow range makes it suitable both for analytical and preparative chromatography.

2. Injection loop: A segment of tubing of known volume which is filled with the sample solution before it is injected into the column. Loop volume can range from a few microliters to 50ml or more.

3. Injection valve: A motorized valve which links the mixer and sample loop to the column. Typically the valve has three positions for loading the sample loop, for injecting the sample from the loop into the column, and for connecting the pumps directly to the waste line to wash them or change buffer solutions. The injection valve has a sample loading port through which the sample can be loaded into the injection loop, usually from a hypodermic syringe using a Luer-lock connection.

4. Column: The column is a glass or plastic cylinder packed with beads of resin and filled with buffer solution. It is normally mounted vertically with the buffer flowing downward from top to bottom. A glass frit at the bottom of the column retains the resin beads in the column while allowing the buffer and dissolved proteins to exit.

5. Flow Cells: The effluent from the column passes through one or more flow cells to measure the concentration of protein in the effluent (by UV light absorption at 280nm). The conductivity cell measures the buffer conductivity, usually in millisiemens/cm, which indicates the concentration of salt in the buffer. A flow cell which measures pH of the buffer is also commonly included. Usually each flow cell is connected to a separate electronics module which provides power and amplifies the signal.

6. Monitor/Recorder: The flow cells are connected to a display and/or recorder. On

older systems this was a simple chart recorder, on modern systems a computer with hardware interface and display is used. This permits the experimenter to identify when peaks in protein concentration occur, indicating that specific components of the mixture are being eluted.

7. Fraction collector: The fraction collector is typically a rotating rack that can be filled with test tubes or similar containers. It allows samples to be collected in fixed volumes, or can be controlled to direct specific fractions detected as peaks of protein concentration, into separate containers.

Many systems include various optional components. A filter may be added between the mixer and column to minimize clogging. In large FPLC columns the sample may be loaded into the column directly using a small peristaltic pump rather than an injection loop. When the buffer contains dissolved gas, bubbles may form as pressure drops where the buffer exits the column; these bubbles create artifacts if they pass through the flow cells. This may be prevented by degassing the buffers, e.g. with a degasser, or by adding a flow restrictor downstream of the flow cells to maintain a pressure of 1-5 bar in the effluent line.

FPLC Columns

The columns used in FPLC are large [mm id] tubes that contain small [μ] particles or gel beads that are known as stationary phase. The chromatographic bed is composed by the gel beads inside the column and the sample is introduced into the injector and carried into the column by the flowing solvent. As a result of different components adhering to or diffusing through the gel, the sample mixture gets separated.

Columns used with an FPLC can separate macromolecules based on size, charge distribution (ion exchange), hydrophobicity, reverse-phase or biorecognition (as with affinity chromatography). For easy use, a wide range of pre-packed columns for techniques such as ion exchange, gel filtration (size exclusion), hydrophobic interaction, and affinity chromatography are available. FPLC differs from HPLC in that the columns used for FPLC can only be used up to maximum pressure of 3-4 MPa (435-580 psi). Thus, if the pressure of HPLC can be limited, each FPLC column may also be used in an HPLC machine.

Optimizing Protein Purification

Using a combination of chromatographic methods, purification of the target molecule is achieved. The purpose of purifying proteins with FPLC is to deliver quantities of the target at sufficient purity in a biologically active state to suit its further use. The quality of the end product varies depending the type and amount of starting material, efficiency of separation, and selectivity of the purification resin. The ultimate goal of a given purification protocol is to deliver the required yield and purity of the target molecule in the quickest, cheapest, and safest way for acceptable results. The range of purity required can be from that required for basic analysis (SDS-PAGE or ELISA, for example), with only bulk impurities removed, to pure enough for structural analysis (NMR or X-ray crystallography), approaching >99% target molecule. Purity required can also mean pure enough that the biological activity of the target is retained. These demands can be used to determine the amount of starting material required to reach the experimental goal. If the starting material is limited and full optimization of purification protocol cannot be performed, then a safe standard protocol that requires a minimum adjustment and optimization steps are expected. This may not be optimal with respect to experimental time, yield, and economy but it will achieve the experimental goal. On the other hand, if the starting material is enough to develop more complete protocol, the amount of work to reach the separation goal depends on the available sample information and target molecule properties. Limits to development of purification protocols many times depends on the source of the substance to be purified, whether from natural sources (harvested tissues or organisms, for example), recombinant sources (such as using prokaryotic or eukaryotic vectors in their respective expression systems), or totally synthetic sources.

No chromatographic techniques provide 100% yield of active material and overall yields depend on the number of steps in the purification protocol. By optimizing each step for the intended purpose and arranging them that minimizes inter step treatments, the number of steps will be minimized.

A typical multistep purification protocol starts with a preliminary capture step which many times utilizes ion exchange chromatography (IEC). The media (stationary phase) employed range from large bead resins (good for fast flow rates and little to no sample clarification at the expense of resolution) to small bead resins (for best possible resolution with all other factors being equal). Short and wide column geometries are amenable to high flow rates also at the expense of resolution, typically because of lateral diffusion of sample on the column. For techniques such as size exclusion chromatography to be useful, very long, thin columns and minimal sample volumes (maximum 5% of column volume) are required. Hydrophobic interaction chromatography (HIC) can also be used for first and/ or intermediate steps. Selectivity in HIC is independent of running pH and descending salt gradients are used. For HIC, conditioning involves adding ammonium sulphate to the sample to match the buffer A concentration. If HIC is used before IEC, the ionic strength would have to be lowered to match that of buffer A for IEC step by dilution, dialysis or buffer exchange by gel filtration. This is why IEC is usually performed prior to HIC as the high salt elution conditions for IEC are ideal for binding to HIC resins in the next purification step. Polishing is used to achieve the final level of purification required and is commonly performed on a gel filtration column. An extra intermediate purification step can be added or optimization of the different steps is performed for improving purity. This extra step usually involves another round of IEC under completely different conditions.

Although this is an example of a common purification protocol for proteins, the buffer conditions, flow rates, and resins used to achieve final goals can be chosen to cover a broad range of target proteins. This flexibility is imperative for a functional purification system as all proteins behave differently and often deviate from predictions.

Countercurrent Chromatography

Countercurrent chromatography (CCC, also counter-current chromatography) is a form of liquid–liquid chromatography that uses a liquid stationary phase that is held in place by centrifugal force and is used to separate, identify, and quantify the chemical components of a mixture. In its broadest sense, countercurrent chromatography encompasses a collection of related liquid chromatography techniques that employ two immiscible liquid phases without a solid support. The two liquid phases come in contact with each other as at least one phase is pumped through a column, a hollow tube or a series of chambers connected with channels, which contains both phases. The resulting dynamic mixing and settling action allows the components to be separated by their respective solubilities in the two phases. A wide variety of two phase solvent systems consisting of at least two immiscible liquids may be employed to provide the proper selectivity for the desired separation.

Some types of countercurrent chromatography, such as dual flow CCC, feature a true countercurrent process where the two immiscible phases flow past each other and exit at opposite ends of the column. More often, however, one liquid acts as the stationary phase and is retained in the column while the mobile phase is pumped through it. The liquid stationary phase is held in place by gravity or by centrifugal force. An example of a gravity method is called droplet counter current chromatography (DCCC). There are two modes by which the stationary phase is retained by centrifugal force: hydrostatic and hydrodynamic. In the hydrostatic method, the column, a series of chambers connected by channels, is rotated about a central axis. Hydrostatic instruments are marketed under the name centrifugal partition chromatography (CPC). Hydrodynamic instruments are often marketed as high-speed or high-performance countercurrent chromatography (HSCCC and HPCCC respectively) instruments which rely on the Archimedes' screw force in a helical coil to retain the stationary phase in the column.

The components of a CCC system are similar to most liquid chromatography configurations such as high-performance liquid chromatography. One or more pumps deliver the phases to the column which is the CCC instrument itself. Samples are introduced into the column through a sample loop filled with an automated or manual syringe. The outflow is monitored with various detectors such as ultraviolet–visible spectroscopy or mass spectrometry. The operation of the pumps, CCC instrument, sample injection, and detection may be controlled manually or with a microprocessor.

History

The predecessor of modern countercurrent chromatography theory and practice was countercurrent distribution (CCD). The theory of CCD was described in the 1930s by Randall and Longtin. Archer Martin and Richard Laurence Millington Synge developed the methodology further during the 1940s. Finally, Lyman C. Craig introduced the Craig countercurrent distribution apparatus in 1944 which made CCD practical for laboratory work. CCD was used to separate a wide variety of useful compounds for several decades.

Support-free Liquid Chromatography

Standard column chromatography consists of a solid stationary phase and a liquid mobile phase, while gas chromatography (GC) uses a solid or liquid stationary phase on a solid support and a gaseous mobile phase. By contrast, in liquid-liquid chromatography, both the mobile and stationary phases are liquid. The contrast is, however not as stark as it first appears. In reversed-phase chromatography, for example, the stationary phase can be regarded as a liquid which is immobilized by chemical bonding to a micro-porous silica solid support. In countercurrent chromatography centrifugal or gravitational forces immobilize the stationary liquid layer. By eliminating solid supports, permanent adsorption of the analyte onto the column is avoided, and a high recovery of the analyte can be achieved. The countercurrent chromatography instrument is easily switched between normal phase chromatography and reversed-phase chromatography simply by changing the mobile and stationary phases. With column chromatography, the separation potential is limited by the commercially available stationary phase media and its particular characteristics. Nearly any pair of immiscible solutions can be used in countercurrent chromatography provided that the stationary phase can be successfully retained.

Solvent costs are also generally lower than for high-performance liquid chromatography (HPLC). In comparison to column chromatography, flows and total solvent usage can in most countercurrent chromatography separations may be reduced by half and even up to a tenth. Also, the cost of purchasing and disposing of stationary phase media is eliminated. Another advantage of countercurrent chromatography is that experiments conducted in the laboratory can be scaled to industrial volumes. When gas chromatography or HPLC is carried out with large volumes, resolution is lost due to issues with surface-to-volume ratios and flow dynamics; this is avoided when both phases are liquid.

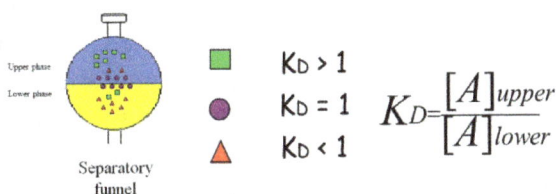

$$K_D = \frac{[A]_{upper}}{[A]_{lower}}$$

Partition coefficient (K_D)

The CCC separation process can be thought of as occurring in three stages: mixing, settling, and separation of the two phases (although they often occur continuously). Vigorous mixing of the phases is critical in order to maximize the interfacial area between them and enhance mass transfer. The analyte will distribute between the phases according to its partition coefficient which is also called the distribution coefficient, distribution constant, or partition ratio and is represented by P, K, D, K_c, or K_D. The partition coefficient for an analyte in a particular biphasic solvent system is independent of the volume of the instrument, flow rate, stationary phase retention volume ratio and the g-force required to immobilize the stationary phase. The degree of stationary phase retention is a crucial parameter. Common factors that influence stationary phase retention are flow rate, solvent composition of the biphasic solvent system, and the g-force. The stationary phase retention is represented by the stationary phase volume retention ratio (Sf) which is the volume of the stationary phase divided by the total volume of the instrument. The settling time is a property of the solvent system and the sample matrix, both of which greatly influence stationary phase retention.

To most process chemists, the term "countercurrent" implies two immiscible liquids moving in opposing directions, as typically occurs in large centrifugal extractor units. With the exception of dual flow CCC, most countercurrent chromatography modes of operation have a stationary phase and a mobile phase. Even in this situation, countercurrent flows occur within the instrument column. Several researchers have proposed renaming both CCC & CPC to liquid-liquid chromatography, but others feel the term "countercurrent" itself is a misnomer.

Unlike column chromatography and high-performance liquid chromatography, countercurrent chromatography operators can inject large volumes relative to column volume. Typically 5 to 10% of coil volume can be injected. In some cases this can be increased to as high as 15 to 20% of the coil volume. Typically, most modern commercial CCC and CPC can inject 5 to 40 g per liter capacity. The range is so large, even for a specific instrument, let alone all instrument options, as the type of target, matrix and available biphasic solvent vary so much. Approximately 10 g per liter would be a more typical value, that the majority of applications could use as a base value.

The countercurrent separation starts with choosing an appropriate biphasic solvent system for the desired separation. A wide array of biphasic solvent mixtures are available to the CCC practitioner including the combination n-hexane (or heptane), ethyl acetate, methanol and water in different proportions. This basic solvent system is sometimes referred to as the HEMWat solvent system. The choice of solvent system may be guided by perusal of the CCC literature. The familiar technique of thin layer chromatography may also be employed to determine an optimal solvent system. The organization of solvent systems into "families" has greatly facilitated the choice of solvent systems as well. A solvent system can be tested with a one-flask partitioning experiment. The measured partition coefficient from the partitioning experiment will indicate the elution behavior of the compound. Typically, it is desirable to choose a solvent system where the target compound(s) have a partition coefficient between 0.25 and 8. Historically, it was thought that no commercial countercurrent chromatograph could cope with the high viscosities of ionic liquids. However, modern instruments that can accommodate 30 to 70+ % ionic liquids (and potentially 100% ionic liquid, if both phases are suitably customized ionic liquids) have become available. Ionic liquids can be customized for polar / non-polar organic, achiral and chiral compounds, bio-molecule, and inorganic separations, as ionic liquids can be customized to have extraordinary solvency and specificity.

After the biphasic solvent system has been chosen a batch of is formulated and equilibrated in a separatory funnel. This step is called pre-equilibration of the solvent system. The two phases are separated. Then the column is filled with stationary with a pump. Next, the column is set an equilibration conditions, such as the desired rotation speed, and the mobile phase is pumped through the column. The mobile phase displaces the a portion of the stationary phase until column equilibration is achieved and the mobile phase elutes from the column. The sample may be introduced into the column at any time during the column equilibration step or after equilibration has been accomplished. After the volume of eluant surpasses the volume of the mobile phase in the column, the sample components will begin to elute. Compounds with a partition coefficient of unity will elute when one column volume of mobile phase has passed through the column since the time of injection. The flow is stopped after the target compound(s) are eluted or the column is extruded by pumping the stationary phase through the column. An example of a major application of countercurrent chromatography is to take an extremely complex matrix such as a plant extract, perform the countercurrent chromatography separation with a carefully selected solvent system, and extrude the column to recover all of the sample. The original complex matrix will have been fractionated into discrete narrow polarity bands, which may then be assayed for chemical composition or bioactivity. Performing one or more countercurrent chromatography separations in conjunction with other chromatographic and non chromatographic techniques has the potential for rapid advances in compositional recognition of extremely complex matrices.

Droplet CCC

Droplet countercurrent chromatography (DCCC) was introduced in 1970 by Tanimura, Pisano, Ito, and Bowman. DCCC uses only gravity to move the mobile phase through the stationary phase which is held in long vertical tubes connected in series. In the descending mode, droplets of the denser mobile phase and sample are allowed to fall through the columns of the lighter stationary phase using only gravity. If a less dense mobile phase is used it will rise through the stationary phase; this is called ascending mode. The eluent from one column is transferred to another; the more columns that are used, the more theoretical plates can be achieved. DCCC enjoyed some success with natural product separations but was largely eclipsed by the rapid development of high-speed countercurrent chromatography. The main limitation of DCCC is that flow rates are low, and poor mixing is achieved for most binary solvent systems.

Hydrodynamic CCC

The modern era of CCC began with the development by Dr. Yoichiro Ito of the planetary centrifuge which was first introduced in 1966 as a closed helical tube which was rotated on a "planetary" axis as is turned on a "sun" axis. A flow-through model was subsequently developed and the new technique was called countercurrent chromatography in 1970. The technique was further developed by employing test mixtures of DNP amino acids in chloroform:glacial acetic acid:0.1 M aqueous hydrochloric acid (2:2:1 v/v) cosolvent system. Much development was needed to engineer the instrument so that required planetary motion could be sustained while the phases were being pumped through the coil(s). Parameters such as the relative rotation of the two axes (synchronous or non-synchronous), the direction of flow through the coil, and the rotor angles were investigated.

High-speed

By 1982 the technology was sufficiently advanced for the technique to be called "high-speed" countercurrent chromatography (HSCCC). Peter Carmeci initially commercialized the PC Inc. Ito Multilayer Coil Separator/Extractor which utilized a single bobbin (onto which the coil is wound) and a counterbalance, plus a set of "flying leads" which are tubing that connect the bobbins. Dr. Walter Conway & others later evolved the bobbin design such that multiple coils, even coils of different tubing sizes, could be placed on the single bobbin. Edward Chou later evolved and commercialized a triple bobbin design as the Pharmatech CCC which had a de-twist mechanism for leads between the three bobbins. The Quattro CCC released in 1993 further evolved the commercially available instruments by utilizing a novel mirror image, twin bobbin design that did not need the de-twist mechanism of the Pharmatech between the multiple bobbins, so could still accommodate multiple bobbins on the same instrument. Hydrodynamic CCC are now available with up to 4 coils per instrument. These coils can be in PTFE, PEEK, PVDF, or stainless steel tubing. The 2, 3 or 4 coils can all be of the same bore to facilitate "2D" CCC. The coils may be connected in series to lengthen the coil and increase the capacity, or the coils may be linked in parallel so that 2, 3, or 4 separations may be done simultaneously. The coils can also be of different sizes, on one instrument, ranging from 1 to 2 to 3.7 to 6 mm on one instrument, thus allowing a single instrument to optimize from mg to kilos per day. More recently instrument derivatives have started to be offered with rotating seals for various hydrodynamic CCC designs, instead of flying leads, either as custom options or as standard.

High-performance

The operating principle of CCC equipment requires a column consisting of a tube coiled around a bobbin. The bobbin is rotated in a double-axis gyratory motion (a cardioid), which causes a variable g-force to act on the column during each rotation. This motion causes the column to see one partitioning step per revolution and components of the sample separate in the column due to their partitioning coefficient between the two immiscible liquid phases. "High-performance" countercurrent chromatography (HPCCC) works in much the same way as HSCCC. A seven-year R&D process produced HPCCC instruments that generated 240 g's, compared to the 80 g's of the HSCCC machines. This increase in g-force and larger bore of the column has enabled a tenfold increase in throughput, due to improved mobile phase flow rates and a higher stationary phase retention. Countercurrent chromatography is a preparative liquid chromatography technique, however with the advent of the higher-g HPCCC instruments it is now possible to operate instruments with sample loadings as low as a few milligrams, whereas in the past 100s of milligrams had been necessary. Major application areas for this technique include natural products purification and also drug development.

Hydrostatic

Hydrostatic CCC or centrifugal partition chromatography (CPC) was invented in the 1980s by the Japanese company Sanki Engineering Ltd, whose president was Kanichi Nunogaki. CPC has been extensively developed in France starting from the late 1990s. In France, they initially optimized the stacked disc concept initiated by Sanki. More recently, in France and UK, non-stacked disc CPC configurations have been developed with PTFE, stainless steel or titanium rotors. These have been designed to overcome possible leakages between the stacked discs of the original concept, and to allow steam cleaning for Good manufacturing practice. The volumes ranging from a 100 ml to 12 liters are available in different rotor materials. The 25 liter rotor CPC has a titanium rotor. This technique is sometimes sold under the name "fast" CPC or "high-performance" CPC.

Principle

CPC is based on the principles of countercurrent chromatography: the stationary phase of a two-phase system is maintained in the instrument under a centrifugal force while the mobile phase is pumped through a series of chambers or cells. The stationary phase is retained inside the rotor by the centrifugal force generated by rotation around a single axis. Discrete chambers connected by channels are the sites of mixing and settling action that allows the compounds being separated to distribute between the two phases. As the mobile phase exits the columns it is collected in test tubes for further analysis. Chemical compounds subjected to CPC are separated based on the partition coefficient of each compound in this two-phase system.

Realization

The centrifugal partition chromatograph instrument is constituted with a unique rotor which contains the column. This rotor rotates on its central axis (while HSCCC column rotates on its planetary axis and simultaneously rotates eccentrically about another solar axis). With less vibrations and noise, the CPC offers a typical rotation speed range from 500 to 2000 rpm. Contrary to hydrodynamic CCC, the rotation speed is not directly proportional to the retention volume ratio of the stationary phase. Like DCCC, CPC can be operated in either descending or ascending mode, where the direction is relative to the force generated by the rotor rather than gravity. A redesigned CPC column with larger chambers and channels has been named centrifugal partition extraction (CPE). In the CPE design, faster flow rates and increased column loading can be achieved.

Advantages

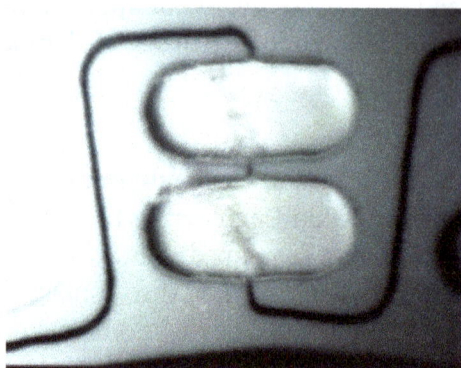

Visualization of mixing and settling in a CPC twin cell

The CPC offers now the direct scale-up from the analytical apparatuses (few milliliters) to industrial apparatuses (some liters) for fast batch production. The CPC seems particularly suited to accommodate aqueous two-phase solvent systems. Generally, CPC instruments can retain solvent systems that are not well-retained in a hydrodynamic instrument due to small differences in density between the phases. It has been very helpful for the development of CPC instrumentation to visualize the flow patterns which give rise to the mixing and settling in the CPC chamber with an asynchronous camera and a stroboscope triggered by the CPC rotor.

Modes of Operation

The aforementioned hydrodynamic and hydrostatic instruments may be employed in a variety of ways, or modes of operation, in order to address the particular separation needs of the scientist. Many modes of operation have been devised to take advantage of the strengths and potentialities of the countercurrent chromatography technique. Generally, the following modes may be performed with commercially available instruments.

Normal-phase

Normal phase elution is achieved by pumping the non-aqueous or phase of a biphasic solvent system through the column as the mobile phase, with a more polar stationary phase being retained in the column. The cause of original nomenclature of is relevant. As original stationary phases of paper chromatography were superseded by more efficient materials such as diatomaceous earths (natural micro-porous silica) and followed by modern silica gel, the thin-layer chromatography stationary phase was polar (hydroxy groups attached to silica) and maximum retention was achieved with non-polar solvents such as n-hexane. Progressively more polar eluents were then used to move polar compounds up the plate. Various alkane bonded phases were tried with C18 becoming the most popular. Alkane chains were chemically bonded to the silica, and a reversal of the elution trend occurred. Thus a polar stationary became "normal" phase chromatography, and the non-polar stationary phase chromatography became "reversed" phase chromatography.

Reversed-phase

In reversed-phase (e.g. aqueous mobile phase) elution, the aqueous phase is used as the mobile phase with a less polar stationary phase. In countercurrent chromatography the same solvent system may be used in either normal or reversed phase mode simply by switching the direction of mobile phase flow through the column.

Elution-extrusion

The extrusion of stationary phase from the column at the end of a separation experiment by stopping rotation and pumping solvent or gas through the column was used by CCC practitioners before the term EECCC was suggested. In elution-extrusion mode (EECCC), The mobile phase is extruded after a certain point by switching the phase being pumped into the system whilst maintaining rotation. For example, if the separation has been initiated with the aqueous phase as the mobile phase at a certain point the organic phase is pumped through the column which effectively pushes out both phases that are present in the column at the time of switching. The complete sample is eluted in the order of polarity (either normal or reversed) without loss of resolution by diffusion. It requires only one column volume of solvent phase and leaves the column full of fresh stationary phase for the subsequent separation.

Gradient

The use of a solvent gradient is very well developed in column chromatography but is less common in CCC. A solvent gradient is produced by increasing (or decreasing) the polarity of the mobile phase during the separation to achieve optimal resolution across a wider range of polarities. For example, a methanol-water mobile phase gradient may be employed using heptane as the stationary phase. This is not possible with all biphasic solvent systems, due to excessive loss of stationary phase created by disruption the equilibrium conditions within the column. Gradients may either be produced in steps, or continuously.

Dual-mode

In dual-mode, the mobile and stationary phases are reversed part way through the separation experiment. This requires changing the phase being pumped through the column as well as the direction of flow. Dual-mode operation is likely to elute the entire sample from the column but the order of elution is disrupted by switching the phase and direction of flow.

Dual-flow

Dual-flow, also known as dual, countercurrent chromatography occurs when both phases are flowing in opposite directions inside the column. Instruments are available for dual-flow operation for both Hydrodynamic and hydrostatic CCC. Dual-flow countercurrent chromatography was first described by Yoichiro Ito in 1985 for foam CCC where gas-liquid separations were performed. Liquid–liquid separations soon followed. The countercurrent chromatography instrument must be modified so that both ends of the column have both inlet and outlet capabilities. This mode may accommodate continuous or sequential separations with the sample being introduced in the middle of the column or between two bobbins in a hydrodynamic instrument. A technique called intermittent countercurrent extraction (ICcE) is a quasi-continuous method where the flow of the phases is alternated "intermittently" between normal and reversed-phase elution so that the stationary phase also alternates.

Recycling and Sequential

Recycling chromatography is mode practiced in both HPLC and CCC. In recycling chromatography, the target compounds are reintroduced into the column after they elute. Each pass through the column increases the number of theoretical plates the compounds experience and enhances chromatographic resolution. Direct recycling must be done with an isocratic solvent system. With this mode, the eluant can be selectively re-chromatographed on the same or a different column in order to facilitate the separation. This process of selective recycling has been termed a "heart-cut" and is especially effective in purifying selected target compounds with some sacrificial loss of recovery. The process of re-separating selected fractions from one chromatography experiment with another chromatographic method has long been practiced by scientists. Recycling and sequential chromatography is a streamlined version of this process. In CCC, the separation characteristics of the column may be modified simply by changing the composition of the biphasic solvent system.

Ion-exchange and pH-Zone-refining

In an conventional CCC experiment the biphasic solvent system is pre-equilibrated before the instrument is filled with the stationary phase and equilibrated with the mobile phase. An ion-exchange mode has been created by modifying both of the phases after pre-equilibration. Generally, an ionic displacer (or eluter) is added to mobile phase and an ionic retainer is added to the stationary phase. For example, the aqueous mobile phase may contain NaI as a displacer and the organic stationary phase may be modified with the quaternary ammonium salt called Aliquat 336 as a retainer. The mode known a pH-zone-refining is a type of ion-exchange mode that utilizes acids and/or bases as solvent modifiers. Typically, the analytes are eluted in an order determined by their pKa values. For example, 6 oxindole alkaloids were isolated from a 4.5g sample of *Gelsemium elegans* stem extract with a biphasic solvent system composed of hexane–ethyl acetate–methanol–water (3:7:1:9, v/v) where 10 mM triethylamine (TEA) was added to the upper organic stationary phase as a retainer and 10 mM hydrochloric acid (HCl) to the aqueous mobile phase as an eluter. Ion-exchange modes such as pH-zone-refining have tremendous potential because high sample loads can be achieved without sacrificing separation power. It works best with ionizable compounds such as nitrogen containing alkaloids or carboxylic acid containing fatty acids.

Applications

Countercurrent chromatography and related liquid-liquid separation techniques have been used on both industrial and laboratory scale to purify a wide variety of chemical substances. Separation realizations include proteins, DNA, antibiotics, vitamins, natural products, pharmaceuticals, metal ions, pesticides, enantiomers, polyaromatic hydrocarbons from environmental samples, active enzymes, and carbon nanotubes. Countercurrent chromatography is known for its high dynamic range of scalability: milligram to kilogram quantities purified chemical components may be obtained with this technique. It also has the advantage of accommodating chemically complex samples with undissolved particulates.

High-performance Liquid Chromatography

A modern self-contained HPLC.

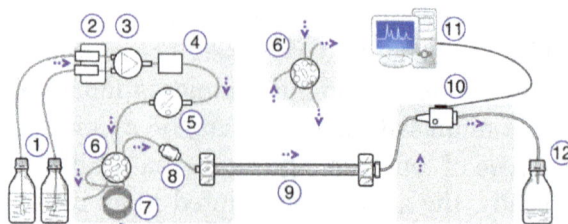

Schematic representation of an HPLC unit. (1) Solvent reservoirs, (2) Solvent degasser, (3) Gradient valve, (4) Mixing vessel for delivery of the mobile phase, (5) High-pressure pump, (6) Switching valve in "inject position", (6') Switching valve in "load position", (7) Sample injection loop, (8) Pre-column (guard column), (9) Analytical column, (10) Detector (i.e. IR, UV), (11) Data acquisition, (12) Waste or fraction collector.

High-performance liquid chromatography (HPLC; formerly referred to as high-pressure liquid chromatography), is a technique in analytical chemistry used to separate, identify, and quantify each component in a mixture. It relies on pumps to pass a pressurized liquid solvent containing the sample mixture through a column filled with a solid adsorbent material. Each component in the sample interacts slightly differently with the adsorbent material, causing different flow rates for the different components and leading to the separation of the components as they flow out the column.

HPLC has been used for manufacturing (e.g. during the production process of pharmaceutical and biological products), legal (e.g. detecting performance enhancement drugs in urine), research (e.g. separating the components of a complex biological sample, or of similar synthetic chemicals from each other), and medical (e.g. detecting vitamin D levels in blood serum) purposes.

Chromatography can be described as a mass transfer process involving adsorption. HPLC relies on pumps to pass a pressurized liquid and a sample mixture through a column filled with adsorbent, leading to the separation of the sample components. The active component of the column, the adsorbent, is typically a granular material made of solid particles (e.g. silica, polymers, etc.), 2–50 micrometers in size. The components of the sample mixture are separated from each other due to their different degrees of interaction with the adsorbent particles. The pressurized liquid is typically a mixture of solvents (e.g. water, acetonitrile and/or methanol) and is referred to as a "mobile phase". Its composition and temperature play a major role in the separation process by influencing the interactions taking place between sample components and adsorbent. These interactions are physical in nature, such as hydrophobic (dispersive), dipole–dipole and ionic, most often a combination.

HPLC is distinguished from traditional ("low pressure") liquid chromatography because operational pressures are significantly higher (50–350 bar), while ordinary liquid chromatography typically relies on the force of gravity to pass the mobile phase through the column. Due to the small sample amount separated in analytical HPLC, typical column dimensions are 2.1–4.6 mm diameter, and 30–250 mm length. Also HPLC columns are made with smaller sorbent particles (2–50 micrometer in average particle size). This gives HPLC superior resolving power (the ability to distinguish between compounds) when separating mixtures, which makes it a popular chromatographic technique.

The schematic of an HPLC instrument typically includes a sampler, pumps, and a detector. The sampler brings the sample mixture into the mobile phase stream which carries it into the column. The pumps deliver the desired flow and composition of the mobile phase through the column. The detector generates a signal proportional to the amount of sample component emerging from the column, hence allowing for quantitative analysis of the sample components. A digital microprocessor and user software control the HPLC instrument and provide data analysis. Some models of mechanical pumps in a HPLC instrument can mix multiple solvents together in ratios changing in time, generating a composition gradient in the mobile phase. Various detectors are in common use, such as UV/Vis, photodiode array (PDA) or based on mass spectrometry. Most HPLC instruments also have a column oven that allows for adjusting the temperature at which the separation is performed.

Operation

The sample mixture to be separated and analyzed is introduced, in a discrete small volume (typically microliters), into the stream of mobile phase percolating through the column. The components of the sample move through the column at different velocities, which are a function of specific physical interactions with the adsorbent (also called stationary phase). The velocity of each component depends on its chemical nature, on the nature of the stationary phase (column) and on the composition of the mobile phase. The time at which a specific analyte elutes (emerges from the column) is called its retention time. The retention time measured under particular conditions is an identifying characteristic of a given analyte.

Many different types of columns are available, filled with adsorbents varying in particle size, and in the nature of their surface ("surface chemistry"). The use of smaller particle size packing materials requires the use of higher operational pressure ("backpressure") and typically improves chromatographic resolution (i.e. the degree of separation between consecutive analytes emerging from the column). Sorbent particles may be hydrophobic or polar in nature.

Common mobile phases used include any miscible combination of water with various organic solvents (the most common are acetonitrile and methanol). Some HPLC techniques use water-free mobile phases. The aqueous component of the mobile phase may contain acids (such as formic, phosphoric or trifluoroacetic acid) or salts to assist in the separation of the sample components. The composition of the mobile phase may be kept constant ("isocratic elution mode") or varied ("gradient elution mode") during the chromatographic analysis. Isocratic elution is typically effective in the separation of sample components that are not very different in their affinity for the stationary phase. In gradient elution the composition of the mobile phase is varied typically from low to high eluting strength. The eluting strength of the mobile phase is reflected by analyte retention times with high eluting strength producing fast elution (=short retention times). A typical gradient profile in reversed phase chromatography might start at 5% acetonitrile (in water or aqueous buffer) and progress linearly to 95% acetonitrile over 5–25 minutes. Periods of constant mobile phase composition may be part of any gradient profile. For example, the mobile phase composition may be kept constant at 5% acetonitrile for 1–3 min, followed by a linear change up to 95% acetonitrile.

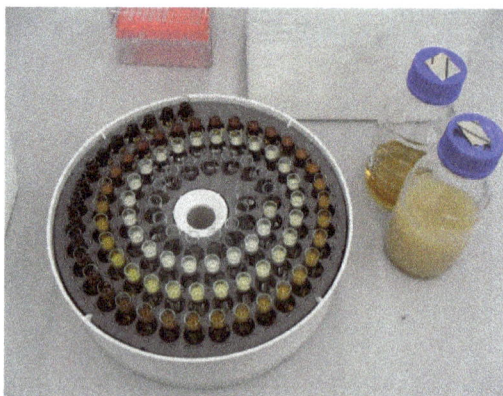

A rotary fraction collector collecting HPLC output. The system is being used to isolate a fraction containing Complex I from *E. coli* plasma membranes. About 50 litres of bacteria were needed to isolate this amount.

The chosen composition of the mobile phase (also called eluent) depends on the intensity of interactions between various sample components ("analytes") and stationary phase (e.g. hydrophobic interactions in reversed-phase HPLC). Depending on their affinity for the stationary and mobile phases analytes partition between the two during the separation process taking place in the column. This partitioning process is similar to that which occurs during a liquid–liquid extraction but is continuous, not step-wise. In this example, using a water/acetonitrile gradient, more hydrophobic components will elute (come off the column) late, once the mobile phase gets more concentrated in acetonitrile (i.e. in a mobile phase of higher eluting strength).

The choice of mobile phase components, additives (such as salts or acids) and gradient conditions depends on the nature of the column and sample components. Often a series of trial runs is performed with the sample in order to find the HPLC method which gives adequate separation.

History and Development

Prior to HPLC scientists used standard liquid chromatographic techniques. Liquid chromatographic systems were largely inefficient due to the flow rate of solvents being dependent on gravity. Separations took many hours, and sometimes days to complete.

Gas chromatography (GC) at the time was more powerful than liquid chromatography (LC), however, it was believed that gas phase separation and analysis of very polar high molecular weight biopolymers was impossible. GC was ineffective for many biochemists because of the thermal instability of the solutes. As a result, alternative methods were hypothesized which would soon result in the development of HPLC.

Following on the seminal work of Martin and Synge in 1941, it was predicted by Cal Giddings, Josef Huber, and others in the 1960s that LC could be operated in the high-efficiency mode by reducing the packing-particle diameter substantially below the typical LC (and GC) level of 150 μm and using pressure to increase the mobile phase velocity. These predictions underwent extensive experimentation and refinement throughout the 60s into the 70s. Early developmental research began to improve LC particles, and the invention of Zipax, a superficially porous particle, was promising for HPLC technology.

The 1970s brought about many developments in hardware and instrumentation. Researchers began using pumps and injectors to make a rudimentary design of an HPLC system. Gas amplifier pumps were ideal because they operated at constant pressure and did not require leak free seals or check valves for steady flow and good quantitation. Hardware milestones were made at Dupont IPD such as a low-dwell-volume gradient device being utilized as well as replacing the septum injector with a loop injection valve.

While instrumentational developments were important, the history of HPLC is primarily about the history and evolution of particle technology. After the introduction of porous layer particles, there has been a steady trend to reduced particle size to improve efficiency. However, by decreasing particle size, new problems arose. The practical disadvantages stem from the excessive pressure drop needed to force mobile fluid through the column and the difficulty of preparing a uniform packing of extremely fine materials. Every time particle size is reduced significantly, another round of instrument development usually must occur to handle the pressure.

Types

Partition Chromatography

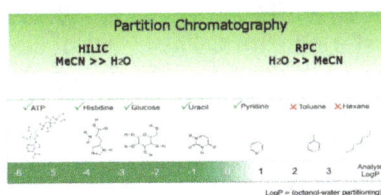

HILIC Partition Technique Useful Range

Partition chromatography was one of the first kinds of chromatography that chemists developed. The partition coefficient principle has been applied in paper chromatography, thin layer chromatography, gas phase and liquid–liquid separation applications. The 1952 Nobel Prize in chemistry was earned by Archer John Porter Martin and Richard Laurence Millington Synge for their development of the technique, which was used for their separation of amino acids. Partition chromatography uses a retained solvent, on the surface or within the grains or fibers of an "inert" solid supporting matrix as with paper chromatography; or takes advantage of some coulombic and/or hydrogen donor interaction with the stationary phase. Analyte molecules partition between a liquid stationary phase and the eluent. Just as in Hydrophilic Interaction Chromatography (HILIC; a sub-technique within HPLC), this method separates analytes based on differences in their polarity. HILIC most often uses a bonded polar stationary phase and a mobile phase made primarily of acetonitrile with water as the strong component. Partition HPLC has been used historically on unbonded silica or alumina supports. Each works effectively for separating analytes by relative polar differences. HILIC bonded phases have the advantage of separating acidic, basic and neutral solutes in a single chromatographic run.

The polar analytes diffuse into a stationary water layer associated with the polar stationary phase and are thus retained. The stronger the interactions between the polar analyte and the polar stationary phase (relative to the mobile phase) the longer the elution time. The interaction strength depends on the functional groups part of the analyte molecular structure, with more polarized groups (e.g. hydroxyl-) and groups capable of hydrogen bonding inducing more retention. Coulombic (electrostatic) interactions can also increase retention. Use of more polar solvents in the mobile phase will decrease the retention time of the analytes, whereas more hydrophobic solvents tend to increase retention times.

Normal–phase Chromatography

Normal–phase chromatography was one of the first kinds of HPLC that chemists developed. Also known as normal-phase HPLC (NP-HPLC) this method separates analytes based on their affinity for a polar stationary surface such as silica, hence it is based on analyte ability to engage in polar interactions (such as hydrogen-bonding or dipole-dipole type of interactions) with the sorbent surface. NP-HPLC uses a non-polar, non-aqueous mobile phase (e.g. Chloroform), and works effectively for separating analytes readily soluble in non-polar solvents. The analyte associates with and is retained by the polar stationary phase. Adsorption strengths increase with increased analyte polarity. The interaction strength depends not only on the functional groups present in the structure of the analyte molecule, but also on steric factors. The effect of steric hindrance on interaction strength allows this method to resolve (separate) structural isomers.

The use of more polar solvents in the mobile phase will decrease the retention time of analytes, whereas more hydrophobic solvents tend to induce slower elution (increased retention times). Very polar solvents such as traces of water in the mobile phase tend to

adsorb to the solid surface of the stationary phase forming a stationary bound (water) layer which is considered to play an active role in retention. This behavior is somewhat peculiar to normal phase chromatography because it is governed almost exclusively by an adsorptive mechanism (i.e. analytes interact with a solid surface rather than with the solvated layer of a ligand attached to the sorbent surface;). Adsorption chromatography is still widely used for structural isomer separations in both column and thin-layer chromatography formats on activated (dried) silica or alumina supports.

Partition- and NP-HPLC fell out of favor in the 1970s with the development of reversed-phase HPLC because of poor reproducibility of retention times due to the presence of a water or protic organic solvent layer on the surface of the silica or alumina chromatographic media. This layer changes with any changes in the composition of the mobile phase (e.g. moisture level) causing drifting retention times.

Recently, partition chromatography has become popular again with the development of Hilic bonded phases which demonstrate improved reproducibility, and due to a better understanding of the range of usefulness of the technique.

Displacement Chromatography

The basic principle of displacement chromatography is: A molecule with a high affinity for the chromatography matrix (the displacer) will compete effectively for binding sites, and thus displace all molecules with lesser affinities. There are distinct differences between displacement and elution chromatography. In elution mode, substances typically emerge from a column in narrow, Gaussian peaks. Wide separation of peaks, preferably to baseline, is desired in order to achieve maximum purification. The speed at which any component of a mixture travels down the column in elution mode depends on many factors. But for two substances to travel at different speeds, and thereby be resolved, there must be substantial differences in some interaction between the biomolecules and the chromatography matrix. Operating parameters are adjusted to maximize the effect of this difference. In many cases, baseline separation of the peaks can be achieved only with gradient elution and low column loadings. Thus, two drawbacks to elution mode chromatography, especially at the preparative scale, are operational complexity, due to gradient solvent pumping, and low throughput, due to low column loadings. Displacement chromatography has advantages over elution chromatography in that components are resolved into consecutive zones of pure substances rather than "peaks". Because the process takes advantage of the nonlinearity of the isotherms, a larger column feed can be separated on a given column with the purified components recovered at significantly higher concentration.

Reversed-phase Chromatography (RPC)

Reversed phase HPLC (RP-HPLC) has a non-polar stationary phase and an aqueous, moderately polar mobile phase. One common stationary phase is a silica which has been surface-modified with RMe_2SiCl, where R is a straight chain alkyl group such as $C_{18}H_{37}$ or C_8H_{17}. With such stationary phases, retention time is longer for molecules which are less polar, while polar molecules elute more readily (early in the analysis). An investigator can increase retention times by adding more water to the mobile phase; thereby making the affinity of the hydrophobic analyte for the hydrophobic stationary phase stronger relative to the now more hydrophilic mobile phase. Similarly, an investigator can decrease retention time by adding more organic solvent to the eluent. RP-HPLC is so commonly used that it is often incorrectly referred to as "HPLC" without further specification. The pharmaceutical industry regularly employs RP-HPLC to qualify drugs before their release.

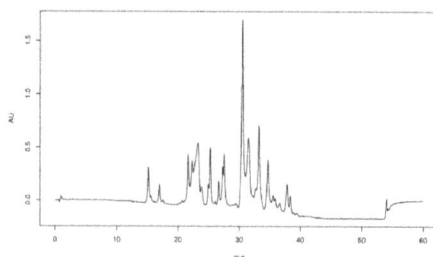

A chromatogram of complex mixture (perfume water) obtained by reversed phase HPLC

RP-HPLC operates on the principle of hydrophobic interactions, which originates from the high symmetry in the dipolar water structure and plays the most important role in all processes in life science. RP-HPLC allows the measurement of these interactive forces. The binding of the analyte to the stationary phase is proportional to the contact surface area around the non-polar segment of the analyte molecule upon association with the ligand on the stationary phase. This solvophobic effect is dominated by the force of water for "cavity-reduction" around the analyte and the C_{18}-chain versus the complex of both. The energy released in this process is proportional to the surface tension of the eluent (water: 7.3×10^{-6} J/cm2, methanol: 2.2×10^{-6} J/cm2) and to the hydrophobic surface of the analyte and the ligand respectively. The retention can be decreased by adding a less polar solvent (methanol, acetonitrile) into the mobile phase to reduce the surface tension of water. Gradient elution uses this effect by automatically reducing the polarity and the surface tension of the aqueous mobile phase during the course of the analysis.

Structural properties of the analyte molecule play an important role in its retention characteristics. In general, an analyte with a larger hydrophobic surface area (C–H, C–C, and generally non-polar atomic bonds, such as S-S and others) is retained longer because it is non-interacting with the water structure. On the other hand, analytes with higher polar surface area (conferred by the presence of polar groups, such as -OH, $-NH_2$, COO^- or $-NH_3^+$ in their structure) are less retained as they are better integrated into water. Such interactions are subject to steric effects in that very large molecules may have only restricted access to the pores of the stationary phase, where the interactions with surface ligands (alkyl chains) take place. Such surface hindrance typically results in less retention.

Retention time increases with hydrophobic (non-polar) surface area. Branched chain compounds elute more rapidly than their corresponding linear isomers because the overall surface area is decreased. Similarly organic compounds with single C–C bonds elute later than those with a C=C or C–C triple bond, as the double or triple bond is shorter than a single C–C bond.

Aside from mobile phase surface tension (organizational strength in eluent structure), other mobile phase modifiers can affect analyte retention. For example, the addition of inorganic salts causes a moderate linear increase in the surface tension of aqueous solutions (ca. 1.5×10^{-7} J/cm^2 per Mol for NaCl, 2.5×10^{-7} J/cm^2 per Mol for $(NH_4)_2SO_4$), and because the entropy of the analyte-solvent interface is controlled by surface tension, the addition of salts tend to increase the retention time. This technique is used for mild separation and recovery of proteins and protection of their biological activity in protein analysis (hydrophobic interaction chromatography, HIC).

Another important factor is the mobile phase pH since it can change the hydrophobic character of the analyte. For this reason most methods use a buffering agent, such as sodium phosphate, to control the pH. Buffers serve multiple purposes: control of pH, neutralize the charge on the silica surface of the stationary phase and act as ion pairing agents to neutralize analyte charge. Ammonium formate is commonly added in mass spectrometry to improve detection of certain analytes by the formation of analyte-ammonium adducts. A volatile organic acid such as acetic acid, or most commonly formic acid, is often added to the mobile phase if mass spectrometry is used to analyze the column effluent. Trifluoroacetic acid is used infrequently in mass spectrometry applications due to its persistence in the detector and solvent delivery system, but can be effective in improving retention of analytes such as carboxylic acids in applications utilizing other detectors, as it is a fairly strong organic acid. The effects of acids and buffers vary by application but generally improve chromatographic resolution.

Reversed phase columns are quite difficult to damage compared with normal silica columns; however, many reversed phase columns consist of alkyl derivatized silica particles and should never be used with aqueous bases as these will destroy the underlying silica particle. They can be used with aqueous acid, but the column should not be exposed to the acid for too long, as it can corrode the metal parts of the HPLC equipment. RP-HPLC columns should be flushed with clean solvent after use to remove residual acids or buffers, and stored in an appropriate composition of solvent. The metal content of HPLC columns must be kept low if the best possible ability to separate substances is to be retained. A good test for the metal content of a column is to inject a sample which is a mixture of 2,2'- and 4,4'- bipyridine. Because the 2,2'-bipy can chelate the metal, the shape of the peak for the 2,2'-bipy will be distorted (tailed) when metal ions are present on the surface of the silica.

Size-exclusion Chromatography

Size-exclusion chromatography (SEC), also known as *gel permeation chromatography* or *gel filtration chromatography*, separates particles on the basis of molecular size (actually by a particle's Stokes radius). It is generally a low resolution chromatography and thus it is often reserved for the final, "polishing" step of the purification. It is also useful for determining the tertiary structure and quaternary structure of purified proteins. SEC is used primarily for the analysis of large molecules such as proteins or polymers. SEC works by trapping these smaller molecules in the pores of a particle. The larger molecules simply pass by the pores as they are too large to enter the pores. Larger molecules therefore flow through the column quicker than smaller molecules, that is, the smaller the molecule, the longer the retention time.

This technique is widely used for the molecular weight determination of polysaccharides. SEC is the official technique (suggested by European pharmacopeia) for the molecular weight comparison of different commercially available low-molecular weight heparins.

Ion-exchange Chromatography

In ion-exchange chromatography (IC), retention is based on the attraction between solute ions and charged sites bound to the stationary phase. Solute ions of the same charge as the charged sites on the column are excluded from binding, while solute ions of the opposite charge of the charged sites of the column are retained on the column. Solute ions that are retained on the column can be eluted from the column by changing the solvent conditions (e.g. increasing the ion effect of the solvent system by increasing the salt concentration of the solution, increasing the column temperature, changing the pH of the solvent, etc...).

Types of ion exchangers include:

- Polystyrene resins – These allow cross linkage which increases the stability of the chain. Higher cross linkage reduces swerving, which increases the equilibration time and ultimately improves selectivity.

- Cellulose and dextran ion exchangers (gels) – These possess larger pore sizes and low charge densities making them suitable for protein separation.

- Controlled-pore glass or porous silica

In general, ion exchangers favor the binding of ions of higher charge and smaller radius.

An increase in counter ion (with respect to the functional groups in resins) concentration reduces the retention time. A decrease in pH reduces the retention time in cation exchange while an increase in pH reduces the retention time in anion exchange. By lowering the pH of the solvent in a cation exchange column, for instance, more hydrogen ions are available to compete for positions on the anionic stationary phase, thereby eluting weakly bound cations.

This form of chromatography is widely used in the following applications: water purification, preconcentration of trace components, ligand-exchange chromatography, ion-exchange chromatography of proteins, high-pH anion-exchange chromatography of carbohydrates and oligosaccharides, and others.

Bioaffinity Chromatography

This chromatographic process relies on the property of biologically active substances to form stable, specific, and reversible complexes. The formation of these complexes involves the participation of common molecular forces such as the Van der Waals interaction, electrostatic interaction, dipole-dipole interaction, hydrophobic interaction, and the hydrogen bond. An efficient, biospecific bond is formed by a simultaneous and concerted action of several of these forces in the complementary binding sites.

Aqueous Normal-phase Chromatography

Aqueous normal-phase chromatography (ANP) is a chromatographic technique which encompasses the mobile phase region between reversed-phase chromatography (RP) and organic normal phase chromatography (ONP). This technique is used to achieve unique selectivity for hydrophilic compounds, showing normal phase elution using reversed-phase solvents.

Isocratic and Gradient Elution

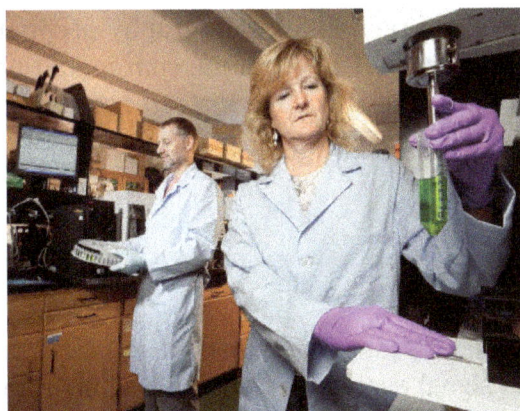

At the ARS Natural Products Utilization Research Unit in Oxford, MS., a support scientist (r) extracts plant pigments that will be analyzed by a plant physiologist (l) using an HPLC system.

A separation in which the mobile phase composition remains constant throughout the procedure is termed *isocratic* (meaning *constant composition*). The word was coined by Csaba Horvath who was one of the pioneers of HPLC.,

The mobile phase composition does not have to remain constant. A separation in which the mobile phase composition is changed during the separation process is described as a *gradient elution*. One example is a gradient starting at 10% methanol and ending at 90% methanol after 20 minutes. The two components of the mobile phase are typically termed "A" and "B"; *A* is the "weak" solvent which allows the solute to elute only slowly, while *B* is the "strong" solvent which rapidly elutes the solutes from the column. In reversed-phase chromatography, solvent *A* is often water or an aqueous buffer, while *B* is an organic solvent miscible with water, such as acetonitrile, methanol, THF, or isopropanol.

In isocratic elution, peak width increases with retention time linearly according to the equation for N, the number of theoretical plates. This leads to the disadvantage that late-eluting peaks get very flat and broad. Their shape and width may keep them from being recognized as peaks.

Gradient elution decreases the retention of the later-eluting components so that they elute faster, giving narrower (and taller) peaks for most components. This also improves the peak shape for tailed peaks, as the increasing concentration of the organic eluent pushes the tailing part of a peak forward. This also increases the peak height (the peak looks "sharper"), which is important in trace analysis. The gradient program may include sudden "step" increases in the percentage of the organic component, or different slopes at different times – all according to the desire for optimum separation in minimum time.

In isocratic elution, the selectivity does not change if the column dimensions (length and inner diameter) change – that is, the peaks elute in the same order. In gradient elution, the elution order may change as the dimensions or flow rate change.

The driving force in reversed phase chromatography originates in the high order of the water structure. The role of the *organic component of the mobile phase* is to reduce this high order and thus *reduce the retarding strength of the aqueous component.*

Parameters

Theoretical

HPLC separations have theoretical parameters and equations to describe the separation of components into signal peaks when detected by instrumentation such as by a UV detector or a mass spectrometer. The parameters are largely derived from two sets of chromatagraphic theory: plate theory (as part of Partition chromatography), and the rate theory of chromatography / *Van Deemter equation*. Of course, they can be put in practice through analysis of HPLC chromatograms, although rate theory is considered the more accurate theory.

They are analogous to the calculation of retention factor for a paper chromatography separation, but describes how well HPLC separates a mixture into two or more components that are detected as peaks (bands) on a chromatogram. The HPLC parameters are the: efficiency factor(N), the retention factor (kappa prime), and the separation factor (alpha). Together the factors are variables in a resolution equation, which describes how well two components' peaks separated or overlapped each other. These parameters are mostly only used for describing HPLC reversed phase and HPLC normal phase separations, since those separations tend to be more subtle than other HPLC modes (e.g. ion exchange and size exclusion).

- Void volume is the amount of space in a column that is occupied by solvent. It is the space within the column that is outside of the column's internal packing material. Void volume is measured on a chromatogram as the first component peak detected, which is usually the solvent that was present in the sample mixture; ideally the sample solvent flows through the column without interacting with the column, but is still detectable as distinct from the HPLC solvent. The void volume is used as a correction factor.

- Efficiency factor (N) practically measures how sharp component peaks on the chromatogram are, as ratio of the component peak's area ("retention time") relative to the width of the peaks at their widest point (at the baseline). Peaks that are tall, sharp, and relatively narrow indicate that separation method efficiently removed a component from a mixture; high efficiency. Efficiency is very dependent upon the HPLC column and the HPLC method used. Efficiency factor is synonymous with plate number, and the 'number of theoretical plates'.

- Retention factor (*kappa prime*) measures how long a component of the mixture stuck to the column, measured by the area under the curve of its peak in a chromatogram (since HPLC chromatograms are a function of time). Each chromatogram peak will have its own retention factor (e.g. *kappa*1 for the retention factor of the first peak). This factor may be corrected for by the void volume of the column.

- Separation factor (*alpha*) is a relative comparison on how well two neighboring components of the mixture were separated (i.e. two neighboring bands on a chromatogram). This factor is defined in terms of a ratio of the retention factors of a pair of neighboring chromatogram peaks, and may also be corrected for by the void volume of the column. The greater the separation factor value is over 1.0, the better the separation, until about 2.0 beyond which an HPLC method is probably not needed for separation.

Resolution equations relate the three factors such that high efficiency and separation factors improve the resolution of component peaks in a HPLC separation.

Internal Diameter

Tubing on a nano-liquid chromatography (nano-LC) system, used for very low flow capacities.

The internal diameter (ID) of an HPLC column is an important parameter that influences the detection sensitivity and separation selectivity in gradient elution. It also determines the quantity of analyte that can be loaded onto the column. Larger columns are usually seen in industrial applications, such as the purification of a drug product for later use. Low-ID columns have improved sensitivity and lower solvent consumption at the expense of loading capacity.

- Larger ID columns (over 10 mm) are used to purify usable amounts of material because of their large loading capacity.

- Analytical scale columns (4.6 mm) have been the most common type of columns, though smaller columns are rapidly gaining in popularity. They are used in traditional quantitative analysis of samples and often use a UV-Vis absorbance detector.

- Narrow-bore columns (1–2 mm) are used for applications when more sensitivity is desired either with special UV-vis detectors, fluorescence detection or with other detection methods like liquid chromatography-mass spectrometry

- Capillary columns (under 0.3 mm) are used almost exclusively with alternative detection means such as mass spectrometry. They are usually made from fused silica capillaries, rather than the stainless steel tubing that larger columns employ.

Particle Size

Most traditional HPLC is performed with the stationary phase attached to the outside of small spherical silica particles (very small beads). These particles come in a variety of sizes with 5 μm beads being the most common. Smaller particles generally provide more surface area and better separations, but the pressure required for optimum linear velocity increases by the inverse of the particle diameter squared.

This means that changing to particles that are half as big, keeping the size of the column the same, will double the performance, but increase the required pressure by a factor of four. Larger particles are used in preparative HPLC (column diameters 5 cm up to >30 cm) and for non-HPLC applications such as solid-phase extraction.

Pore Size

Many stationary phases are porous to provide greater surface area. Small pores provide greater surface area while larger pore size has better kinetics, especially for larger analytes. For example, a protein which is only slightly smaller than a pore might enter the pore but does not easily leave once inside.

Pump Pressure

Pumps vary in pressure capacity, but their performance is measured on their ability to yield a consistent and reproducible flow rate. Pressure may reach as high as 60 MPa (6000 lbf/in²), or about 600 atmospheres. Modern HPLC systems have been improved to work at much higher pressures, and therefore are able to use much smaller particle sizes in the columns (<2 μm). These "Ultra High Performance Liquid Chromatography" systems or UHPLCs can work at up to 120 MPa (17,405 lbf/in²), or about 1200 atmospheres. The term "UPLC" is a trademark of the Waters Corporation, but is sometimes used to refer to the more general technique of UHPLC.

Detectors

HPLC most commonly uses a UV-Vis absorbance detector, however, a wide range of other chromatography detectors can be used. A kind of commonly utilized detector includes refractive index detectors, which provide readings by measuring the changes in the refractive index of the effluent as it moves through the flow cell. In certain cases, it is possible to use multiple detectors, for example LCMS normally combines UV-Vis with a mass spectrometer.

Applications

Manufacturing

As briefly mentioned, HPLC has many applications in both laboratory and clinical science. It is a common technique used in pharmaceutical development as it is a dependable way to obtain and ensure product purity. While HPLC can produce extremely high quality (pure) products, it is not always the primary method used in the production of bulk drug materials. According to the European pharmacopoeia, HPLC is used in only 15.5% of syntheses. However, it plays a role in 44% of syntheses in the United States pharmacopoeia. This could possibly be due to differences in monetary and time constraints, as HPLC on a large scale can be an expensive technique. An increase in specificity, precision, and accuracy that occurs with HPLC unfortunately corresponds to an increase in cost.

Legal

This technique is also used for detection of illicit drugs in urine. The most common method of drug detection is an immunoassay. This method is much more convenient. However, convenience comes at the cost of specificity and coverage of a wide range of drugs. As HPLC is a method of determining (and possibly increasing) purity, using HPLC alone in evaluating concentrations of drugs is somewhat insufficient. With this, HPLC in this context is often performed in conjunction with mass spectrometry. Using liquid chromatography instead of gas chromatography in conjunction with MS circumvents the necessity for derivitizing with acetylating or alkylation agents, which can be a burdensome extra step. This technique has been used to detect a variety of agents like doping agents, drug metabolites, glucuronide conjugates, amphetamines, opioids, cocaine, BZDs, ketamine, LSD, cannabis, and pesticides Performing HPLC in conjunction with Mass spectrometry reduces the absolute need for standardizing HPLC experimental runs.

Research

Similar assays can be performed for research purposes, detecting concentrations of potential clinical candidates like anti-fungal and asthma drugs. This technique is obviously useful in observing multiple species in collected samples, as well, but requires the use of standard solutions when information about species identity is sought out. It is used as a method to confirm results of synthesis reactions, as purity is essential in this type of research. However, mass spectrometry is still the more reliable way to identify species.

Medical

Medical use of HPLC can include drug analysis, but falls more closely under the category of nutrient analysis. While urine is the most common medium for analyzing drug concentrations, blood serum is the sample collected for most medical analyses with HPLC. Other methods of detection of molecules that are useful for clinical studies have been tested against HPLC, namely immunoassays. In one example of this, competitive protein binding assays (CPBA) and HPLC were compared for sensitivity in detection of vitamin D. Useful for diagnosing vitamin D deficiencies in children, it was found that sensitivity and specificity of this CPBA reached only 40% and 60%, respectively, of the capacity of HPLC. While an expensive tool, the accuracy of HPLC is nearly unparalleled.

Supercritical Fluid Chromatography

Supercritical Fluid Chromatography (SFC) is a form of normal phase chromatography, first used in 1962, that is used for the analysis and purification of low to moderate molecular weight, thermally labile molecules. It can also be used for the separation of chiral compounds. Principles are similar to those of high performance liquid chromatography (HPLC), however SFC typically utilizes carbon dioxide as the mobile phase; therefore the entire chromatographic flow path must be pressurized. Because the supercritical phase represents a state in which liquid and gas properties converge, supercritical fluid chromatography is sometimes called "convergence chromatography."

Applications

SFC is used in industry primarily for separation of chiral molecules, and uses the same columns as standard HPLC systems. SFC is now commonly used for achiral separations and purifications in the pharmaceutical industry.

Apparatus

SFC with CO_2 utilizes carbon dioxide pumps that require that the incoming CO_2 and pump heads be kept cold in order to maintain the carbon dioxide at a temperature and pressure that keeps it in a liquid state where it can be effectively metered at some specified flow rate. The CO_2 subsequently becomes supercritical post the injector and in the column oven when the temperature and pressure it is subjected to are raised above the triple point of the liquid and the supercritical state is achieved. SFC as a chromatographic process has been likened to a process having the combined properties of the power of a liquid to dissolve a matrix, with the chromatographic interactions and kinetics of a gas. The result is that you can get a lot of mass on column per injection, and still maintain a high chromatographic efficiency. Typically, gradient elution is employed in analytical SFC using a polar co-solvent such as methanol, possibly with a weak acid or base at low concentrations ~1%. The effective plate counts per analysis can be observed to exceed 500K plates per metre routinely with 5 um material. The operator uses software to set mobile phase flow rate, co-solvent composition, system back pressure and column oven temperature which must exceed 40 °C for supercritical conditions to be achieved with CO_2. In addition, SFC provides an additional control parameter - pressure - by using an automated back pressure regulator. From an operational standpoint, SFC is as simple and robust as HPLC but fraction collection is more convenient because the primary mobile phase evaporates leaving only the analyte and a small volume of polar co-solvent. If the outlet CO_2 is captured, it can be recompressed and recycled, allowing for >90% reuse of CO_2.

Similar to HPLC, SFC uses a variety of detection methods including UV/VIS, mass spectrometry, FID (unlike HPLC) and evaporative light scattering.

Sample Preparation

A rule-of-thumb is that any molecule that will dissolve in methanol or a less polar solvent is compatible with SFC, including polar solutes. CO_2 has polarity similar to n-heptane at its critical point, but the solvent strength can be increased by increasing density or using a polar cosolvent. In practice, when the fraction of cosolvent is high, the mobile phase is not truly supercritical, but this terminology is used regardless.

Mobile Phase

The mobile phase is composed primarily of supercritical carbon dioxide, but since CO_2 on its own is too non-polar to effectively elute many analytes, cosolvents are added to modify the mobile phase polarity. Cosolvents are typically simple alcohols like methanol, ethanol, or isopropyl alcohol. Other solvents such as acetonitrile, chloroform, or ethyl acetate can be used as modifiers. For food-grade materials, the selected cosolvent is often ethanol or ethyl acetate, both of which are generally recognized as safe (GRAS). The solvent limitations are system and column based.

Drawbacks

There have been a few technical issues that have limited adoption of SFC technology, first of which is the high pressure operating conditions. High-pressure vessels are expensive and bulky, and special materials are often needed to avoid dissolving gaskets and O-rings in the supercritical fluid. A second drawback is difficulty in maintaining pressure (backpressure regulation). Whereas liquids are nearly incompressible, so their densities are constant regardless of pressure, supercritical fluids are highly compressible and their physical properties change with pressure - such as the pressure drop across a packed-bed column. Currently, automated backpressure regulators can maintain a constant pressure in the column even if flow rate varies, mitigating this problem. A third drawback is difficulty in gas/liquid separation during collection of product. Upon depressurization, the CO_2 rapidly turns into gas and aerosolizes any dissolved analyte in the process. Cyclone separators have lessened difficulties in gas/liquid separations.

Column Chromatography

Column chromatography in chemistry is a method used to purify individual chemical compounds from mixtures of compounds. It is often used for preparative applications on scales from micrograms up to kilograms. The main advantage of column chromatography is the relatively low cost and disposability of the stationary phase used in the process. The latter prevents cross-contamination and stationary phase degradation due to recycling.

The classical preparative chromatography column is a glass tube with a diameter from 5 mm to 50 mm and a height of 5 cm to 1 m with a tap and some kind of a filter (a glass frit or glass wool plug – to prevent the loss of the stationary phase) at the bottom. Two methods are generally used to prepare a column: the dry method and the wet method.

- For the dry method, the column is first filled with dry stationary phase powder, followed by the addition of mobile phase, which is flushed through the column until it is completely wet, and from this point is never allowed to run dry.

- For the wet method, a slurry is prepared of the eluent with the stationary phase powder and then carefully poured into the column. Care must be taken to avoid air bubbles. A solution of the organic material is pipetted on top of the stationary phase. This layer is usually topped with a small layer of sand or with cotton or glass wool to protect the shape of the organic layer from the velocity of newly added eluent. Eluent is slowly passed through the column to advance the organic material. Often a spherical eluent reservoir or an eluent-filled and stoppered separating funnel is put on top of the column.

A chemist in the 1950s using column chromatography. The Erlenmeyer receptacles are on the floor.

The individual components are retained by the stationary phase differently and separate from each other while they are running at different speeds through the column with the eluent. At the end of the column they elute one at a time. During the entire chromatography process the eluent is collected in a series of fractions. Fractions can be collected automatically by means of fraction collectors. The productivity of chromatography can be increased by running several columns at a time. In this case multi stream collectors are used. The composition of the eluent flow can be monitored and each fraction is analyzed for dissolved compounds, e.g. by analytical chromatography, UV absorption spectra, or fluorescence. Colored compounds (or fluorescent compounds with the aid of a UV lamp) can be seen through the glass wall as moving bands.

Stationary Phase

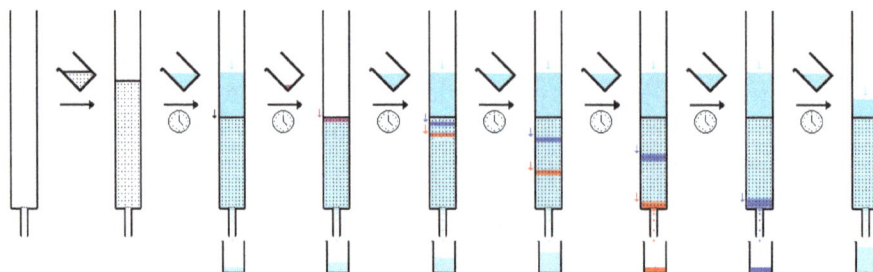
Column chromatography proceeds by a series of steps.

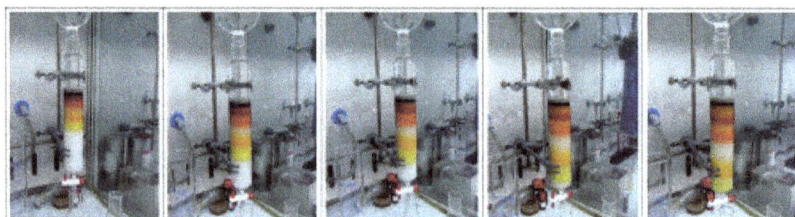
Photographic sequence of a column chromatography

The *stationary phase* or *adsorbent* in column chromatography is a solid. The most common stationary phase for column chromatography is silica gel, the next most common being alumina. Cellulose powder has often been used in the past. Also possible are ion exchange chromatography, reversed-phase chromatography (RP), affinity chromatography or expanded bed adsorption (EBA). The stationary phases are usually finely ground powders or gels and/or are microporous for an increased surface, though in EBA a fluidized bed is used. There is an important ratio between the stationary phase weight and the dry weight of the analyte mixture that can be applied onto the column. For silica column chromatography, this ratio lies within 20:1 to 100:1, depending on how close to each other the analyte components are being eluted.

Mobile Phase (Eluent)

The *mobile phase* or *eluent* is either a pure solvent or a mixture of different solvents. It is chosen so that the retention factor value of the compound of interest is roughly around 0.2 - 0.3 in order to minimize the time and the amount of eluent to run the chromatography. The eluent has also been chosen so that the different compounds can be separated effectively. The eluent is optimized in small scale pretests, often using thin layer chromatography (TLC) with the same stationary phase.

There is an optimum flow rate for each particular separation. A faster flow rate of the eluent minimizes the time required to run a column and thereby minimizes diffusion, resulting in a better separation. However, the maximum flow rate is limited because a finite time is required for the analyte to equilibrate between the stationary phase and mobile phase, A simple laboratory column runs by gravity flow. The flow rate of such a column can be increased by extending the fresh eluent filled column above the top of the stationary phase or decreased by the tap controls. Faster flow rates can be achieved by using a pump or by using compressed gas (e.g. air, nitrogen, or argon) to push the solvent through the column (flash column chromatography).

The particle size of the stationary phase is generally finer in flash column chromatography than in gravity column chromatography. For example, one of the most widely used silica gel grades in the former technique is mesh 230 – 400 (40 – 63 μm), while the latter technique typically requires mesh 70 – 230 (63 – 200 μm) silica gel.

A spreadsheet that assists in the successful development of flash columns has been developed. The spreadsheet estimates the retention volume and band volume of analytes, the fraction numbers expected to contain each analyte, and the resolution between adjacent peaks. This information allows users to select optimal parameters for preparative-scale separations before the flash column itself is attempted.

Automated Systems

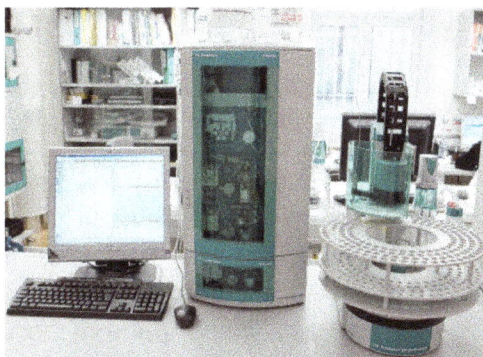

An automated ion chromatography system.

Column chromatography is an extremely time consuming stage in any lab and can quickly become the bottleneck for any process lab. Therefore, several manufacturers like Biotage, Buchi, Interchim and Teledyne Isco have developed automated flash chromatography systems (typically referred to as LPLC, low pressure liquid chromatography, around 350–525 kPa or 50.8–76.1 psi) that minimize human involvement in the purification process. Automated systems will include components normally found on more expensive high performance liquid chromatography (HPLC) systems such as a gradient pump, sample injection ports, a UV detector and a fraction collector to collect the eluent. Typically these automated systems can separate samples from a few milligrams up to an industrial many kilogram scale and offer a much cheaper and quicker solution to doing multiple injections on prep-HPLC systems.

The resolution (or the ability to separate a mixture) on an LPLC system will always be lower compared to HPLC, as the packing material in an HPLC column can be much smaller, typically only 5 micrometre thus increasing stationary phase surface area, increasing surface interactions and giving better separation. However, the use of this small packing media causes the high back pressure and is why it is termed high pressure liquid chromatography. The LPLC columns are typically packed with silica of around 50 micrometres, thus reducing back pressure and resolution, but it also removes the need for expensive high pressure pumps. Manufacturers are now starting to move into higher pressure flash chromatography systems and have termed these as medium pressure liquid chromatography (MPLC) systems which operate above 1 MPa (150 psi).

Column Chromatogram Resolution Calculation

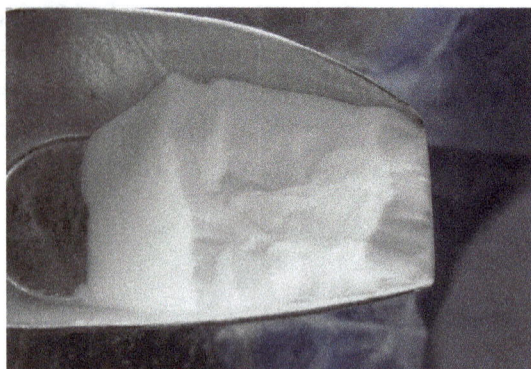

Powdery silica gel for column chromatography

Typically, column chromatography is set up with peristaltic pumps, flowing buffers and the solution sample through the top of the column. The solutions and buffers pass through the column where a fraction collector at the end of the column setup collects the eluted samples. Prior to the fraction collection, the samples that are eluted from the column pass through a detector such as a spectrophotometer or mass spectrometer so that the concentration of the separated samples in the sample solution mixture can be determined.

For example, if you were to separate two different proteins with different binding capacities to the column from a solution sample, a good type of detector would be a spectrophotometer using a wavelength of 280 nm. The higher the concentration of protein that passes through the eluted solution through the column, the higher the absorbance of that wavelength.

Because the column chromatography has a constant flow of eluted solution passing through the detector at varying concentrations, the detector must plot the concentration of the eluted sample over a course of time. This plot of sample concentration versus time is called a chromatogram.

The ultimate goal of chromatography is to separate different components from a solution mixture. The resolution expresses the extent of separation between the components from the mixture. The higher the resolution of the chromatogram, the better the extent of separation of the samples the column gives. This data is a good way of deter-

mining the column's separation properties of that particular sample. The resolution can be calculated from the chromatogram.

The separate curves in the diagram represent different sample elution concentration profiles over time based on their affinity to the column resin. To calculate resolution, the retention time and curve width are required.

Retention time: The time from the start of signal detection by the detector to the peak height of the elution concentration profile of each different sample.

Curve width: The width of the concentration profile curve of the different samples in the chromatogram in units of time.

A simplified method of calculating chromatogram resolution is to use the plate model. The plate model assumes that the column can be divided into a certain number of sections, or plates and the mass balance can be calculated for each individual plate. This approach approximates a typical chromatogram curve as a Gaussian distribution curve. By doing this, the curve width is estimated as 4 times the standard deviation of the curve, 4σ. The retention time is the time from the start of signal detection to the time of the peak height of the Gaussian curve.

From the variables in the figure above, the resolution, plate number, and plate height of the column plate model can be calculated using the equations:

Resolution (R_s)

$R_s = 2(t_{RB} - t_{RA})/(w_B + w_A)$ Where: t_{RB} = retention time of solute B t_{RA} = retention time of solute A w_B = Gaussian curve width of solute B w_A = Gaussian curve width of solute A Plate Number (N): $N = (t_R)^2/(w/4)^2$ Plate Height (H): $H = L/N$ Where L is the length of the column.

Column Adsorption Equilibrium

For an adsorption column, the column resin (the stationary phase) is composed of microbeads. Even smaller particles such as proteins, carbohydrates, metal ions, or other chemical compounds are conjugated onto the microbeads. Each binding particle that is attached to the microbead can be assumed to bind in a 1:1 ratio with the solute sample sent through the column that needs to be purified or separated.

Binding between the target molecule to be separated and the binding molecule on the column beads can be modeled using a simple equilibrium reaction $K_{eq} = [CS]/([C][S])$ where K_{eq} is the equilibrium constant, [C] and [S] are the concentrations of the target molecule and the binding molecule on the column resin, respectively. [CS] is the concentration of the complex of the target molecule bound to the column resin.

Using this as a basis, three different isotherms can be used to describe the binding dynamics of a column chromatography: linear, Langmuir, and Freundlich.

The linear isotherm occurs when the solute concentration needed to be purified is very small relative to the binding molecule. Thus, the equilibrium can be defined as:

$[CS] = K_{eq}[C]$.

For industrial scale uses, the total binding molecules on the column resin beads must be factored in because unoccupied sites must be taken into account. The Langmuir isotherm and Freundlich isotherm are useful in describing this equilibrium. Langmuir Isotherm: $[CS] = (K_{eq}S_{tot}[C])/(1 + K_{eq}[C])$, where S_{tot} is the total binding molecules on the beads.

Freundlich Isotherm:

$[CS] = K_{eq}[C]^{1/n}$

The Freundlich isotherm is used when the column can bind to many different samples in the solution that needs to be purified. Because the many different samples have different binding constants to the beads, there are many different Keq's. Therefore, the Langmuir isotherm is not a good model for binding in this case.

Gel Electrophoresis

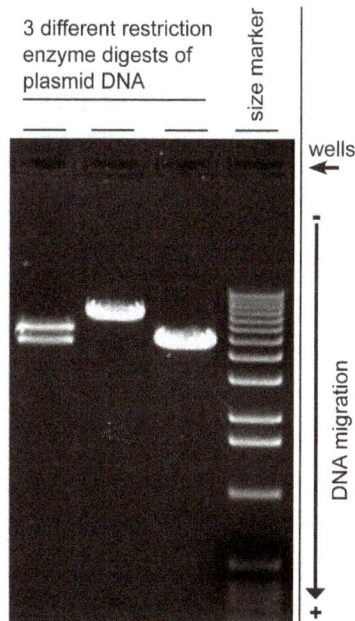

Digital image of 3 plasmid restriction digests run on a 1% w/v agarose gel, 3 volt/cm, stained with ethidium bromide. The DNA size marker is a commercial 1 kbp ladder. The position of the wells and direction of DNA migration is noted.

Gel electrophoresis is a method for separation and analysis of macromolecules (DNA, RNA and proteins) and their fragments, based on their size and charge. It is used in clinical chemistry to separate proteins by charge and/or size (IEF agarose, essentially size independent) and in biochemistry and molecular biology to separate a mixed population of DNA and RNA fragments by length, to estimate the size of DNA and RNA fragments or to separate proteins by charge.

Nucleic acid molecules are separated by applying an electric field to move the negatively charged molecules through a matrix of agarose or other substances. Shorter molecules move faster and migrate farther than longer ones because shorter molecules migrate more easily through the pores of the gel. This phenomenon is called sieving. Proteins are separated by charge in agarose because the pores of the gel are too large to sieve proteins. Gel electrophoresis can also be used for separation of nanoparticles.

Gel electrophoresis uses a gel as an anticonvective medium and/or sieving medium during electrophoresis, the movement of a charged particle in an electrical field. Gels suppress the thermal convection caused by application of the electric field, and can also act as a sieving medium, retarding the passage of molecules; gels can also simply serve to maintain the finished separation, so that a post electrophoresis stain can be applied. DNA Gel electrophoresis is usually performed for analytical purposes, often after amplification of DNA via PCR, but may be used as a preparative technique prior to use of other methods such as mass spectrometry, RFLP, PCR, cloning, DNA sequencing, or Southern blotting for further characterization.

Physical Basis

In simple terms, electrophoresis is a process which enables the sorting of molecules based on size. Using an electric field, molecules (such as DNA) can be made to move through a gel made of agar or polyacrylamide. The electric field consists of a negative charge at one end which pushes the molecules through the gel, and a positive charge at the other end that pulls the molecules through the gel. The molecules being sorted are dispensed into a well in the gel material. The gel is placed in an electrophoresis chamber, which is then connected to a power source. When the electric current is applied, the larger molecules move more slowly through the gel while the smaller molecules move faster. The different sized molecules form distinct bands on the gel.

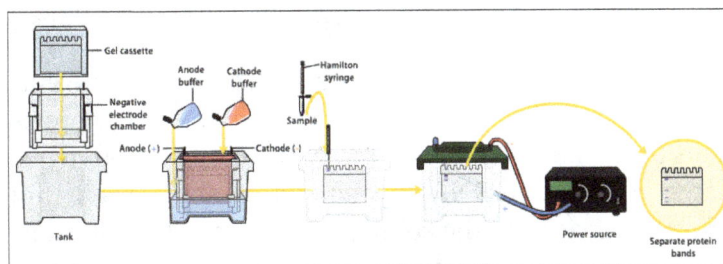

Overview of Gel Electrophoresis.

The term "gel" in this instance refers to the matrix used to contain, then separate the target molecules. In most cases, the gel is a crosslinked polymer whose composition and porosity is chosen based on the specific weight and composition of the target to be analyzed. When separating proteins or small nucleic acids (DNA, RNA, or oligonucleotides) the gel is usually composed of different concentrations of acrylamide and a cross-linker, producing different sized mesh networks of polyacrylamide. When separating larger nucleic acids (greater than a few hundred bases), the preferred matrix is purified agarose. In both cases, the gel forms a solid, yet porous matrix. Acrylamide, in contrast to polyacrylamide, is a neurotoxin and must be handled using appropriate safety precautions to avoid poisoning. Agarose is composed of long unbranched chains of uncharged carbohydrate without cross links resulting in a gel with large pores allowing for the separation of macromolecules and macromolecular complexes.

"Electrophoresis" refers to the electromotive force (EMF) that is used to move the molecules through the gel matrix. By placing the molecules in wells in the gel and applying an electric field, the molecules will move through the matrix at different rates, determined largely by their mass when the charge to mass ratio (Z) of all species is uniform. However, when charges are not all uniform then, the electrical field generated by the electrophoresis procedure will affect the species that have different charges and therefore will attract the species according to their charges being the opposite. Species that are positively charged will migrate towards the cathode which is negatively charged (because this is an electrolytic rather than galvanic cell). If the species are negatively charged they will migrate towards the positively charged anode.

If several samples have been loaded into adjacent wells in the gel, they will run parallel in individual lanes. Depending on the number of different molecules, each lane shows separation of the components from the original mixture as one or more distinct bands, one band per component. Incomplete separation of the components can lead to overlapping bands, or to indistinguishable smears representing multiple unresolved components. Bands in different lanes that end up at the same distance from the top contain molecules that passed through the gel with the same speed, which usually means they are approximately the same size. There are molecular weight size markers available that contain a mixture of molecules of known sizes. If such a marker was run on one lane in the gel parallel to the unknown samples, the bands observed can be compared to those of the unknown in order to determine their size. The distance a band travels is approximately inversely proportional to the logarithm of the size of the molecule.

There are limits to electrophoretic techniques. Since passing current through a gel causes heating, gels may melt during electrophoresis. Electrophoresis is performed in buffer solutions to reduce pH changes due to the electric field, which is important because the charge of DNA and RNA depends on pH, but running for too long can exhaust the buffering capacity of the solution. Further, different preparations of genetic material may not migrate consistently with each other, for morphological or other reasons.

Types of Gel

The types of gel most typically used are agarose and polyacrylamide gels. Each type of gel is well-suited to different types and sizes of analyte. Polyacrylamide gels are usually used for proteins, and have very high resolving power for small fragments of DNA (5-500 bp). Agarose gels on the other hand have lower resolving power for DNA but have greater range of separation, and are therefore used for DNA fragments of usually 50-20,000 bp in size, but resolution of over 6 Mb is possible with pulsed field gel electrophoresis (PFGE). Polyacrylamide gels are run in a vertical configuration while agarose gels are typically run horizontally in a submarine mode. They also differ in their casting methodology, as agarose sets thermally, while polyacrylamide forms in a chemical polymerization reaction.

Agarose

Agarose gels are made from the natural polysaccharide polymers extracted from seaweed. Agarose gels are easily cast and handled compared to other matrices, because the gel setting is a physical rather than chemical change. Samples are also easily recovered. After the experiment is finished, the resulting gel can be stored in a plastic bag in a refrigerator.

Agarose gels do not have a uniform pore size, but are optimal for electrophoresis of proteins that are larger than 200 kDa. Agarose gel electrophoresis can also be used for the separation of DNA fragments ranging from 50 base pair to several megabases (millions of bases), the largest of which require specialized apparatus. The distance between DNA bands of different lengths is influenced by the percent agarose in the gel, with higher percentages requiring longer run times, sometimes days. Instead high percentage agarose gels should be run with a pulsed field electrophoresis (PFE), or field inversion electrophoresis.

"Most agarose gels are made with between 0.7% (good separation or resolution of large 5–10kb DNA fragments) and 2% (good resolution for small 0.2–1kb fragments) agarose dissolved in electrophoresis buffer. Up to 3% can be used for separating very tiny fragments but a vertical polyacrylamide gel is more appropriate in this case. Low percentage gels are very weak and may break when you try to lift them. High percentage gels are often brittle and do not set evenly. 1% gels are common for many applications."

Polyacrylamide

Polyacrylamide gel electrophoresis (PAGE) is used for separating proteins ranging in size from 5 to 2,000 kDa due to the uniform pore size provided by the polyacrylamide gel. Pore size is controlled by modulating the concentrations of acrylamide and bis-acrylamide powder used in creating a gel. Care must be used when creating this type of gel, as acrylamide is a potent neurotoxin in its liquid and powdered forms.

Traditional DNA sequencing techniques such as Maxam-Gilbert or Sanger methods used polyacrylamide gels to separate DNA fragments differing by a single base-pair in length so the sequence could be read. Most modern DNA separation methods now use agarose gels, except for particularly small DNA fragments. It is currently most often used in the field of immunology and protein analysis, often used to separate different proteins or isoforms of the same protein into separate bands. These can be transferred onto a nitrocellulose or PVDF membrane to be probed with antibodies and corresponding markers, such as in a western blot.

Typically resolving gels are made in 6%, 8%, 10%, 12% or 15%. Stacking gel (5%) is poured on top of the resolving gel and a gel comb (which forms the wells and defines the lanes where proteins, sample buffer and ladders will be placed) is inserted. The percentage chosen depends on the size of the protein that one wishes to identify or probe in the sample. The smaller the known weight, the higher the percentage that should be used. Changes on the buffer system of the gel can help to further resolve proteins of very small sizes.

Starch

Partially hydrolysed potato starch makes for another non-toxic medium for protein electrophoresis. The gels are slightly more opaque than acrylamide or agarose. Non-denatured proteins can be separated according to charge and size. They are visualised using Napthal Black or Amido Black staining. Typical starch gel concentrations are 5% to 10%.

Gel Conditions

Denaturing

TTGE profiles representing the bifidobacterial diversity of fecal samples from two healthy volunteers (A and B) before and after AMC (Oral Amoxicillin-Clavulanic Acid) treatment

Denaturing gels are run under conditions that disrupt the natural structure of the analyte, causing it to unfold into a linear chain. Thus, the mobility of each macromolecule depends only on its linear length and its mass-to-charge ratio. Thus, the secondary, tertiary, and quaternary levels of biomolecular structure are disrupted, leaving only the primary structure to be analyzed.

Nucleic acids are often denatured by including urea in the buffer, while proteins are denatured using sodium dodecyl sulfate, usually as part of the SDS-PAGE process. For full denaturation of proteins, it is also necessary to reduce the covalent disulfide bonds that stabilize their tertiary and quaternary structure, a method called reducing PAGE. Reducing conditions are usually maintained by the addition of beta-mercaptoethanol or dithiothreitol. For general analysis of protein samples, reducing PAGE is the most common form of protein electrophoresis.

Denaturing conditions are necessary for proper estimation of molecular weight of RNA. RNA is able to form more intramolecular interactions than DNA which may result in change of its electrophoretic mobility. Urea, DMSO and glyoxal are the most often used denaturing agents to disrupt RNA structure. Originally, highly toxic methylmercury hydroxide was often used in denaturing RNA electrophoresis, but it may be method of choice for some samples.

Denaturing gel electrophoresis is used in the DNA and RNA banding pattern-based methods DGGE (denaturing gradient gel electrophoresis), TGGE (temperature gradient gel electrophoresis), and TTGE (temporal temperature gradient electrophoresis).

Native

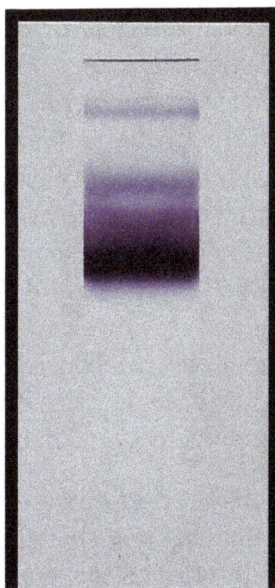

Specific enzyme-linked staining: Glucose-6-Phosphate Dehydrogenase isoenzymes in *Plasmodium falciparum* infected Red blood cells

Native gels are run in non-denaturing conditions, so that the analyte's natural structure is maintained. This allows the physical size of the folded or assembled complex to affect the mobility, allowing for analysis of all four levels of the biomolecular structure. For biological samples, detergents are used only to the extent that they are necessary to lyse lipid membranes in the cell. Complexes remain—for the most part—associated and folded as they would be in the cell. One downside, however, is that complexes may not separate cleanly or predictably, as it is difficult to predict how the molecule's shape and size will affect its mobility. Addressing and solving this problem is a major aim of quantitative native PAGE.

Unlike denaturing methods, native gel electrophoresis does not use a charged denaturing agent. The molecules being separated (usually proteins or nucleic acids) therefore differ not only in molecular mass and intrinsic charge, but also the cross-sectional area, and thus experience different electrophoretic forces dependent on the shape of the overall structure. For proteins, since they remain in the native state they may be visualised not only by general protein staining reagents but also by specific enzyme-linked staining.

Native gel electrophoresis is typically used in proteomics and metallomics. However, native PAGE is also used to scan genes (DNA) for unknown mutations as in Single-strand conformation polymorphism.

Buffers

Buffers in gel electrophoresis are used to provide ions that carry a current and to maintain the pH at a relatively constant value. There are a number of buffers used for electrophoresis. The most common being, for nucleic acids Tris/Acetate/EDTA (TAE), Tris/Borate/EDTA (TBE). Many other buffers have been proposed, e.g. lithium borate, which is almost never used, based on Pubmed citations (LB), iso electric histidine, pK matched goods buffers, etc.; in most cases the purported rationale is lower current (less heat) and or matched ion mobilities, which leads to longer buffer life. Borate is problematic; Borate can polymerize, and/or interact with cis diols such as those found in RNA. TAE has the lowest buffering capacity but provides the best resolution for larger DNA. This means a lower voltage and more time, but a better product. LB is relatively new and is ineffective in resolving fragments larger than 5 kbp; However, with its low conductivity, a much higher voltage could be used (up to 35 V/cm), which means a shorter analysis time for routine electrophoresis. As low as one base pair size difference could be resolved in 3% agarose gel with an extremely low conductivity medium (1 mM Lithium borate).

Most SDS-PAGE protein separations are performed using a "discontinuous" (or DISC) buffer system that significantly enhances the sharpness of the bands within the gel. During electrophoresis in a discontinuous gel system, an ion gradient is formed in the early stage of electrophoresis that causes all of the proteins to focus into a single sharp band in a process called isotachophoresis. Separation of the proteins by size is achieved in the lower, "resolving" region of the gel. The resolving gel typically has a much smaller pore size, which leads to a sieving effect that now determines the electrophoretic mobility of the proteins.

Visualization

After the electrophoresis is complete, the molecules in the gel can be stained to make them visible. DNA may be visualized using ethidium bromide which, when intercalated into DNA, fluoresce under ultraviolet light, while protein may be visualised using silver stain or Coomassie Brilliant Blue dye. Other methods may also be used to visualize the separation of the mixture's components on the gel. If the molecules to be separated contain radioactivity, for example in a DNA sequencing gel, an autoradiogram can be recorded of the gel. Photographs can be taken of gels, often using a Gel Doc system.

Downstream Processing

After separation, an additional separation method may then be used, such as isoelectric focusing or SDS-PAGE. The gel will then be physically cut, and the protein complexes extracted from each portion separately. Each extract may then be analysed, such as by peptide mass fingerprinting or de novo peptide sequencing after in-gel digestion. This can provide a great deal of information about the identities of the proteins in a complex.

Applications

- Estimation of the size of DNA molecules following restriction enzyme digestion, e.g. in restriction mapping of cloned DNA.

- Analysis of PCR products, e.g. in molecular genetic diagnosis or genetic finger-printing

- Separation of restricted genomic DNA prior to Southern transfer, or of RNA prior to Northern transfer.

Gel electrophoresis is used in forensics, molecular biology, genetics, microbiology and biochemistry. The results can be analyzed quantitatively by visualizing the gel with UV light and a gel imaging device. The image is recorded with a computer operated camera, and the intensity of the band or spot of interest is measured and compared against standard or markers loaded on the same gel. The measurement and analysis are mostly done with specialized software.

Depending on the type of analysis being performed, other techniques are often implemented in conjunction with the results of gel electrophoresis, providing a wide range of field-specific applications.

Nucleic Acids

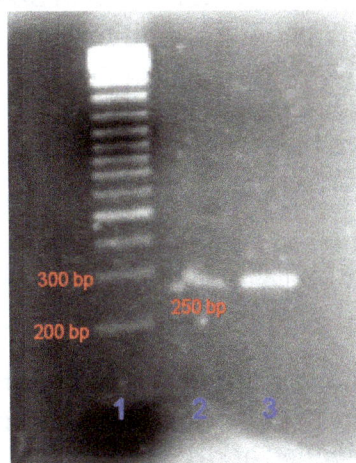

An agarose gel of a PCR product compared to a DNA ladder.

In the case of nucleic acids, the direction of migration, from negative to positive electrodes, is due to the naturally occurring negative charge carried by their sugar-phosphate backbone.

Double-stranded DNA fragments naturally behave as long rods, so their migration through the gel is relative to their size or, for cyclic fragments, their radius of gyration. Circular DNA such as plasmids, however, may show multiple bands, the speed of migration may depend on whether it is relaxed or supercoiled. Single-stranded DNA or RNA tend to fold up into molecules with complex shapes and migrate through the gel in a complicated manner based on their tertiary structure. Therefore, agents that disrupt the hydrogen bonds, such as sodium hydroxide or formamide, are used to denature the nucleic acids and cause them to behave as long rods again.

Gel electrophoresis of large DNA or RNA is usually done by agarose gel electrophoresis. Characterization through ligand interaction of nucleic acids or fragments may be performed by mobility shift affinity electrophoresis.

Electrophoresis of RNA samples can be used to check for genomic DNA contamination and also for RNA degradation. RNA from eukaryotic organisms shows distinct bands of 28s and 18s rRNA, the 28s band being approximately twice as intense as the 18s band. Degraded RNA has less sharply defined bands, has a smeared appearance, and intensity ratio is less than 2:1.

Proteins

Proteins, unlike nucleic acids, can have varying charges and complex shapes, therefore they may not migrate into the polyacrylamide gel at similar rates, or at all, when placing a negative to positive EMF on the sample. Proteins therefore, are usually denatured in the presence of a detergent such as sodium dodecyl sulfate (SDS) that coats the proteins with a negative charge. Generally, the amount of SDS bound is relative to the size of the protein (usually 1.4g SDS per gram of protein), so that the resulting denatured proteins have an overall negative charge, and all the proteins have a similar charge to mass ratio. Since denatured proteins act like long rods instead of having a complex tertiary shape, the rate at which the resulting SDS coated proteins migrate in the gel is relative only to its size and not its charge or shape.

Proteins are usually analyzed by sodium dodecyl sulfate polyacrylamide gel electrophoresis (SDS-PAGE), by native gel electrophoresis, by preparative gel electrophoresis (QPNC-PAGE), or by 2-D electrophoresis.

Characterization through ligand interaction may be performed by electroblotting or by affinity electrophoresis in agarose or by capillary electrophoresis as for estimation of binding constants and determination of structural features like glycan content through lectin binding.

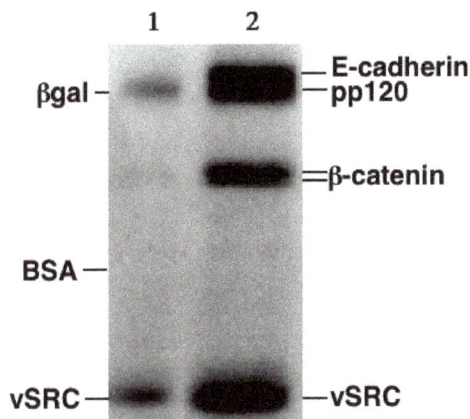

SDS-PAGE autoradiography – The indicated proteins are present in different concentrations in the two samples.

History

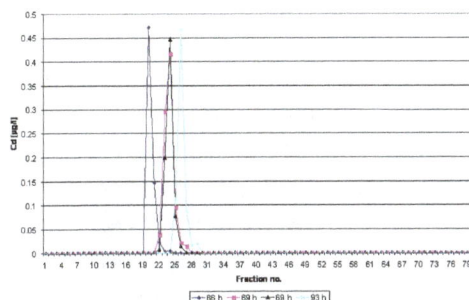

Electropherogram showing four PAGE runs of a native chromatographically prepurified high molecular weight cadmium protein ($\approx$ 200 ku) as a function of time of polymerization of the gel (4% T, 2.67% C).

- 1930s – first reports of the use of sucrose for gel electrophoresis

- 1955 – introduction of starch gels, mediocre separation (Smithies)

- 1959 – introduction of acrylamide gels; disc electrophoresis (Ornstein and Davis); accurate control of parameters such as pore size and stability; and (Raymond and Weintraub)

- 1966 – first use of agar gels

- 1969 – introduction of denaturing agents especially SDS separation of protein subunit (Weber and Osborn)

- 1970 – Laemmli separated 28 components of T4 phage using a stacking gel and SDS

- 1972 – agarose gels with ethidium bromide stain

- 1975 – 2-dimensional gels (O'Farrell); isoelectric focusing then SDS gel electrophoresis

- 1977 – sequencing gels

- 1983 – pulsed field gel electrophoresis enables separation of large DNA molecules

- 1983 – introduction of capillary electrophoresis

- 2004 – standardized time of polymerization of acrylamide gels enables clean and predictable separation of native proteins (Kastenholz)

A 1959 book on electrophoresis by Milan Bier cites references from the 1800s. However, Oliver Smithies made significant contributions. Bier states: "The method of Smithies ... is finding wide application because of its unique separatory power." Taken in context, Bier clearly implies that Smithies' method is an improvement.

Agarose Gel Electrophoresis

Agarose gel electrophoresis is a method of gel electrophoresis used in biochemistry, molecular biology, genetics, and clinical chemistry to separate a mixed population of macromolecules such as DNA or proteins in a matrix of agarose. The proteins may be separated by charge and/or size (isoelectric focusing agarose electrophoresis is essentially size independent), and the DNA and RNA fragments by length. Biomolecules are separated by applying an electric field to move the charged molecules through an agarose matrix, and the biomolecules are separated by size in the agarose gel matrix.

Agarose gel is easy to cast, has relatively fewer charged groups, and is particularly suitable for separating DNA of size range most often encountered in laboratories, which accounts for the popularity of its use. The separated DNA may be viewed with stain, most commonly under UV light, and the DNA fragments can be extracted from the gel with relative ease. Most agarose gels used are between 0.7 - 2% dissolved in a suitable electrophoresis buffer.

Properties of Agarose Gel

Agarose gel is a three-dimensional matrix formed of helical agarose molecules in supercoiled bundles that are aggregated into three-dimensional structures with channels and pores through which biomolecules can pass. The 3-D structure is held together with hydrogen bonds and can therefore be disrupted by heating back to a liquid state. The melting temperature is different from the gelling temperature, depending on the sources, agarose gel has a gelling temperature of 35-42 °C and a melting temperature of 85-95 °C. Low-melting and low-gelling agaroses made through chemical modifications are also available.

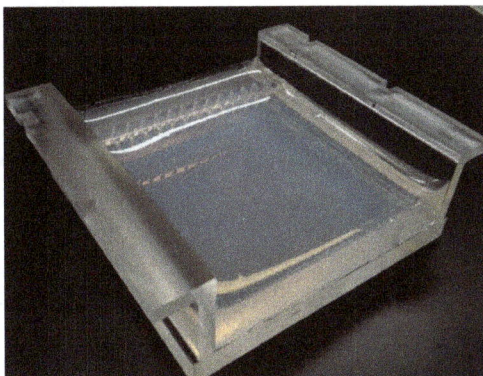

An agarose gel cast in tray used for gel electrophoresis

Agarose gel has large pore size and good gel strength, making it suitable as an anticonvection medium for the electrophoresis of DNA and large protein molecules. The pore size of a 1% gel has been estimated from 100 nm to 200-500 nm, and its gel strength allows gels as dilute as 0.15% to form a slab for gel electrophoresis. Low-concentration gels (0.1 - 0.2%) however are fragile and therefore hard to handle. Agarose gel has lower resolving power than polyacrylamide gel for DNA but has a greater range of separation, and is therefore used for DNA fragments of usually 50-20,000 bp in size. The limit of resolution for standard agarose gel electrophoresis is around 750 kb, but resolution of over 6 Mb is possible with pulsed field gel electrophoresis (PFGE). It can also be used to separate large proteins, and it is the preferred matrix for the gel electrophoresis of particles with effective radii larger than 5-10 nm. A 0.9% agarose gel has pores large enough for the entry of bacteriophage T4.

The agarose polymer contains charged groups, in particular pyruvate and sulphate. These negatively charged groups create a flow of water in the opposite direction to the movement of DNA in a process called electroendosmosis (EEO), and can therefore retard the movement of DNA and cause blurring of bands. Higher concentration gel would have higher electroosmotic flow. Low EEO agarose is therefore generally preferred for use in agarose gel electrophoresis of nucleic acids, but high EEO agarose may be used for other purposes. The lower sulphate content of low EEO agarose, particularly low-melting point (LMP) agarose, is also beneficial in cases where the DNA extracted from gel is to be used for further manipulation as the presence of contaminating sulphates may affect some subsequent procedures, such as ligation and PCR. Zero EEO agaroses however are undesirable for some applications as they may be made by adding positively charged groups and such groups can affect subsequent enzyme reactions. Electroendosmosis is a reason agarose is used in preference to agar as the agaropectin component in agar contains a significant amount of negatively charged sulphate and carboxyl groups. The removal of agaropectin in agarose substantially reduce the EEO, as well as reducing the non-specific adsorption of biomolecules to the gel matrix. However, for some applications such as the electrophoresis of serum proteins, a high EEO may be desirable, and agaropeptin may be added in the gel used.

Migration of Nucleic Acids in Agarose Gel

Factors Affecting Migration of Nucleic Acid in Gel

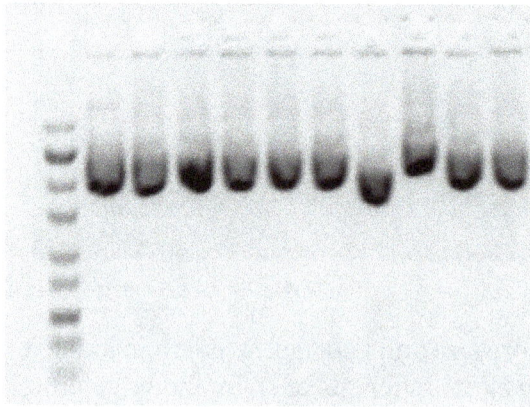

Gels of plasmid preparations usually show a major band of supercoiled DNA with other fainter bands in the same lane. Note that by convention DNA gel is displayed with smaller DNA fragments nearer to the bottom of the gel. This is because historically DNA gels were run vertically and the smaller DNA fragments move downwards faster.

A number of factors can affect the migration of nucleic acids: the dimension of the gel pores (gel concentration), size of DNA being electrophoresed, the voltage used, the ionic strength of the buffer, and the concentration of intercalating dye such as ethidium bromide if used during electrophoresis.

Smaller molecules travel faster than larger molecules in gel, and double-stranded DNA moves at a rate that is inversely proportional to the $\log_{10}$ of the number of base pairs. This relationship however breaks down with very large DNA fragments, and separation of very large DNA fragments requires the use of pulsed field gel electrophoresis (PFGE).

For standard agarose gel electrophoresis, larger molecules are resolved better using a low concentration gel while smaller molecules separate better at high concentration gel. High concentrations gel however requires longer run times (sometimes days).

The movement of the DNA may be affected by the conformation of the DNA molecule, for example, supercoiled DNA usually moves faster than relaxed DNA because it is tightly coiled and hence more compact. In a normal plasmid DNA preparation, multiple forms of DNA may be present. Gel electrophoresis of the plasmids would normally show the negatively supercoiled form as the main band, while nicked DNA (open circular form) and the relaxed closed circular form appears as minor bands. The rate at which the various forms move however can change using different electrophoresis conditions, and the mobility of larger circular DNA may be more strongly affected than linear DNA by the pore size of the gel.

Ethidium bromide which intercalates into circular DNA can change the charge, length, as well as the superhelicity of the DNA molecule, therefore its presence in gel during electrophoresis can affect its movement. Agarose gel electrophoresis can be used to resolve circular DNA with different supercoiling topology.

DNA damage due to increased cross-linking will also reduce electrophoretic DNA migration in a dose-dependent way.

The rate of migration of the DNA is proportional to the voltage applied, i.e. the higher the voltage, the faster the DNA moves. The resolution of large DNA fragments however is lower at high voltage. The mobility of DNA may also change in an unsteady field - in a field that is periodically reversed, the mobility of DNA of a particular size may drop significantly at a particular cycling frequency. This phenomenon can result in band inversion in field inversion gel electrophoresis (FIGE), whereby larger DNA fragments move faster than smaller ones.

Migration Anomalies

- "Smiley" gels - this edge effect is caused when the voltage applied is too high for the gel concentration used.

- Overloading of DNA - overloading of DNA slows down the migration of DNA fragments.

- Contamination - presence of impurities, such as salts or proteins can affect the movement of the DNA.

Mechanism of Migration and Separation

The negative charge of its phosphate backbone moves the DNA towards the positively charged anode during electrophoresis. However, the migration of DNA molecules in solution, in the absence of a gel matrix, is independent of molecular weight during electrophoresis. The gel matrix is therefore responsible for the separation of DNA by size during electrophoresis, and a number of models exist to explain the mechanism of separation of biomolecules in gel matrix. A widely accepted one is the Ogston model which treats the polymer matrix as a sieve. A globular protein or a random coil DNA moves through the interconnected pores, and the movement of larger molecules is more likely to be impeded and slowed down by collisions with the gel matrix, and the molecules of different sizes can therefore be separated in this sieving process.

The Ogston model however breaks down for large molecules whereby the pores are significantly smaller than size of the molecule. For DNA molecules of size greater than 1 kb, a reptation model (or its variants) is most commonly used. This model assumes that the DNA can crawl in a "snake-like" fashion (hence "reptation") through the pores as an elongated molecule. A biased reptation model applies at higher electric field strength, whereby the leading end of the molecule become strongly biased in the forward direction and pulls the rest of the molecule along. Real-time fluorescence microscopy of stained molecules, however, showed more subtle dynamics during electrophoresis, with the DNA showing considerable elasticity as it alternately stretching in the direction of the applied field and then contracting into a ball, or becoming hooked into a U-shape when it gets caught on the polymer fibres.

General Procedure

The details of an agarose gel electrophoresis experiment may vary depending on methods, but most follow a general procedure.

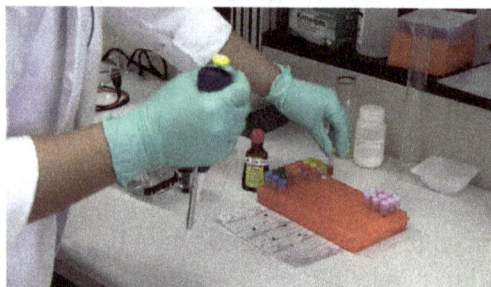

Video showing assembly of the rig and loading/running of the gel.

Casting of Gel

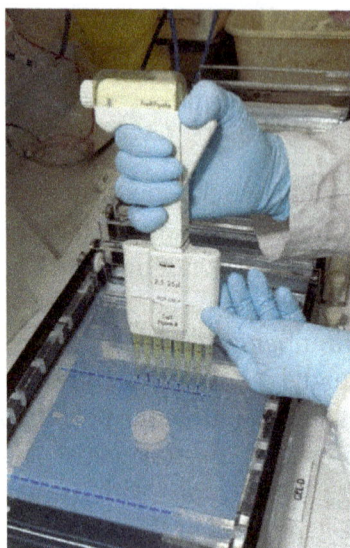

Loading DNA samples into the wells of an agarose gel using a multi-channel pipette.

The gel is prepared by dissolving the agarose powder in an appropriate buffer, such as TAE or TBE, to be used in electrophoresis. The agarose is dispersed in the buffer before heating it to near-boiling point, but avoid boiling. The melted agarose is allowed to cool sufficiently before pouring the solution into a cast as the cast may warp or crack if the agarose solution is too hot. A comb is placed in the cast to create wells for loading sample, and the gel should be completely set before use.

The concentration of gel affects the resolution of DNA separation. For a standard agarose gel electrophoresis, a 0.8% gives good separation or resolution of large 5–10kb DNA fragments, while 2% gel gives good resolution for small 0.2–1kb fragments. 1% gels is often used for a standard electrophoresis. The concentration is measured in weight of agarose over volume of buffer used (g/ml). High percentage gels are often brittle and may not set evenly, while low percentage gels (0.1-0.2%) are fragile and not easy to handle. Low-melting-point (LMP) agarose gels are also more fragile than normal agarose gel. Low-melting point agarose may be used on its own or simultaneously with standard agarose for the separation and isolation of DNA. PFGE and FIGE are often done with high percentage agarose gels.

Loading of Samples

Once the gel has set, the comb is removed, leaving wells where DNA samples can be loaded. Loading buffer is mixed with the DNA sample before the mixture is loaded into the wells. The loading buffer contains a dense compound, which may be glycerol, sucrose, or Ficoll, that raises the density of the sample so that the DNA sample may sink to the bottom of the well. If the DNA sample contains residual ethanol after its preparation, it may float out of the well. The loading buffer also include colored dyes such as xylene cyanol and bromophenol blue used to monitor the progress of the electrophoresis. The DNA samples are loaded using a pipette.

Electrophoresis

Agarose gel slab in electrophoresis tank with bands of dyes indicating progress of the electrophoresis. The DNA moves towards anode.

Agarose gel electrophoresis is most commonly done horizontally in a submarine mode whereby the slab gel is completely submerged in buffer during electrophoresis. It is also possible, but less common, to perform the electrophoresis vertically, as well as horizontally with the gel raised on agarose legs using an appropriate apparatus. The buffer used in the gel is the same as the running buffer in the electrophoresis tank, which is why electrophoresis in the submarine mode is possible with agarose gel.

For optimal resolution of DNA greater than 2 kb in size in standard gel electrophoresis,

5 to 8 V/cm is recommended (the distance in cm refers to the distance between electrodes, therefore this recommended voltage would be 5 to 8 multiplied by the distance between the electrodes in cm). Voltage may also be limited by the fact that it heats the gel and may cause the gel to melt if it is run at high voltage for a prolonged period, especially if the gel used is LMP agarose gel. Too high a voltage may also reduce resolution, as well as causing band streaking for large DNA molecules. Too low a voltage may lead to broadening of band for small DNA fragments due to dispersion and diffusion.

Since DNA is not visible in natural light, the progress of the electrophoresis is monitored using colored dyes. Xylene cyanol (light blue color) comigrates large DNA fragments, while Bromophenol blue (dark blue) comigrates with the smaller fragments. Less commonly used dyes include Cresol Red and Orange G which migrate ahead of bromophenol blue. A DNA marker is also run together for the estimation of the molecular weight of the DNA fragments. Note however that the size of a circular DNA like plasmids cannot be accurately gauged using standard markers unless it has been linearized by restriction digest, alternatively a supercoiled DNA marker may be used.

Staining and Visualization

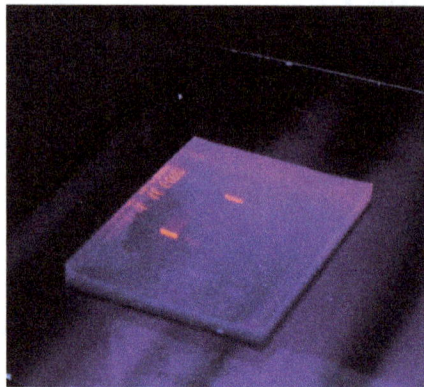

The gel with UV illumination: DNA stained with ethidium bromide appears as glowing orange bands.

DNA as well as RNA are normally visualized by staining with ethidium bromide, which intercalates into the major grooves of the DNA and fluoresces under UV light. The ethidium bromide may be added to the agarose solution before it gels, or the DNA gel may be stained later after electrophoresis. Destaining of the gel is not necessary but may produce better images. Other methods of staining are available; examples are SYBR Green, GelRed, methylene blue, brilliant cresyl blue, Nile blue sulphate, and crystal violet. SYBR Green, GelRed and other similar commercial products are sold as safer alternatives to ethidium bromide as it has been shown to be mutagenic in Ames test, although the carcinogenicity of ethidium bromide has not actually been established. SYBR Green requires the use of a blue-light transilluminator. DNA stained with crystal violet can be viewed under natural light without the use of a UV transilluminator which is an advantage, however it may not produce a strong band.

When stained with ethidium bromide, the gel is viewed with an ultraviolet (UV) transilluminator. Standard transilluminators use wavelengths of 302/312-nm (UV-B), however exposure of DNA to UV radiation for as little as 45 seconds can produce damage to DNA and affect subsequent procedures, for example reducing the efficiency of transformation, *in vitro* transcription, and PCR. Exposure of the DNA to UV radiation therefore should be limited. Using a higher wavelength of 365 nm (UV-A range) causes less damage to the DNA but also produces much weaker fluorescence with ethidium bromide. Where multiple wavelengths can be selected in the transilluminator, the shorter wavelength would be used to capture images, while the longer wavelength should be used if it is necessary to work on the gel for any extended period of time.

The transilluminator apparatus may also contain image capture devices, such as a digital or polaroid camera, that allow an image of the gel to be taken or printed.

Cutting out agarose gel slices. Protective equipment must be worn when using UV transilluminator.

Downstream Procedures

The separated DNA bands are often used for further procedures, and a DNA band may be cut out of the gel as a slice, dissolved and purified. Contaminants however may affect some downstream procedures such as PCR, and low melting point agarose may be preferred in some cases as it contains fewer of the sulphates that can affect some enzymatic reactions. The gels may also be used for blotting techniques.

Buffers

In general, the ideal buffer should have good conductivity, produce less heat and have a long life. There are a number of buffers used for agarose electrophoresis; common ones for nucleic acids include Tris/Acetate/EDTA (TAE) and Tris/Borate/EDTA (TBE). Many other buffers have been proposed, e.g. lithium borate (LB), which is almost never used, based on Pubmed citations, iso electric histidine, pK matched goods buffers, etc.; in most cases the purported rationale is lower current (less heat) and or matched ion mobilities, which leads to longer buffer life. Tris-phosphate buffer has high buffering capacity but cannot be used if DNA extracted is to be used in phosphate sensitive reaction. Borate is problematic; Borate can polymerize, and/or interact with cis diols such as those found in RNA. TAE has the lowest buffering capacity but provides the best resolution for larger DNA. This means a lower voltage and more time, but a better product. LB is relatively new and is ineffective in resolving fragments larger than 5 kbp; However, with its low conductivity, a much higher voltage could be used (up to 35 V/cm), which means a shorter analysis time for routine electrophoresis. As low as one base pair size difference could be resolved in 3% agarose gel with an extremely low conductivity medium (1 mM lithium borate). The buffers used contain EDTA to inactivate many nucleases which require divalent cation for their function.

Applications

- Estimation of the size of DNA molecules following restriction enzyme digestion, e.g. in restriction mapping of cloned DNA.

- Analysis of PCR products, e.g. in molecular genetic diagnosis or genetic fingerprinting

- Separation of DNA fragments for extraction and purification.

- Separation of restricted genomic DNA prior to Southern transfer, or of RNA prior to Northern transfer.

Agarose gels are easily cast and handled compared to other matrices and nucleic acids are not chemically altered during electrophoresis. Samples are also easily recovered. After the experiment is finished, the resulting gel can be stored in a plastic bag in a refrigerator.

Electrophoresis is performed in buffer solutions to reduce pH changes due to the electric field, which is important because the charge of DNA and RNA depends on pH, but running for too long can exhaust the buffering capacity of the solution. Further, different preparations of genetic material may not migrate consistently with each other, for morphological or other reasons.

Polyacrylamide Gel Electrophoresis

Polyacrylamide gel electrophoresis (PAGE), describes a technique widely used in biochemistry, forensics, genetics, molecular biology and biotechnology to separate biological macromolecules, usually proteins or nucleic acids, according to their electrophoretic mobility. Mobility is a function of the length, conformation and charge of the molecule.

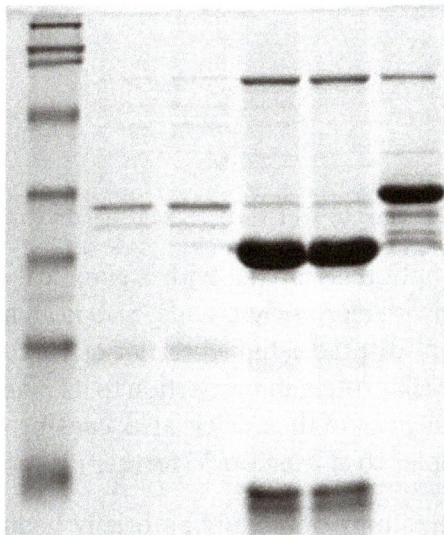

Picture of an SDS-PAGE. The molecular markers (ladder) are in the left lane

As with all forms of gel electrophoresis, molecules may be run in their native state, preserving the molecules' higher-order structure, or a chemical denaturant may be added to remove this structure and turn the molecule into an unstructured linear chain whose mobility depends only on its length and mass-to-charge ratio. For nucleic acids, urea is the most commonly used denaturant. For proteins, sodium dodecyl sulfate (SDS) is an anionic detergent applied to protein samples to linearize proteins and to impart a negative charge to linearized proteins. This procedure is called SDS-PAGE. In most proteins, the binding of SDS to the polypeptide chain imparts an even distribution of charge per unit mass, thereby resulting in a fractionation by approximate size during electrophoresis. Proteins that have a greater hydrophobic content, for instance many membrane proteins, and those that interact with surfactants in their native environment, are intrinsically harder to treat accurately using this method, due to the greater variability in the ratio of bound SDS.

Procedure

Sample Preparation

Samples may be any material containing proteins or nucleic acids. These may be biologically derived, for example from prokaryotic or eukaryotic cells, tissues, viruses, environmental samples, or purified proteins. In the case of solid tissues or cells, these are often first broken down mechanically using a blender (for larger sample volumes), using a homogenizer (smaller volumes), by sonicator or by using cycling of high pressure, and a combination of biochemical and mechanical techniques – including various types of filtration and centrifugation – may be used to separate different cell compartments and organelles prior to electrophoresis. Synthetic biomolecules such as oligonucleotides may also be used as analytes.

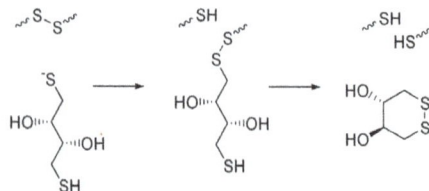

Reduction of a typical disulfide bond by DTT via two sequential thiol-disulfide exchange reactions.

The sample to analyze is optionally mixed with a chemical denaturant if so desired, usually SDS for proteins or urea for nucleic acids. SDS is an anionic detergent that denatures secondary and non–disulfide–linked tertiary structures, and additionally applies a negative charge to each protein in proportion to its mass. Urea breaks the hydrogen bonds between the base pairs of the nucleic acid, causing the constituent strands to separate. Heating the samples to at least 60 °C further promotes denaturation.

In addition to SDS, proteins may optionally be briefly heated to near boiling in the presence of a reducing agent, such as dithiothreitol (DTT) or 2-mercaptoethanol (beta-mercaptoethanol/BME), which further denatures the proteins by reducing disulfide linkages, thus overcoming some forms of tertiary protein folding, and breaking up quaternary protein structure (oligomeric subunits). This is known as reducing SDS-PAGE.

A tracking dye may be added to the solution. This typically has a higher electrophoretic mobility than the analytes to allow the experimenter to track the progress of the solution through the gel during the electrophoretic run.

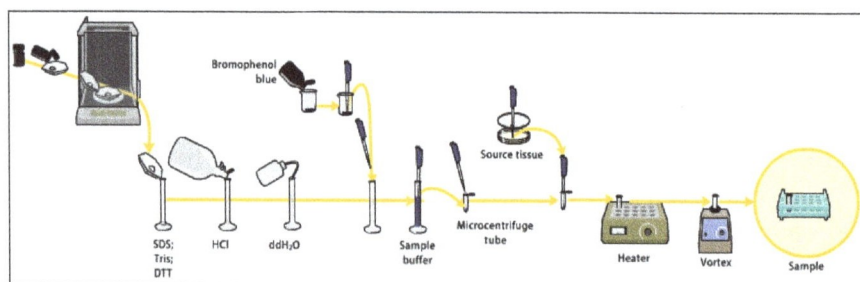

Preparing Acrylamide Gels

The gels typically consist of acrylamide, bisacrylamide, the optional denaturant (SDS or urea), and a buffer with an adjusted pH. The solution may be degassed under a vacuum to prevent the formation of air bubbles during polymerization. Alternatively, butanol may be added to the resolving gel (for proteins) after it is poured, as butanol removes bubbles and makes the surface smooth. A source of free radicals and a stabilizer, such as ammonium persulfate and TEMED are added to initiate polymerization. The polymerization reaction creates a gel because of the added bisacrylamide, which can form cross-links between two acrylamide molecules. The ratio of bisacrylamide to acrylamide can be varied for special purposes, but is generally about 1 part in 35. The acrylamide concentration of the gel can also be varied, generally in the range from 5%

to 25%. Lower percentage gels are better for resolving very high molecular weight molecules, while much higher percentages are needed to resolve smaller proteins.

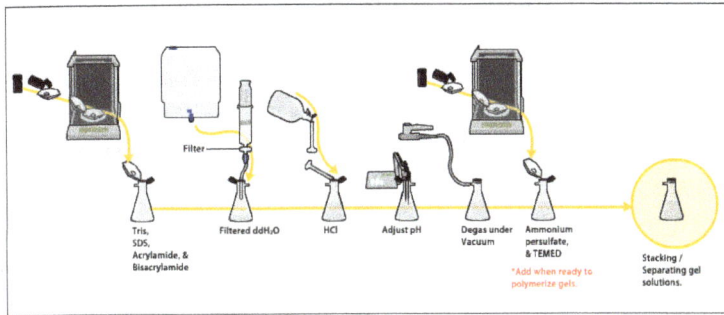

Gels are usually polymerized between two glass plates in a gel caster, with a comb inserted at the top to create the sample wells. After the gel is polymerized the comb can be removed and the gel is ready for electrophoresis.

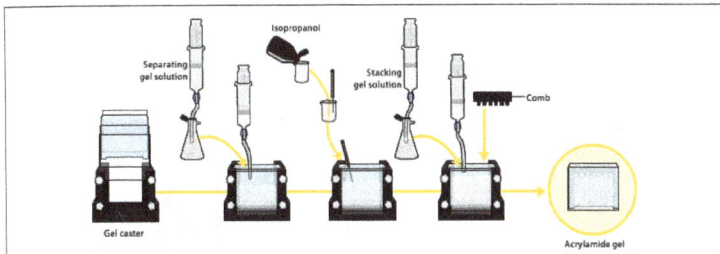

Electrophoresis

Various buffer systems are used in PAGE depending on the nature of the sample and the experimental objective. The buffers used at the anode and cathode may be the same or different.

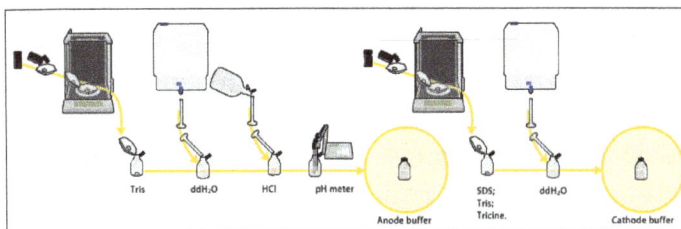

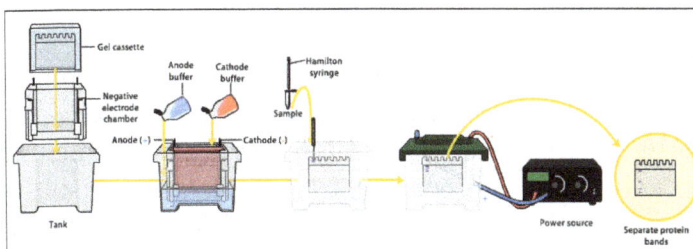

An electric field is applied across the gel, causing the negatively charged proteins or nucleic acids to migrate across the gel away from the negative electrode (which is the cathode being that this is an electrolytic rather than galvanic cell) and towards the positive electrode (the anode). Depending on their size, each biomolecule moves differently through the gel matrix: small molecules more easily fit through the pores in the gel, while larger ones have more difficulty. The gel is run usually for a few hours, though this depends on the voltage applied across the gel; migration occurs more quickly at higher voltages, but these results are typically less accurate than at those at lower voltages. After the set amount of time, the biomolecules have migrated different distances based on their size. Smaller biomolecules travel farther down the gel, while larger ones remain closer to the point of origin. Biomolecules may therefore be separated roughly according to size, which depends mainly on molecular weight under denaturing conditions, but also depends on higher-order conformation under native conditions. However, certain glycoproteins behave anomalously on SDS gels.

Further Processing

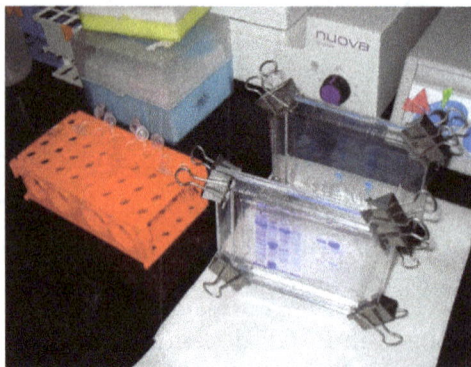

Two SDS-PAGE-gels after a completed run

Following electrophoresis, the gel may be stained (for proteins, most commonly with Coomassie Brilliant Blue R-250; for nucleic acids, ethidium bromide; or for either, silver stain), allowing visualization of the separated proteins, or processed further (e.g. Western blot). After staining, different species biomolecules appear as distinct bands within the gel. It is common to run molecular weight size markers of known molecular weight in a separate lane in the gel to calibrate the gel and determine the approximate molecular mass of unknown biomolecules by comparing the distance traveled relative to the marker.

For proteins, SDS-PAGE is usually the first choice as an assay of purity due to its reliability and ease. The presence of SDS and the denaturing step make proteins separate, approximately based on size, but aberrant migration of some proteins may occur. Different proteins may also stain differently, which interferes with quantification by staining. PAGE may also be used as a preparative technique for the purification of proteins. For example, quantitative preparative native continuous polyacrylamide gel electrophoresis (QPNC-PAGE) is a method for separating native metalloproteins in complex biological matrices.

Chemical Ingredients and Their Roles

Polyacrylamide gel (PAG) had been known as a potential embedding medium for sectioning tissues as early as 1964, and two independent groups employed PAG in electrophoresis in 1959. It possesses several electrophoretically desirable features that make it a versatile medium. It is a synthetic, thermo-stable, transparent, strong, chemically relatively inert gel, and can be prepared with a wide range of average pore sizes. The pore size of a gel is determined by two factors, the total amount of acrylamide present (%T) (T = Total concentration of acrylamide and bisacrylamide monomer) and the amount of cross-linker (%C) (C = bisacrylamide concentration). Pore size decreases with increasing %T; with cross-linking, 5%C gives the smallest pore size. Any increase or decrease in %C from 5% increases the pore size, as pore size with respect to %C is a parabolic function with vertex as 5%C. This appears to be because of non-homogeneous bundling of polymer strands within the gel. This gel material can also withstand high voltage gradients, is amenable to various staining and destaining procedures, and can be digested to extract separated fractions or dried for autoradiography and permanent recording.

Components

- Chemical buffer Stabilizes the pH value to the desired value within the gel itself and in the electrophoresis buffer. The choice of buffer also affects the electrophoretic mobility of the buffer counterions and thereby the resolution of the gel. The buffer should also be unreactive and not modify or react with most proteins. Different buffers may be used as cathode and anode buffers, respectively, depending on the application. Multiple pH values may be used within a single gel, for example in DISC electrophoresis. Common buffers in PAGE include Tris, Bis-Tris, or imidazole.

- Counterion balance the intrinsic charge of the buffer ion and also affect the electric field strength during electrophoresis. Highly charged and mobile ions are often avoided in SDS-PAGE cathode buffers, but may be included in the gel itself, where it migrates ahead of the protein. In applications such as DISC SDS-PAGE the pH values within the gel may vary to change the average charge of the counterions during the run to improve resolution. Popular counterions are glycine and tricine. Glycine has been used as the source of trailing ion or slow ion because its pKa is 9.69 and mobility of glycinate are such that the effective mobility can be set at a value below that of the slowest known proteins of net negative charge in the pH range. The minimum pH of this range is approximately 8.0.

- Acrylamide (C_3H_5NO; mW: 71.08). When dissolved in water, slow, spontaneous autopolymerization of acrylamide takes place, joining molecules together by head on tail fashion to form long single-chain polymers. The presence of a free radical-generating system greatly accelerates polymerization. This kind of reaction is known as Vinyl addition polymerisation. A solution of these polymer chains becomes viscous but does not form a gel, because the chains simply slide over one another. Gel formation requires linking various chains together. Acrylamide is carcinogenic, a neurotoxin, and a reproductive toxin. It is also essential to store acrylamide in a cool dark and dry place to reduce autopolymerisation and hydrolysis.

- Bisacrylamide (**N,N'**-Methylenebisacrylamide) ($C_7H_{10}N_2O_2$; mW: 154.17). Bisacrylamide is the most frequently used cross linking agent for polyacrylamide gels. Chemically it can be thought of as two acrylamide molecules coupled head to head at their non-reactive ends. Bisacrylamide can crosslink two polyacrylamide chains to one another, thereby resulting in a gel.

- Sodium Dodecyl Sulfate (SDS) ($C_{12}H_{25}NaO_4S$; mW: 288.38). (only used in denaturing protein gels) SDS is a strong detergent agent used to denature native proteins to unfolded, individual polypeptides. When a protein mixture is heated to 100 °C in presence of SDS, the detergent wraps around the polypeptide backbone. It binds to polypeptides in a constant weight ratio of 1.4 g SDS/g of polypeptide. In this process, the intrinsic charges of polypeptides become negligible when compared to the negative charges contributed by SDS. Thus polypeptides after treatment become rod-like structures possessing a uniform charge density, that is same net negative charge per unit weight. The electrophoretic mobilities of these proteins is a linear function of the logarithms of their molecular weights.

  Without SDS, different proteins with similar molecular weights would migrate differently due to differences in mass-charge ratio, as each protein has an isoelectric point and molecular weight particular to its primary structure. This is known as Native PAGE. Adding SDS solves this problem, as it binds to and unfolds the protein, giving a near uniform negative charge along the length of the polypeptide.

- Urea ($CO(NH_2)_2$; mW: 60.06). Urea is a chaotropic agent that increases the entropy of the system by interfering with intramolecular interactions mediated by non-covalent forces such as hydrogen bonds and van der Waals forces. Macromolecular structure is dependent on the net effect of these forces, therefore it follows that an increase in chaotropic solutes denatures macromolecules,

- Ammonium persulfate (APS) ($N_2H_8S_2O_8$; mW: 228.2). APS is a source of free radicals and is often used as an initiator for gel formation. An alternative source of free radicals is riboflavin, which generated free radicals in a photochemical reaction.

- TEMED (**N, N, N′, N′**-tetramethylethylenediamine) ($C_6H_{16}N_2$; mW: 116.21). TEMED stabilizes free radicals and improves polymerization. The rate of polymerisation and the properties of the resulting gel depend on the concentrations of free radicals. Increasing the amount of free radicals results in a decrease in the average polymer chain length, an increase in gel turbidity and a decrease in gel elasticity. Decreasing the amount shows the reverse effect. The lowest catalytic concentrations that allow polymerisation in a reasonable period of time should be used. APS and TEMED are typically used at approximately equimolar concentrations in the range of 1 to 10 mM.

Chemicals for Processing and Visualization

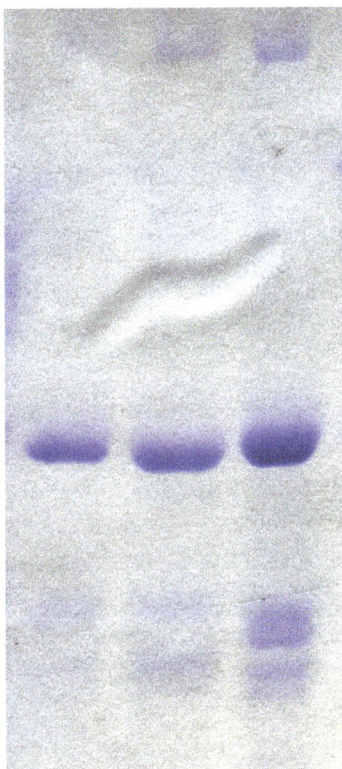

PAGE of rotavirus proteins stained with Coomassie blue

The following chemicals and procedures are used for processing of the gel and the protein samples visualized in it:

- Tracking dye. As proteins and nucleic acids are mostly colorless, their progress through the gel during electrophoresis cannot be easily followed. Anionic dyes of a known electrophoretic mobility are therefore usually included in the PAGE sample buffer. A very common tracking dye is Bromophenol blue (BPB,

3',3",5',5" tetrabromophenolsulfonphthalein). This dye is coloured at alkali and neutral pH and is a small negatively charged molecule that moves towards the anode. Being a highly mobile molecule it moves ahead of most proteins. As it reaches the anodic end of the electrophoresis medium electrophoresis is stopped. It can weakly bind to some proteins and impart a blue colour. Other common tracking dyes are xylene cyanol, which has lower mobility, and Orange G, which has a higher mobility.

- Loading aids. Most PAGE systems are loaded from the top into wells within the gel. To ensure that the sample sinks to the bottom of the gel, sample buffer is supplemented with additives that increase the density of the sample. These additives should be non-ionic and non-reactive towards proteins to avoid interfering with electrophoresis. Common additives are glycerol and sucrose.

- Coomassie Brilliant Blue R-250 (CBB)($C_{45}H_{44}N_3NaO_7S_2$; mW: 825.97). CBB is the most popular protein stain. It is an anionic dye, which non-specifically binds to proteins. The structure of CBB is predominantly non-polar, and it is usually used in methanolic solution acidified with acetic acid. Proteins in the gel are fixed by acetic acid and simultaneously stained. The excess dye incorporated into the gel can be removed by destaining with the same solution without the dye. The proteins are detected as blue bands on a clear background. As SDS is also anionic, it may interfere with staining process. Therefore, large volume of staining solution is recommended, at least ten times the volume of the gel.

- Ethidium bromide (EtBr) is the traditionally most popular nucleic acid stain.

- Silver staining. Silver staining is used when more sensitive method for detection is needed, as classical Coomassie Brilliant Blue staining can usually detect a 50 ng protein band, Silver staining increases the sensitivity typically 50 times. The exact chemical mechanism by which this happens is still largely unknown. Silver staining was introduced by Kerenyi and Gallyas as a sensitive procedure to detect trace amounts of proteins in gels. The technique has been extended to the study of other biological macromolecules that have been separated in a variety of supports. Many variables can influence the colour intensity and every protein has its own staining characteristics; clean glassware, pure reagents and water of highest purity are the key points to successful staining. Silver staining was developed in the 14th century for colouring the surface of glass. It has been used extensively for this purpose since the 16th century. The colour produced by the early silver stains ranged between light yellow and an orange-red. Camillo Golgi perfected the silver staining for the study of the nervous system. Golgi's method stains a limited number of cells at random in their entirety.

- Western blotting is a process by which proteins separated in the acrylamide gel are electrophoretically transferred to a stable, manipulable membrane such as a nitrocellulose, nylon, or PVDF membrane. It is then possible to apply immuno-chemical techniques to visualise the transferred proteins, as well as accurately identify relative increases or decreases of the protein of interest.

Gel Electrophoresis of Proteins

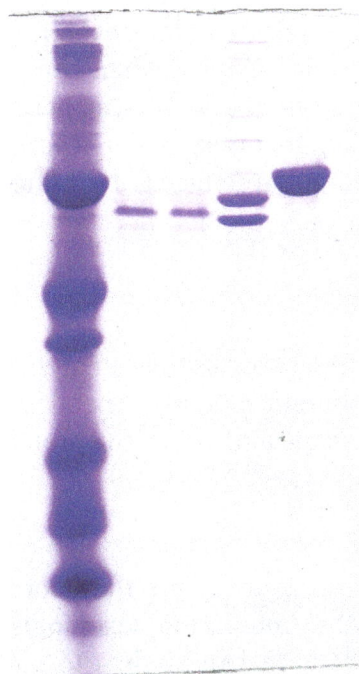

Proteins separated by SDS-PAGE, Coomassie Brilliant Blue staining

Protein electrophoresis is a method for analysing the proteins in a fluid or an extract. The electrophoresis may be performed with a small volume of sample in a number of alternative ways with or without a supporting medium: SDS polyacrylamide gel electrophoresis (in short: gel electrophoresis, PAGE, or SDS-electrophoresis), free-flow electrophoresis, electrofocusing, isotachophoresis, affinity electrophoresis, immuno-electrophoresis, counterelectrophoresis, and capillary electrophoresis. Each method has many variations with individual advantages and limitations. Gel electrophoresis is often performed in combination with electroblotting immunoblotting to give additional information about a specific protein. Because of practical limitations, protein electrophoresis is generally not suited as a preparative method.

Denaturing Gel Methods

SDS-PAGE

SDS-PAGE, sodium dodecyl sulfate polyacrylamide gel electrophoresis, describes a collection of related techniques to separate proteins according to their electrophoretic mobility (a function of the length of a polypeptide chain and its charge) while in the denatured (unfolded) state. In most proteins, the binding of SDS to the polypeptide chain imparts an even distribution of charge per unit mass, thereby resulting in a fractionation by approximate size during electrophoresis.

SDS is a strong detergent agent used to denature native proteins to unfolded, individual polypeptides. When a protein mixture is heated to 100 °C in presence of SDS, the detergent wraps around the polypeptide backbone. In this process, the intrinsic charges of polypeptides becomes negligible when compared to the negative charges contributed by SDS. Thus polypeptides after treatment become rod-like structures possessing a uniform charge density, that is same net negative charge per unit length. The electrophoretic mobilities of these proteins will be a linear function of the logarithms of their molecular weights.

Native Gel Methods

Native gels, also known as non-denaturing gels, analyze proteins that are still in their folded state. Thus, the electrophoretic mobility depends not only on the charge-to-mass ratio, but also on the physical shape and size of the protein.

Blue Native PAGE

BN-PAGE is a native PAGE technique, where the Coomassie Brilliant Blue dye provides the necessary charges to the protein complexes for the electrophoretic separation. The disadvantage of Coomassie is that in binding to proteins it can act like a detergent causing complexes to dissociate. Another drawback is the potential quenching of chemoluminescence (e.g. in subsequent western blot detection or activity assays) or fluorescence of proteins with prosthetic groups (e.g. heme or chlorophyll) or labelled with fluorescent dyes.

Clear Native PAGE

CN-PAGE (commonly referred to as Native PAGE) separates acidic water-soluble and membrane proteins in a polyacrylamide gradient gel. It uses no charged dye so the electrophoretic mobility of proteins in CN-PAGE (in contrast to the charge shift technique BN-PAGE) is related to the intrinsic charge of the proteins. The migration distance depends on the protein charge, its size and the pore size of the gel. In many cases this method has lower resolution than BN-PAGE, but CN-PAGE offers advantages whenever Coomassie dye would interfere with further analytical techniques, for example it has been described as a very efficient microscale separation technique for FRET analyses. Also CN-PAGE is milder than BN-PAGE so it can retain labile supramolecular assemblies of membrane protein complexes that are dissociated under the conditions of BN-PAGE.

Quantitative Native PAGE

The folded protein complexes of interest separate cleanly and predictably due to the specific properties and polymerization time of the polyacrylamide gel. The separated proteins are continuously eluted into a physiological eluent and transported to a fraction collector. In a few specific PAGE fractions metal cofactors can be identified and absolutely quantified by high-resolution ICP-MS. The natural structures of the isolated metalloproteins are elucidated by solution NMR spectroscopy.

Buffer Systems

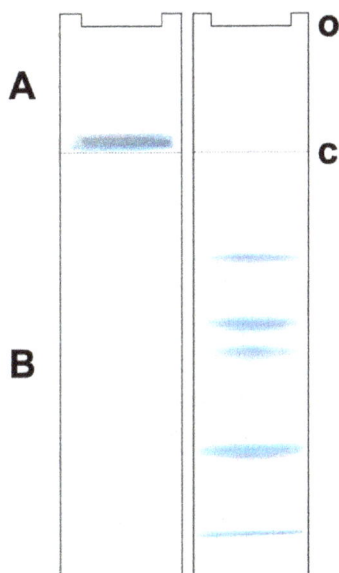

Postulated migration of proteins in a Laemmli gel system A: Stacking gel, B: Resolving gel, o: sample application c: discontinuities in the buffer and electrophoretic matrix

Most protein separations are performed using a "discontinuous" (or DISC) buffer system that significantly enhances the sharpness of the bands within the gel. During electrophoresis in a discontinuous gel system, an ion gradient is formed in the early stage of electrophoresis that causes all of the proteins to focus into a single sharp band. The formation of the ion gradient is achieved by choosing a pH value at which the ions of the buffer are only moderately charged compared to the SDS-coated proteins. These conditions provide an environment in which Kohlrausch's reactions determine the molar conductivity. As a result, SDS-coated proteins are concentrated to several fold in a thin zone of the order of 19 μm within a few minutes. At this stage all proteins migrate at the same migration speed by isotachophoresis. This occurs in a region of the gel that has larger pores so that the gel matrix does not retard the migration during the focusing or "stacking" event. Separation of the proteins by size is achieved in the lower, "resolving" region of the gel. The resolving gel typically has a much smaller pore size, which leads to a sieving effect that now determines the electrophoretic mobility of the proteins. At the same time, the separating part of the gel also has a pH value in which the buffer ions on average carry a greater charge, causing them to "outrun" the SDS-covered proteins and eliminate the ion gradient and thereby the stacking effect.

A very widespread discontinuous buffer system is the tris-glycine or "Laemmli" system that stacks at a pH of 6.8 and resolves at a pH of ~8.3-9.0. A drawback of this system is that these pH values may promote disulfide bond formation between cysteine residues in the proteins because the pKa of cysteine ranges from 8-9 and because reducing agent present in the loading buffer doesn't co-migrate with the proteins. Recent advances in buffering technology alleviate this problem by resolving the proteins at a pH well below the pKa of cysteine (e.g., bis-tris, pH 6.5) and include reducing agents (e.g. sodium bisulfite) that move into the gel ahead of the proteins to maintain a reducing environment. An additional benefit of using buffers with lower pH values is that the acrylamide gel is more stable at lower pH values, so the gels can be stored for long periods of time before use.

SDS Gradient Gel Electrophoresis of Proteins

As voltage is applied, the anions (and negatively charged sample molecules) migrate toward the positive electrode (anode) in the lower chamber, the leading ion is Cl^- (high mobility and high concentration); glycinate is the trailing ion (low mobility and low concentration). SDS-protein particles do not migrate freely at the border between the Cl^- of the gel buffer and the Gly^- of the cathode buffer. Friedrich Kohlrausch found that Ohm's law also applies to dissolved electrolytes. Because of the voltage drop between the Cl^- and Glycine-buffers, proteins are compressed (stacked) into micrometer thin layers. The boundary moves through a pore gradient and the protein stack gradually disperses due to a frictional resistance increase of the gel matrix. Stacking and unstacking occurs continuously in the gradient gel, for every protein at a different position. For a complete protein unstacking the polyacrylamide-gel concentration must exceed 16% T. The two-gel system of "Laemmli" is a simple gradient gel. The pH discontinuity of the buffers is of no significance for the separation quality, and a "stacking-gel" with a different pH is not needed.

Visualization

The most popular protein stain is Coomassie Brilliant Blue. It is an anionic dye, which non-specifically binds to proteins. Proteins in the gel are fixed by acetic acid and simultaneously stained. The excess dye incorporated into the gel can be removed by destaining with the same solution without the dye. The proteins are detected as blue bands on a clear background.

When more sensitive method than staining by Coomassie is needed silver staining is usually used. Silver staining is a sensitive procedure to detect trace amounts of proteins in gels, but can also visualize nucleic acid or polysaccharides.

Similarly as in nucleic acid gel electrophoresis, tracking dye is often used. Anionic dyes of a known electrophoretic mobility are usually included in the sample buffer. A very common tracking dye is Bromophenol blue. This dye is coloured at alkali and neutral pH and is a small negatively charged molecule that moves towards the anode. Being a highly mobile molecule it moves ahead of most proteins.

Medical Applications

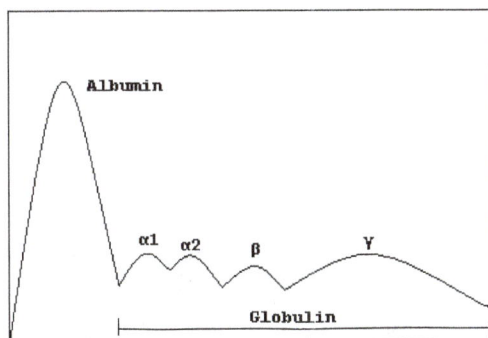

Schematic representation of a protein electrophoresis gel.

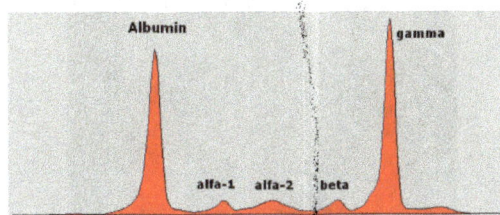

Serum protein electrophoresis showing a paraprotein (peak in the gamma zone) in a patient with multiple myeloma.

In medicine, protein electrophoresis is a method of analysing the proteins mainly in blood serum. Before the widespread use of gel electrophoresis, protein electrophoresis was performed as free-flow electrophoresis (on paper) or as immunoelectrophoresis.

Traditionally, two classes of blood proteins are considered: serum albumin and globulin. They are generally equal in proportion, but albumin as a molecule is much smaller and lightly, negatively-charged, leading to an accumulation of albumin on the electrophoretic gel. A small band before albumin represents transthyretin (also named prealbumin). Some forms of medication or body chemicals can cause their own band, but it usually is small. Abnormal bands (spikes) are seen in monoclonal gammopathy of undetermined significance and multiple myeloma, and are useful in the diagnosis of these conditions.

The globulins are classified by their banding pattern (with their main representatives):

- The *alpha* (α) band consists of two parts, 1 and 2:

- ○ α_1 - α_1-antitrypsin, α_1-acid glycoprotein.

- ○ α_2 - haptoglobin, α_2-macroglobulin, α_2-antiplasmin, ceruloplasmin.

- The *beta* (β) band - transferrin, LDL, complement

- The *gamma* (γ) band - immunoglobulin (IgA, IgD, IgE, IgG and IgM). Paraproteins (in multiple myeloma) usually appear in this band.

Normal present medical procedure involves determination of numerous proteins in plasma including hormones and enzymes, some of them also determined by electrophoresis. However, gel electrophoresis is mainly a research tool, also when the subject is blood proteins.

Gel Electrophoresis of Nucleic Acids

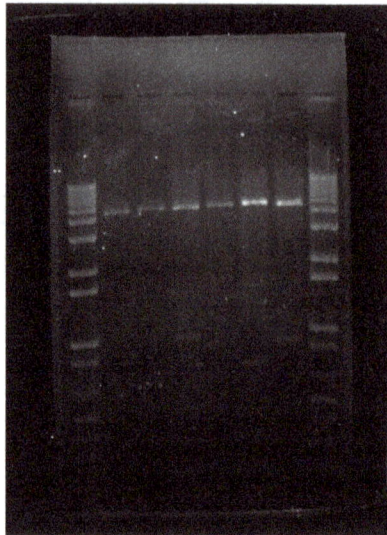

Digital printout of an agarose gel electrophoresis of *cat*-insert plasmid DNA

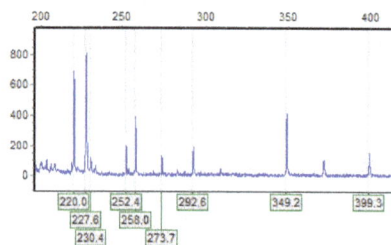

DNA electropherogram trace

Nucleic acid electrophoresis is an analytical technique used to separate DNA or RNA fragments by size and reactivity. Nucleic acid molecules which are to be analyzed are set upon a viscous medium, the gel, where an electric field induces the nucleic acids to migrate toward the anode, due to the net negative charge of the sugar-phosphate backbone of the nucleic acid chain. The separation of these fragments is accomplished by exploiting the mobilities with which different sized molecules are able to pass through the gel. Longer molecules migrate more slowly because they experience more resistance within the gel. Because the size of the molecule affects its mobility, smaller fragments end up nearer to the anode than longer ones in a given period. After some time, the voltage is removed and the fragmentation gradient is analyzed. For larger separations between similar sized fragments, either the voltage or run time can be increased. Extended runs across a low voltage gel yield the most accurate resolution. Voltage is, however, not the sole factor in determining electrophoresis of nucleic acids.

The nucleic acid to be separated can be prepared in several ways before separation by electrophoresis. In the case of large DNA molecules, the DNA is frequently cut into smaller fragments using a DNA restriction endonuclease (or restriction enzyme). In other instances, such as PCR amplified samples, enzymes present in the sample that might affect the separation of the molecules are removed through various means before analysis. Once the nucleic acid is properly prepared, the samples of the nucleic acid solution are placed in the wells of the gel and a voltage is applied across the gel for a specified amount of time.

The DNA fragments of different lengths are visualized using a fluorescent dye specific for DNA, such as ethidium bromide. The gel shows bands corresponding to different nucleic acid molecules populations with different molecular weight. Fragment size is usually reported in "nucleotides", "base pairs" or "kb" (for thousands of base pairs) depending upon whether single- or double-stranded nucleic acid has been separated. Fragment size determination is typically done by comparison to commercially available DNA markers containing linear DNA fragments of known length.

The types of gel most commonly used for nucleic acid electrophoresis are agarose (for relatively long DNA molecules) and polyacrylamide (for high resolution of short DNA molecules, for example in DNA sequencing). Gels have conventionally been run in a "slab" format such as that shown in the figure, but capillary electrophoresis has become important for applications such as high-throughput DNA sequencing. Electrophoresis techniques used in the assessment of DNA damage include alkaline gel electrophoresis and pulsed field gel electrophoresis.

For short DNA segments such as 20 to 60 bp double stranded DNA, running them in Polyacrylamide gel (PAGE) will give better resolution(native condition). Similarly, RNA and single stranded DNA can be run and visualised by PAGE gels containing denaturing agents such as Urea. PAGE gels are widely used in techniques such as DNA foot printing, EMSA and other DNA-protein interaction techniques.

The measurement and analysis are mostly done with a specialized gel analysis software. Capillary electrophoresis results are typically displayed in a trace view called an electropherogram.

Factors Affecting Migration of Nucleic Acids

A number of factors can affect the migration of nucleic acids: the dimension of the gel pores, the voltage used, the ionic strength of the buffer, and the concentration intercalating dye such as ethidium bromide if used during electrophoresis.

Size of DNA

The gel sieves the DNA by the size of the DNA molecule whereby smaller molecules travel faster. Double-stranded DNA moves at a rate that is approximately inversely proportional to the logarithm of the number of base pairs. This relationship however breaks down with very large DNA fragments and it is not possible to separate them using standard agarose gel electrophoresis. The limit of resolution depends on gel composition and field strength. and the mobility of larger circular DNA may be more strongly affected than linear DNA by the pore size of the gel. Separation of very large DNA fragments requires pulse field gel electrophoresis (PFGE). In field inversion gel electrophoresis (FIGE, a kind of PFGE), it is possible to have "band inversion" - where large molecules may move faster than small molecules.

Conformation of DNA

The conformation of the DNA molecule can significantly affect the movement of the DNA, for example, supercoiled DNA usually moves faster than relaxed DNA because it is tightly coiled and hence more compact. In a normal plasmid DNA preparation, multiple forms of DNA may be present, and gel from the electrophoresis of the plasmids would normally show a main band which would be the negatively supercoiled form, while other forms of DNA may appear as minor fainter bands. These minor bands may be nicked DNA (open circular form) and the relaxed closed circular form which normally run slower than supercoiled DNA, and the single-stranded form (which can sometimes appear depending on the preparation methods) may move ahead of the supercoiled DNA. The rate at which the various forms move however can change using different electrophoresis conditions, for example linear DNA may run faster or slower than supercoiled DNA depending on conditions, and the mobility of larger circular DNA may be more strongly affected than linear DNA by the pore size of the gel. Unless supercoiled DNA markers are used, the size of a circular DNA like plasmid therefore may be more accurately gauged after it has been linearized by restriction digest.

DNA damage due to increased cross-linking will also reduce electrophoretic DNA migration in a dose-dependent way.

Concentration of Ethidium Bromide

Circular DNA are more strongly affected by ethidium bromide concentration than linear DNA if ethidium bromide is present in the gel during electrophoresis. All naturally occurring DNA circles are underwound, but ethidium bromide which intercalates into circular DNA can change the charge, length, as well as the superhelicity of the DNA molecule, therefore its presence during electrophoresis can affect its movement in gel. Increasing ethidium bromide intercalated into the DNA can change it from a negatively supercoiled molecule into a fully relaxed form, then to positively coiled superhelix at maximum intercalation. Agarose gel electrophoresis can be used to resolve circular DNA with different supercoiling topology.

Gel Concentration

The concentration of the gel determines the pore size of the gel which affect the migration of DNA. The resolution of the DNA changes with the percentage concentration of the gel. Increasing the agarose concentration of a gel reduces the migration speed and improves separation of smaller DNA molecules, while lowering gel concentration permits large DNA molecules to be separated. For a standard agarose gel electrophoresis, a 0.7% gives good separation or resolution of large 5–10kb DNA fragments, while 2% gel gives good resolution for small 0.2–1kb fragments. Up to 3% can be used for separating very tiny fragments but a vertical polyacrylamide gel would be more appropriate for resolving small fragments. High concentrations gel however requires longer run times (sometimes days) and high percentage gels are often brittle and may not set evenly. High percentage agarose gels should be run with PFGE or FIGE. Low percentage gels (0.1–0.2%) are fragile and may break. 1% gels are common for many applications.

Applied Field

At low voltages, the rate of migration of the DNA is proportional to the voltage applied, i.e. the higher the voltage, the faster the DNA moves. However, in increasing electric field strength, the mobility of high-molecular-weight DNA fragments increases differentially, and the effective range of separation decreases and resolution therefore is lower at high voltage. For optimal resolution of DNA > 2kb in size in standard gel electrophoresis, 5 to 8 V/cm is recommended. Voltage is also limited by the fact that it heats the gel and may cause the gel to melt if a gel is run at high voltage for prolonged period, particularly for low-melting point agarose gel.

The mobility of DNA however may change in an unsteady field. In a field that is periodically reversed, the mobility of DNA of a particular size may drop significantly at a particular cycling frequency. This phenomenon can result in band inversion whereby larger DNA fragments move faster than smaller ones in PFGE.

Mechanism of Migration and Separation

The negative charge of its phosphate backbone moves the DNA towards the positively charged anode during electrophoresis. However, the migration of DNA molecules in solution, in the absence of a gel matrix, is independent of molecular weight during electrophoresis, i.e. there is no separation by size without a gel matrix. Hydrodynamic interaction between different parts of the DNA are cut off by streaming counterions moving in the opposite direction, so no mechanism exists to generate a dependence of velocity on length on a scale larger than screening length of about 10 nm. This makes it different from other processes such as sedimentation or diffusion where long-ranged hydrodynamic interaction are important.

The gel matrix is therefore responsible for the separation of DNA by size during electrophoresis, however the precise mechanism responsible the separation is not entirely clear. A number of models exists for the mechanism of separation of biomolecules in gel matrix, a widely accepted one is the Ogston model which treats the polymer matrix as a sieve consisting of randomly distributed network of inter-connected pores. A globular protein or a random coil DNA moves through the connected pores large enough to accommodate its passage, and the movement of larger molecules is more likely to be impeded and slowed down by collisions with the gel matrix, and the molecules of different sizes can therefore be separated in this process of sieving.

The Ogston model however breaks down for large molecules whereby the pores are significantly smaller than size of the molecule. For DNA molecules of size greater than 1 kb, a reptation model (or its variants) is most commonly used. This model assumes that the DNA can crawl in a "snake-like" fashion (hence "reptation") through the pores as an elongated molecule. At higher electric field strength, this turned into a biased reptation model, whereby the leading end of the molecule become strongly biased in the forward direction, and this leading edge pulls the rest of the molecule along. In the fully biased mode, the mobility reached a saturation point and DNA beyond a certain size cannot be separated. Perfect parallel alignment of the chain with the field however is not observed in practice as that would mean the same mobility for long and short molecules. Further refinement of the biased reptation model takes into account of the internal fluctuations of the chain.

The biased reptation model has also been used to explain the mobility of DNA in PFGE. The orientation of the DNA is progressively built up by reptation after the onset of a field, and the time it reached the steady state velocity is dependent on the size of the molecule. When the field is changed, larger molecules take longer to reorientate, it is therefore possible to discriminate between the long chains that cannot reach its steady state velocity from the short ones that travel most of the time in steady velocity. Other models, however, also exist.

Real-time fluorescence microscopy of stained molecules showed more subtle dynamics during electrophoresis, with the DNA showing considerable elasticity as it alternately stretching in the direction of the applied field and then contracting into a ball, or becoming hooked into a U-shape when it gets caught on the polymer fibres. This observation may be termed the "caterpillar" model. Other model proposes that the DNA gets entangled with the polymer matrix, and the larger the molecule, the more likely it is to become entangled and its movement impeded.

Visualization

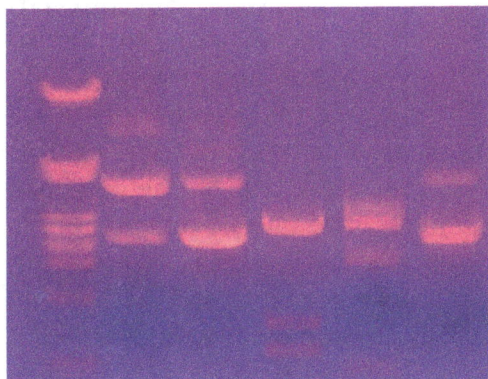

DNA gel electrophoresis

The most common dye used to make DNA or RNA bands visible for agarose gel electrophoresis is ethidium bromide, usually abbreviated as EtBr. It fluoresces under UV light when intercalated into the major groove of DNA (or RNA). By running DNA through an EtBr-treated gel and visualizing it with UV light, any band containing more than ~20 ng DNA becomes distinctly visible. EtBr is a known mutagen, and safer alternatives are available, such as GelRed, produced by Biotium, which binds to the minor groove.

SYBR Green [I] is another dsDNA stain, produced by Invitrogen. It is more expensive, but 25 times more sensitive, and possibly safer than EtBr, though there is no data addressing its mutagenicity or toxicity in humans.

SYBR Safe is a variant of SYBR Green that has been shown to have low enough levels of mutagenicity and toxicity to be deemed nonhazardous waste under U.S. Federal regulations. It has similar sensitivity levels to EtBr, but, like SYBR Green, is significantly more expensive. In countries where safe disposal of hazardous waste is mandatory, the costs of EtBr disposal can easily outstrip the initial price difference, however.

Since EtBr stained DNA is not visible in natural light, scientists mix DNA with negatively charged loading buffers before adding the mixture to the gel. Loading buffers are useful because they are visible in natural light (as opposed to UV light for EtBr stained DNA), and they co-sediment with DNA (meaning they move at the same speed as DNA of a certain length). Xylene cyanol and Bromophenol blue are common dyes found in loading buffers; they run about the same speed as DNA fragments that are 5000 bp and 300 bp in length respectively, but the precise position varies with percentage of the gel. Other less frequently used progress markers are Cresol Red and Orange G which run at about 125 bp and 50 bp, respectively.

Visualization can also be achieved by transferring DNA after SDS-PAGE to a nitrocellulose membrane followed by exposure to a hybridization probe. This process is termed Southern blotting.

For fluorescent dyes, after electrophoresis the gel is illuminated with an ultraviolet lamp (usually by placing it on a light box, while using protective gear to limit exposure to ultraviolet radiation). The illuminator apparatus mostly also contains imaging apparatus that takes an image of the gel, after illumination with UV radiation. The ethidium bromide fluoresces reddish-orange in the presence of DNA, since it has intercalated with the DNA. The DNA band can also be cut out of the gel, and can then be dissolved to retrieve the purified DNA. The gel can then be photographed usually with a digital or polaroid camera. Although the stained nucleic acid fluoresces reddish-orange, images are usually shown in black and white. UV damage to the sample can reduce the efficiency of subsequent manipulation of the sample, such as ligation and cloning. This can be avoided by using a blue light excitation source with a blue-excitable stain such as SYBR Green or GelGreen.

Gel electrophoresis research often takes advantage of software-based image analysis tools, such as ImageJ.

1	2	3
A 1% agarose 'slab' gel under normal light, behind a perspex UV shield. Only the marker dyes can be seen	The gel with UV illumination, the ethidium bromide stained DNA glows orange	Digital photo of the gel. Lane 1. Commercial DNA Markers (1kbplus), Lane 2. empty, Lane 3. a PCR product of just over 500 bases, Lane 4. Restriction digest showing a similar fragment cut from a 4.5 kb plasmid vector

Two-dimensional Gel Electrophoresis

Two-dimensional gel electrophoresis, abbreviated as 2-DE or 2-D electrophoresis, is a form of gel electrophoresis commonly used to analyze proteins. Mixtures of proteins are separated by two properties in two dimensions on 2D gels. 2-DE was first independently introduced by O'Farrell and Klose in 1975.

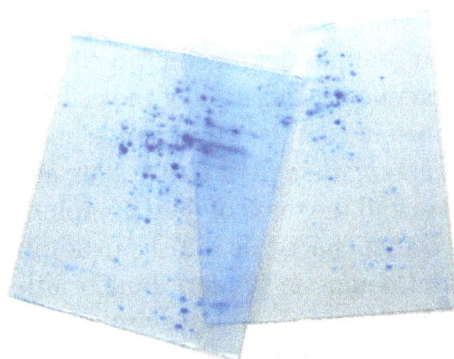

2D-Gels (Coomassie stained)

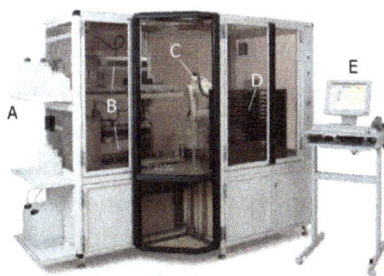

Robots are used for the isolation of protein spots from 2D gels in modern laboratories.

Basis for Separation

2-D electrophoresis begins with electrophoresis in the first dimension and then separates the molecules perpendicularly from the first to create an electropherogram in the second dimension. In electrophoresis in the first dimension, molecules are separated linearly according to their isoelectric point. In the second dimension, the molecules are then separated at 90 degrees from the first electropherogram according to molecular mass. Since it is unlikely that two molecules will be similar in two distinct properties, molecules are more effectively separated in 2-D electrophoresis than in 1-D electrophoresis.

The two dimensions that proteins are separated into using this technique can be isoelectric point, protein complex mass in the native state, and protein mass.

Separation of the proteins by isoelectric point is called isoelectric focusing (IEF). Thereby, a gradient of pH is applied to a gel and an electric potential is applied across the gel, making one end more positive than the other. At all pH values other than their isoelectric point, proteins will be charged. If they are positively charged, they will be pulled towards the more negative end of the gel and if they are negatively charged they will be pulled to the more positive end of the gel. The proteins applied in the first dimension will move along the gel and will accumulate at their isoelectric point; that is, the point at which the overall charge on the protein is 0 (a neutral charge).

For the analysis of the functioning of proteins in a cell, the knowledge of their cooperation is essential. Most often proteins act together in complexes to be fully functional. The analysis of this sub organelle organisation of the cell requires techniques conserving the native state of the protein complexes. In native polyacrylamide gel electrophoresis (native PAGE), proteins remain in their native state and are separated in the electric field following their mass and the mass of their complexes respectively. To obtain a separation by size and not by net charge, as in IEF, an additional charge is transferred to the proteins by the use of Coomassie Brilliant Blue or lithium dodecyl sulfate. After completion of the first dimension the complexes are destroyed by applying the denaturing SDS-PAGE in the second dimension, where the proteins of which the complexes are composed of are separated by their mass.

Before separating the proteins by mass, they are treated with sodium dodecyl sulfate (SDS) along with other reagents (SDS-PAGE in 1-D). This denatures the proteins (that is, it unfolds them into long, straight molecules) and binds a number of SDS molecules roughly proportional to the protein's length. Because a protein's length (when unfolded) is roughly proportional to its mass, this is equivalent to saying that it attaches a number of SDS molecules roughly proportional to the protein's mass. Since the SDS molecules are negatively charged, the result of this is that all of the proteins will have approximately the same mass-to-charge ratio as each other. In addition, proteins will not migrate when they have no charge (a result of the isoelectric focusing step) therefore the coating of the protein in SDS (negatively charged) allows migration of the proteins in the second dimension (SDS-PAGE, it is not compatible for use in the first dimension as it is charged and a nonionic or zwitterionic detergent needs to be used). In the second dimension, an electric potential is again applied, but at a 90 degree angle from the first field. The proteins will be attracted to the more positive side of the gel (because SDS is negatively charged) proportionally to their mass-to-charge ratio. As previously explained, this ratio will be nearly the same for all proteins. The proteins' progress will be slowed by frictional forces. The gel therefore acts like a molecular sieve when the current is applied, separating the proteins on the basis of their molecular weight with larger proteins being retained higher in the gel and smaller proteins being able to pass through the sieve and reach lower regions of the gel.

Detecting Proteins

The result of this is a gel with proteins spread out on its surface. These proteins can then be detected by a variety of means, but the most commonly used stains are silver and Coomassie Brilliant Blue staining. In the former case, a silver colloid is applied to the gel. The silver binds to cysteine groups within the protein. The silver is darkened by exposure to ultra-violet light. The amount of silver can be related to the darkness, and therefore the amount of protein at a given location on the gel. This measurement can only give approximate amounts, but is adequate for most purposes. Silver staining is 100x more sensitive than Coomassie Brilliant Blue with a 40-fold range of linearity.

Molecules other than proteins can be separated by 2D electrophoresis. In supercoiling assays, coiled DNA is separated in the first dimension and denatured by a DNA intercalator (such as ethidium bromide or the less carcinogenic chloroquine) in the second. This is comparable to the combination of native PAGE /SDS-PAGE in protein separation.

Common Techniques

IPG-DALT

A common technique is to use an Immobilized pH gradient (IPG) in the first dimension. This technique is referred to as IPG-DALT. The sample is first separated onto IPG gel (which is commercially available) then the gel is cut into slices for each sample which is then equilibrated in SDS-mercaptoethanol and applied to an SDS-PAGE gel for resolution in the second dimension. Typically IPG-DALT is not used for quantification of proteins due to the loss of low molecular weight components during the transfer to the SDS-PAGE gel.

IEF SDS-PAGE

2D Gel Analysis Software

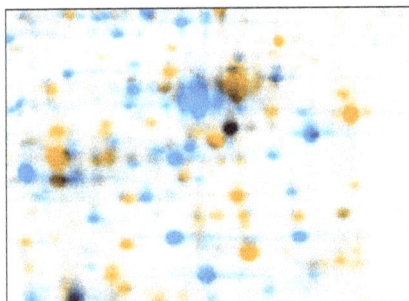

Warping: Images of two 2D electrophoresis gels, overlaid with Delta2D. First image is colored in orange, second one colored in blue. Due to running differences, corresponding spots do not overlap.

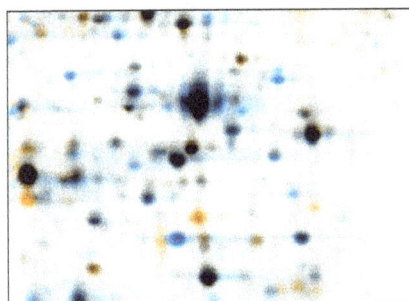

Warping: Images of two 2D electrophoresis gels after warping. First image is colored in orange, second one colored in blue. Corresponding spots overlap after warping. Common spots are colored black, orange spots are only present (or much stronger) on the first image, blue spots are only present (or much stronger) on the second image.

In quantitative proteomics, these tools primarily analyze bio-markers by quantifying individual proteins, and showing the separation between one or more protein "spots" on a scanned image of a 2-DE gel. Additionally, these tools match spots between gels of similar samples to show, for example, proteomic differences between early and advanced stages of an illness. Software packages include BioNumerics 2D, Delta2D, ImageMaster , Melanie, PDQuest, Progenesis and REDFIN – among others. While this technology is widely utilized, the intelligence has not been perfected. For example, while PDQuest and Progenesis tend to agree on the quantification and analysis of well-defined well-separated protein spots, they deliver different results and analysis tendencies with less-defined less-separated spots.

Challenges for automatic software-based analysis include incompletely separated (overlapping) spots (less-defined and/or separated), weak spots / noise (e.g., "ghost spots"), running differences between gels (e.g., protein migrates to different positions on different gels), unmatched/undetected spots, leading to missing values, mismatched spots , errors in quantification (several distinct spots may be erroneously detected as a single spot by the software and/or parts of a spot may be excluded from quantification), and differences in software algorithms and therefore analysis tendencies

Generated picking lists can be used for the automated in-gel digestion of protein spots, and subsequent identification of the proteins by mass spectrometry.

References

- Sheehan, David; O'Sullivan, Siobhan (2003). "Protein Purification Protocols". 244: 253. doi:10.1385/1-59259-655-X:253. ISBN 1-59259-655-X.

- High-speed countercurrent chromatography. Chemical analysis. Yoichiro Ito, Walter D. Conway (eds.). New York: J. Wiley. 1996. ISBN 978-0-471-63749-3.

- Foucault, Alain P. (1994). Centrifugal Partition Chromatography. Chromatographic Science Series, Vol. 68. CRC Press. ISBN 978-0-8247-9257-2.

- Conway, Walter D. (1990). Countercurrent Chromatography: Apparatus, Theory and Applications. New York: VCH Publishers. ISBN 978-0-89573-331-3.

- Berthod, A. (2002). Countercurrent Chromatography: The support-free liquid stationary phase. Wilson & Wilson's Comprehensive Analytical Chemistry Vol. 38. Boston: Elsevier Science Ltd. pp. 1–397. ISBN 978-0-444-50737-2.

- Conway, Walter D.; Petroski, Richard J. (1995). Modern Countercurrent Chromatography. ACS Symposium Series #593. ACS Publications. doi:10.1021/bk-1995-0593. ISBN 978-0-8412-3167-2.

- Ito, Yoichiro; Conway, Walter D. (1995). High-speed Countercurrent Chromatography. Chemical Analysis: A Series of Monographs on Analytical Chemistry and Its Applications (Book 198). New York: John Wiley and Sons. ISBN 978-0-471-63749-3.

- Mandava, N. Bhushan; Ito, Yoichiro (1988). Countercurrent Chromatography: Theory and Practice. Chromatographic Science Series, Vol. 44. New York: Marcel Dekker Inc. ISBN 978-0-8247-7815-6.

- Menet, Jean-Michel; Thiebaut, Didier (1999). Countercurrent Chromatography. Chromatographic Science Series Vol 82. New York: CRC Press. ISBN 978-0-8247-9992-2.

- Snyder, Lloyd R.; Dolan, John W. (2006). High-Performance Gradient Elution: The Practical Application of the Linear-Solvent-Strength Model. Wiley Interscience. ISBN 0470055510.

- Laurence M. Harwood, Christopher J. Moody (13 Jun 1989). Experimental organic chemistry: Principles and Practice (Illustrated ed.). pp. 180–185. ISBN 978-0-632-02017-1.

- Tom Maniatis; E. F. Fritsch; Joseph Sambrook. "Chapter 5, protocol 1". Molecular Cloning - A Laboratory Manual. 1 (3rd ed.). p. 5.2–5.3. ISBN 978-0879691363.

5

Gas Chromatography: An Overview

Chromatography can be divided into preparative and analytic chromatography. Gas chromatography is a type of analytic chromatography which is used in separating compounds and then further analyzing them. The following content also focuses on inverse gas chromatography, unresolved complex mixture, comprehensive two-dimensional gas chromatography, gas chromatography-mass spectrometry, methanizers etc.

Gas Chromatography

Gas chromatography (GC) is a common type of chromatography used in analytical chemistry for separating and analyzing compounds that can be vaporized without decomposition. Typical uses of GC include testing the purity of a particular substance, or separating the different components of a mixture (the relative amounts of such components can also be determined). In some situations, GC may help in identifying a compound. In preparative chromatography, GC can be used to prepare pure compounds from a mixture.

In gas chromatography, the *mobile phase* (or "moving phase") is a carrier gas, usually an inert gas such as helium or an unreactive gas such as nitrogen. Helium remains the most commonly used carrier gas in about 90% of instruments although hydrogen is preferred for improved separations. The *stationary phase* is a microscopic layer of liquid or polymer on an inert solid support, inside a piece of glass or metal tubing called a column (an homage to the fractionating column used in distillation). The instrument used to perform gas chromatography is called a *gas chromatograph* (or "aerograph", "gas separator").

The gaseous compounds being analyzed interact with the walls of the column, which is coated with a stationary phase. This causes each compound to elute at a different time, known as the *retention time* of the compound. The comparison of retention times is what gives GC its analytical usefulness.

Gas chromatography is in principle similar to column chromatography (as well as other forms of chromatography, such as HPLC, TLC), but has several notable differences. First, the process of separating the compounds in a mixture is carried out between a liquid stationary phase and a gas mobile phase, whereas in column chromatography the stationary phase is a solid and the mobile phase is a liquid. (Hence the full name of the procedure is "Gas–liquid chromatography", referring to the mobile and stationary phases, respectively.) Second, the column through which the gas phase passes is located in an oven where the temperature of the gas can be controlled, whereas column chromatography (typically) has no such temperature control. Finally, the concentration of a compound in the gas phase is solely a function of the vapor pressure of the gas.

Gas chromatography is also similar to fractional distillation, since both processes separate the components of a mixture primarily based on boiling point (or vapor pressure) differences. However, fractional distillation is typically used to separate components of a mixture on a large scale, whereas GC can be used on a much smaller scale (i.e. microscale).

Gas chromatography is also sometimes known as vapor-phase chromatography (VPC), or gas–liquid partition chromatography (GLPC). These alternative names, as well as their respective abbreviations, are frequently used in scientific literature. Strictly speaking, GLPC is the most correct terminology, and is thus preferred by many authors.

History

Chromatography dates to 1903 in the work of the Russian scientist, Mikhail Semenovich Tswett. German graduate student Fritz Prior developed solid state gas chromatography in 1947. Archer John Porter Martin, who was awarded the Nobel Prize for his work in developing liquid–liquid (1941) and paper (1944) chromatography, laid the foundation for the development of gas chromatography and he later produced liquid-gas chromatography (1950). Erika Cremer laid the groundwork, and oversaw much of Prior's work.

GC Analysis

A gas chromatograph is a chemical analysis instrument for separating chemicals in a complex sample. A gas chromatograph uses a flow-through narrow tube known as the *column*, through which different chemical constituents of a sample pass in a gas stream (carrier gas, *mobile phase*) at different rates depending on their various chemical and physical properties and their interaction with a specific column filling, called the *stationary phase*. As the chemicals exit the end of the column, they are detected and identified electronically. The function of the stationary phase in the column is to separate different components, causing each one to exit the column at a different time (*retention time*). Other parameters that can be used to alter the order or time of retention are the carrier gas flow rate, column length and the temperature.

In a GC analysis, a known volume of gaseous or liquid analyte is injected into the "entrance" (head) of the column, usually using a microsyringe (or, solid phase microextraction fibers, or a gas source switching system). As the carrier gas sweeps the analyte molecules through the column, this motion is inhibited by the adsorption of the analyte molecules either onto the column walls or onto packing materials in the column. The rate at which the molecules progress along the column depends on the strength of adsorption, which in turn depends on the type of molecule and on the stationary phase materials. Since each type of molecule has a different rate of progression, the various components of the analyte mixture are separated as they progress along the column and reach the end of the column at different times (retention time). A detector is used to monitor the outlet stream from the column; thus, the time at which each component reaches the outlet and the amount of that component can be determined. Generally, substances are identified (qualitatively) by the order in which they emerge (elute) from the column and by the retention time of the analyte in the column.

Physical Components

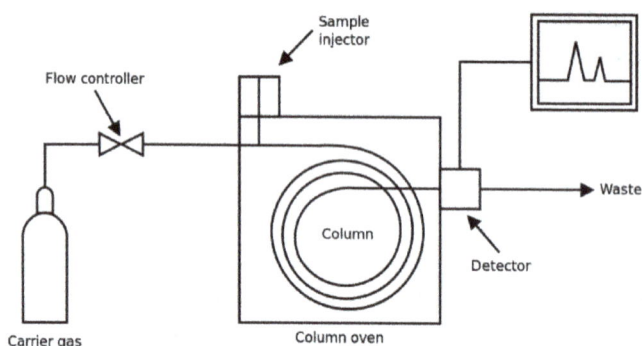

Diagram of a gas chromatograph.

Autosamplers

The autosampler provides the means to introduce a sample automatically into the inlets. Manual insertion of the sample is possible but is no longer common. Automatic insertion provides better reproducibility and time-optimization.

Different kinds of autosamplers exist. Autosamplers can be classified in relation to sample capacity (auto-injectors vs. autosamplers, where auto-injectors can work a small number of samples), to robotic technologies (XYZ robot vs. rotating robot – the most common), or to analysis:

- Liquid

- Static head-space by syringe technology

- Dynamic head-space by transfer-line technology

- Solid phase microextraction (SPME)

Traditionally autosampler manufacturers are different from GC manufacturers and currently no GC manufacturer offers a complete range of autosamplers. Historically, the countries most active in autosampler technology development are the United States, Italy, Switzerland, and the United Kingdom.

Inlets

The column inlet (or injector) provides the means to introduce a sample into a continuous flow of carrier gas. The inlet is a piece of hardware attached to the column head.

Common inlet types are:

- S/SL (split/splitless) injector; a sample is introduced into a heated small chamber via a syringe through a septum – the heat facilitates volatilization of the sample and sample matrix. The carrier gas then either sweeps the entirety (splitless mode) or a portion (split mode) of the sample into the column. In split mode, a part of the sample/carrier gas mixture in the injection chamber is exhausted through the split vent. Split injection is preferred when working with samples with high analyte concentrations (>0.1%) whereas splitless injection is best suited for trace analysis with low amounts of analytes (<0.01%). In splitless mode the split valve opens after a pre-set amount of time to purge heavier elements that would otherwise contaminate the system. This pre-set (splitless) time should be optimized, the shorter time (e.g., 0.2 min) ensures less tailing but loss in response, the longer time (2 min) increases tailing but also signal.

- On-column inlet; the sample is here introduced directly into the column in its entirety without heat, or at a temperature below the boiling point of the solvent. The low temperature condenses the sample into a narrow zone. The column and inlet can then be heated, releasing the sample into the gas phase. This ensures the lowest possible temperature for chromatography and keeps samples from decomposing above their boiling point.

- PTV injector; Temperature-programmed sample introduction was first described by Vogt in 1979. Originally Vogt developed the technique as a method for the introduction of large sample volumes (up to 250 μL) in capillary GC. Vogt introduced the sample into the liner at a controlled injection rate. The temperature of the liner was chosen slightly below the boiling point of the solvent. The low-boiling solvent was continuously evaporated and vented through the split line. Based on this technique, Poy developed the programmed temperature vaporising injector; PTV. By introducing the sample at a low initial liner temperature many of the disadvantages of the classic hot injection techniques could be circumvented.

- Gas source inlet or gas switching valve; gaseous samples in collection bottles are connected to what is most commonly a six-port switching valve. The carrier gas flow is not interrupted while a sample can be expanded into a previously evacuated sample loop. Upon switching, the contents of the sample loop are inserted into the carrier gas stream.

- P/T (Purge-and-Trap) system; An inert gas is bubbled through an aqueous sample causing insoluble volatile chemicals to be purged from the matrix. The volatiles are 'trapped' on an absorbent column (known as a trap or concentrator) at ambient temperature. The trap is then heated and the volatiles are directed into the carrier gas stream. Samples requiring preconcentration or purification can be introduced via such a system, usually hooked up to the S/SL port.

The choice of carrier gas (mobile phase) is important. Hydrogen has a range of flow rates that are comparable to helium in efficiency. However, helium may be more efficient and provide the best separation if flow rates are optimized. Helium is non-flammable and works with a greater number of detectors and older instruments. Therefore, helium is the most common carrier gas used. However, the price of helium has gone up considerably over recent years, causing an increasing number of chromatographers to switch to hydrogen gas. Historical use, rather than rational consideration, may contribute to the continued preferential use of helium.

Detectors

The most commonly used detectors are the flame ionization detector (FID) and the thermal conductivity detector (TCD). Both are sensitive to a wide range of components, and both work over a wide range of concentrations. While TCDs are essentially universal and can be used to detect any component other than the carrier gas (as long as their thermal conductivities are different from that of the carrier gas, at detector temperature), FIDs are sensitive primarily to hydrocarbons, and are more sensitive to them than TCD. However, a FID cannot detect water. Both detectors are also quite robust. Since TCD is non-destructive, it can be operated in-series before a FID (destructive), thus providing complementary detection of the same analytes.

Other detectors are sensitive only to specific types of substances, or work well only in narrower ranges of concentrations. They include:

- Thermal Conductivity detector (TCD), this common detector relies on the thermal conductivity of matter passing around a tungsten -rhenium filament with a current traveling through it. In this set up helium or nitrogen serve as the carrier gas because of their relatively high thermal conductivity which keep the filament cool and maintain uniform resistivity and electrical efficiency of the filament. However, when analyte molecules elute from the column, mixed with carrier gas, the thermal conductivity decreases and this causes a detector response. The response is due to the decreased thermal conductivity causing an increase in filament temperature and resistivity resulting in fluctuations in voltage. Detector sensitivity is proportional to filament current while it is inversely proportional to the immediate environmental temperature of that detector as well as flow rate of the carrier gas.

- Flame Ionization detector (FID), in this common detector electrodes are placed adjacent to a flame fueled by hydrogen / air near the exit of the column, and when carbon containing compounds exit the column they are pyrolyzed by the flame. This detector works only for organic / hydrocarbon containing compounds due to the ability of the carbons to form cations and electrons upon pyrolysis which generates a current between the electrodes. The increase in current is translated and appears as a peak in a chromatogram. FIDs have low detection limits (a few picograms per second) but they are unable to generate ions from carbonyl containing carbons. FID compatible carrier gasses include helium, hydrogen, nitrogen, and argon.

- Catalytic combustion detector (CCD), which measures combustible hydrocarbons and hydrogen.

- Discharge ionization detector (DID), which uses a high-voltage electric discharge to produce ions.

- Dry electrolytic conductivity detector (DELCD), which uses an air phase and high temperature (v. Coulsen) to measure chlorinated compounds.

- Electron capture detector (ECD), which uses a radioactive beta particle (electron) source to measure the degree of electron capture. ECD are used for the detection of molecules containing electronegative / withdrawing elements and functional groups like halogens, carbonyl, nitriles, nitro groups, and organometalics. In this type of detector either nitrogen or 5% methane in argon is used as the mobile phase carrier gas. The carrier gas passes between two electrodes placed at the end of the column, and adjacent to the anode (negative electrode) resides a radioactive foil such as 63Ni. The radioactive foil emits a beta particle (electron) which collides with and ionizes the carrier gas to generate more ions resulting in a current. When analyte molecules with electronegative / withdrawing elements or functional groups electrons are captured which results in a decrease in current generating a detector response.

- Flame photometric detector (FPD), which uses a photomultiplier tube to detect spectral lines of the compounds as they are burned in a flame. Compounds eluting off the column are carried into a hydrogen fueled flame which excites specific elements in the molecules, and the excited elements (P,S, Halogens, Some Metals) emit light of specific characteristic wavelengths. The emitted light is filtered and detected by a photomultiplier tube. In particular, phosphorus emission is around 510–536 nm and sulfur emission os at 394 nm.

- Atomic Emission Detector (AED), a sample eluting from a column enters a chamber which is energized by microwaves that induce a plasma. The plasma causes the analyte sample to decompose and certain elements generate an atomic emission spectra. The atomic emission spectra is diffracted by a diffraction grating and detected by a series of photomultiplier tubes or photo diodes.

- Hall electrolytic conductivity detector (ElCD)

- Helium ionization detector (HID)

- Nitrogen–phosphorus detector (NPD), a form of thermionic detector where nitrogen and phosphorus alter the work function on a specially coated bead and a resulting current is measured.

- Alkali Flame Detector, AFD or Alkali Flame Ionization Detector, AFID. AFD has high sensitivity to nitrogen and phosphorus, similar to NPD. However, the alkaline metal ions are supplied with the hydrogen gas, rather than a bead above the flame. For this reason AFD does not suffer the "fatigue" of he NPD, but provides a constant sensitivity over long period of time. In addition, when alkali ions are not added to the flame, AFD operates like a standard FID.

- Infrared detector (IRD)

- Mass spectrometer (MS), also called GC-MS; highly effective and sensitive, even in a small quantity of sample.

- Photo-ionization detector (PID)

- The Polyarc reactor is an add-on to new or existing GC-FID instruments that converts all organic compounds to methane molecules prior to their detection by the FID. This technique can be used to improve the response of the FID and allow for the detection of many more carbon-containing compounds. The complete conversion of compounds to methane and the now equivalent response in the detector also eliminates the need for calibrations and standards because response factors are all equivalent to those of methane. This allows for the rapid analysis of complex mixtures that contain molecules where standards are not available.

- Pulsed discharge ionization detector (PDD)

- Thermionic ionization detector (TID)

- Vacuum Ultraviolet (VUV) represents the most recent development in Gas Chromatography detectors. Most chemical species absorb and have unique gas phase absorption cross sections in the approximately 120–240 nm VUV wavelength range monitored. Where absorption cross sections are known for analytes, the VUV detector is capable of absolute determination (without calibration) of the number of molecules present in the flow cell in the absence of chemical interferences.

Some gas chromatographs are connected to a mass spectrometer which acts as the detector. The combination is known as GC-MS. Some GC-MS are connected to an NMR spectrometer which acts as a backup detector. This combination is known as GC-MS-NMR. Some GC-MS-NMR are connected to an infrared spectrophotometer which acts as a backup detector. This combination is known as GC-MS-NMR-IR. It must, however, be stressed this is very rare as most analyses needed can be concluded via purely GC-MS.

Methods

This image above shows the interior of a GeoStrata Technologies Eclipse Gas Chromatograph that runs continuously in three-minute cycles. Two valves are used to switch the test gas into the sample loop. After filling the sample loop with test gas, the valves are switched again applying carrier gas pressure to the sample loop and forcing the sample through the column for separation.

The method is the collection of conditions in which the GC operates for a given analysis. Method development is the process of determining what conditions are adequate and/ or ideal for the analysis required.

Conditions which can be varied to accommodate a required analysis include inlet temperature, detector temperature, column temperature and temperature program, carrier gas and carrier gas flow rates, the column's stationary phase, diameter and length, inlet type and flow rates, sample size and injection technique. Depending on the detector(s) installed on the GC, there may be a number of detector conditions that can also be varied. Some GCs also include valves which can change the route of sample and carrier flow. The timing of the opening and closing of these valves can be important to method development.

Carrier Gas Selection and Flow Rates

Typical carrier gases include helium, nitrogen, argon, hydrogen and air. Which gas to use is usually determined by the detector being used, for example, a DID requires helium as the carrier gas. When analyzing gas samples, however, the carrier is sometimes selected based on the sample's matrix, for example, when analyzing a mixture in argon, an argon carrier is preferred, because the argon in the sample does not show up on the chromatogram. Safety and availability can also influence carrier selection, for example, hydrogen is flammable, and high-purity helium can be difficult to obtain in some areas of the world. As a result of helium becoming more scarce, hydrogen is often being substituted for helium as a carrier gas in several applications.

The purity of the carrier gas is also frequently determined by the detector, though the level of sensitivity needed can also play a significant role. Typically, purities of 99.995% or higher are used. The most common purity grades required by modern instruments for the majority of sensitivities are 5.0 grades, or 99.999% pure meaning that there is a total of 10ppm of impurities in the carrier gas that could affect the results. The highest purity grades in common use are 6.0 grades, but the need for detection at very low levels in some forensic and environmental applications has driven the need for carrier gases at 7.0 grade purity and these are now commercially available. Trade names for typical purities include "Zero Grade," "Ultra-High Purity (UHP) Grade," "4.5 Grade" and "5.0 Grade."

The carrier gas linear velocity affects the analysis in the same way that temperature does. The higher the linear velocity the faster the analysis, but the lower the separation between analytes. Selecting the linear velocity is therefore the same compromise between the level of separation and length of analysis as selecting the column temperature. The linear velocity will be implemented by means of the carrier gas flow rate, with regards to the inner diameter of the column.

With GCs made before the 1990s, carrier flow rate was controlled indirectly by controlling the carrier inlet pressure, or "column head pressure." The actual flow rate was measured at the outlet of the column or the detector with an electronic flow meter, or a bubble flow meter, and could be an involved, time consuming, and frustrating process. The pressure setting was not able to be varied during the run, and thus the flow was essentially constant during the analysis. The relation between flow rate and inlet pressure is calculated with Poiseuille's equation for compressible fluids.

Many modern GCs, however, electronically measure the flow rate, and electronically control the carrier gas pressure to set the flow rate. Consequently, carrier pressures and flow rates can be adjusted during the run, creating pressure/flow programs similar to temperature programs.

Stationary Compound Selection

The polarity of the solute is crucial for the choice of stationary compound, which in an optimal case would have a similar polarity as the solute. Common stationary phases in open tubular columns are cyanopropylphenyl dimethyl polysiloxane, carbowax polyethyleneglycol, biscyanopropyl cyanopropylphenyl polysiloxane and diphenyl dimethyl polysiloxane. For packed columns more options are available.

Inlet Types and Flow Rates

The choice of inlet type and injection technique depends on if the sample is in liquid, gas, adsorbed, or solid form, and on whether a solvent matrix is present that has to be vaporized. Dissolved samples can be introduced directly onto the column via a COC injector, if the conditions are well known; if a solvent matrix has to be vaporized and partially removed, a S/SL injector is used (most common injection technique); gaseous samples (e.g., air cylinders) are usually injected using a gas switching valve system; adsorbed samples (e.g., on adsorbent tubes) are introduced using either an external (online or off-line) desorption apparatus such as a purge-and-trap system, or are desorbed in the injector (SPME applications).

Sample Size and Injection Technique

Sample Injection

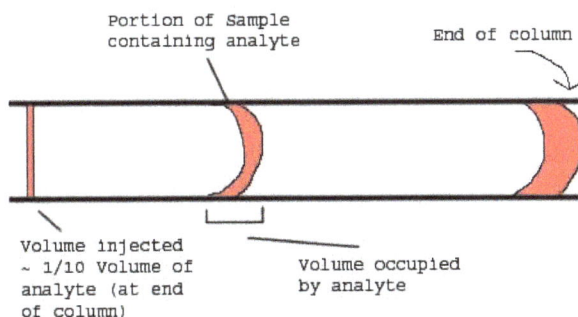

The rule of ten in gas chromatography

The real chromatographic analysis starts with the introduction of the sample onto the column. The development of capillary gas chromatography resulted in many practical problems with the injection technique. The technique of on-column injection, often used with packed columns, is usually not possible with capillary columns. The injection system in the capillary gas chromatograph should fulfil the following two requirements:

1. The amount injected should not overload the column.

2. The width of the injected plug should be small compared to the spreading due to the chromatographic process. Failure to comply with this requirement will reduce the separation capability of the column. As a general rule, the volume injected, V_{inj}, and the volume of the detector cell, V_{det}, should be about 1/10 of the volume occupied by the portion of sample containing the molecules of interest (analytes) when they exit the column.

Some general requirements which a good injection technique should fulfill are:

- It should be possible to obtain the column's optimum separation efficiency.

- It should allow accurate and reproducible injections of small amounts of representative samples.

- It should induce no change in sample composition. It should not exhibit discrimination based on differences in boiling point, polarity, concentration or thermal/catalytic stability.

- It should be applicable for trace analysis as well as for undiluted samples.

However, there are a number of problems inherent in the use of syringes for injection, even when they are not damaged:

- Even the best syringes claim an accuracy of only 3%, and in unskilled hands, errors are much larger

- The needle may cut small pieces of rubber from the septum as it injects sample through it. These can block the needle and prevent the syringe filling the next time it is used. It may not be obvious of what happened.

- A fraction of the sample may get trapped in the rubber, to be released during subsequent injections. This can give rise to ghost peaks in the chromatogram.

- There may be selective loss of the more volatile components of the sample by evaporation from the tip of the needle.

Column Selection

The choice of column depends on the sample and the active measured. The main chemical attribute regarded when choosing a column is the polarity of the mixture, but functional groups can play a large part in column selection. The polarity of the sample must closely match the polarity of the column stationary phase to increase resolution and separation while reducing run time. The separation and run time also depends on the film thickness (of the stationary phase), the column diameter and the column length.

Column Temperature and Temperature Program

A gas chromatography oven, open to show a capillary column

The column(s) in a GC are contained in an oven, the temperature of which is precisely controlled electronically. (When discussing the "temperature of the column," an analyst is technically referring to the temperature of the column oven. The distinction, however, is not important and will not subsequently be made in this article.)

The rate at which a sample passes through the column is directly proportional to the temperature of the column. The higher the column temperature, the faster the sample moves through the column. However, the faster a sample moves through the column, the less it interacts with the stationary phase, and the less the analytes are separated.

In general, the column temperature is selected to compromise between the length of the analysis and the level of separation.

A method which holds the column at the same temperature for the entire analysis is called "isothermal." Most methods, however, increase the column temperature during the analysis, the initial temperature, rate of temperature increase (the temperature "ramp"), and final temperature are called the "temperature program."

A temperature program allows analytes that elute early in the analysis to separate adequately, while shortening the time it takes for late-eluting analytes to pass through the column.

Data Reduction and Analysis

Qualitative Analysis

Generally chromatographic data is presented as a graph of detector response (y-axis) against retention time (x-axis), which is called a chromatogram. This provides a spectrum of peaks for a sample representing the analytes present in a sample eluting from the column at different times. Retention time can be used to identify analytes if the method conditions are constant. Also, the pattern of peaks will be constant for a sample under constant conditions and can identify complex mixtures of analytes. However, in most modern applications, the GC is connected to a mass spectrometer or similar detector that is capable of identifying the analytes represented by the peaks.

Quantitative Analysis

The area under a peak is proportional to the amount of analyte present in the chromatogram. By calculating the area of the peak using the mathematical function of integration, the concentration of an analyte in the original sample can be determined. Concentration can be calculated using a calibration curve created by finding the response for a series of concentrations of analyte, or by determining the relative response factor of an analyte. The relative response factor is the expected ratio of an analyte to an internal standard (or external standard) and is calculated by finding the response of a known amount of analyte and a constant amount of internal standard (a chemical added to the sample at a constant concentration, with a distinct retention time to the analyte).

In most modern GC-MS systems, computer software is used to draw and integrate peaks, and match MS spectra to library spectra.

Applications

In general, substances that vaporize below 300 °C (and therefore are stable up to that temperature) can be measured quantitatively. The samples are also required to be salt-free; they should not contain ions. Very minute amounts of a substance can be measured, but it is often required that the sample must be measured in comparison to a sample containing the pure, suspected substance known as a reference standard.

Various temperature programs can be used to make the readings more meaningful; for example to differentiate between substances that behave similarly during the GC process.

Professionals working with GC analyze the content of a chemical product, for example in assuring the quality of products in the chemical industry; or measuring toxic substances in soil, air or water. GC is very accurate if used properly and can measure picomoles of a substance in a 1 ml liquid sample, or parts-per-billion concentrations in gaseous samples.

In practical courses at colleges, students sometimes get acquainted to the GC by studying the contents of Lavender oil or measuring the ethylene that is secreted by *Nicotiana benthamiana* plants after artificially injuring their leaves. These GC analyse hydrocarbons (C2-C40+). In a typical experiment, a packed column is used to separate the light gases, which are then detected with a TCD. The hydrocarbons are separated using a capillary column and detected with a FID. A complication with light gas analyses that include H_2 is that He, which is the most common and most sensitive inert carrier (sensitivity is proportional to molecular mass) has an almost identical thermal conductivity to hydrogen (it is the difference in thermal conductivity between two separate filaments in a Wheatstone Bridge type arrangement that shows when a component has been eluted). For this reason, dual TCD instruments used with a separate channel for hydrogen that uses nitrogen as a carrier are common. Argon is often used when analysing gas phase chemistry reactions such as F-T synthesis so that a single carrier gas can be used rather than two separate ones. The sensitivity is less, but this is a trade off for simplicity in the gas supply.

Gas Chromatography is used extensively in forensic science. Disciplines as diverse as solid drug dose (pre-consumption form) identification and quantification, arson investigation, paint chip analysis, and toxicology cases, employ GC to identify and quantify various biological specimens and crime-scene evidence.

GCs in Popular Culture

Movies, books and TV shows tend to misrepresent the capabilities of gas chromatography and the work done with these instruments.

In the U.S. TV show CSI, for example, GCs are used to rapidly identify unknown samples. For example, an analyst may say fifteen minutes after receiving the sample: "This is gasoline bought at a Chevron station in the past two weeks."

In fact, a typical GC analysis takes much more time; sometimes a single sample must be run more than an hour according to the chosen program; and even more time is needed to "heat out" the column so it is free from the first sample and can be used for the next. Equally, several runs are needed to confirm the results of a study – a GC analysis of a single sample may simply yield a result per chance.

Also, GC does not positively identify most samples; and not all substances in a sample will necessarily be detected. All a GC truly tells you is at which relative time a component eluted from the column and that the detector was sensitive to it. To make results meaningful, analysts need to know which components at which concentrations are to be expected; and even then a small amount of a substance can hide itself behind a substance having both a higher concentration and the same relative elution time. Last but not least it is often needed to check the results of the sample against a GC analysis of a reference sample containing only the suspected substance.

A GC-MS can remove much of this ambiguity, since the mass spectrometer will identify the component's molecular weight. But this still takes time and skill to do properly.

Similarly, most GC analyses are not push-button operations. You cannot simply drop a sample vial into an auto-sampler's tray, push a button and have a computer tell you everything you need to know about the sample. The operating program must be carefully chosen according to the expected sample composition.

A push-button operation can exist for running similar samples repeatedly, such as in a chemical production environment or for comparing 20 samples from the same experiment to calculate the mean content of the same substance. However, for the kind of investigative work portrayed in books, movies and TV shows this is clearly not the case.

Gas Chromatography–mass Spectrometry

Gas chromatography–mass spectrometry (GC-MS) is an analytical method that combines the features of gas-chromatography and mass spectrometry to identify different substances within a test sample. Applications of GC-MS include drug detection, fire investigation, environmental analysis, explosives investigation, and identification of unknown samples, including that of material samples obtained from planet Mars during probe missions as early as the 1970s. GC-MS can also be used in airport security to detect substances in luggage or on human beings. Additionally, it can identify trace elements in materials that were previously thought to have disintegrated beyond identification. Like liquid chromatography–mass spectrometry, it allows analysis and detection even of tiny amounts of a substance.

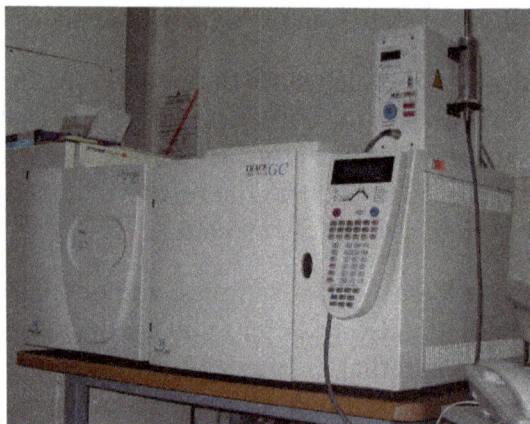

Example of a GC-MS instrument

GC-MS has been widely heralded as a "gold standard" for forensic substance identification because it is used to perform a 100% specific test, which positively identifies the presence of a particular substance. A nonspecific test merely indicates that any of several in a category of substances is present. Although a nonspecific test could statistically suggest the identity of the substance, this could lead to false positive identification.

History

The use of a mass spectrometer as the detector in gas chromatography was developed during the 1950s after being originated by James and Martin in 1952. These comparatively sensitive devices were originally limited to laboratory settings.

The development of affordable and miniaturized computers has helped in the simplification of the use of this instrument, as well as allowed great improvements in the amount of time it takes to analyze a sample. In 1964, Electronic Associates, Inc. (EAI), a leading U.S. supplier of analog computers, began development of a computer controlled quadrupole mass spectrometer under the direction of Robert E. Finnigan. By 1966 Finnigan and collaborator Mike Uthe's EAI division had sold over 500 quadrupole residual gas-analyzer instruments. In 1967, Finnigan left EAI to form the Finnigan Instrument Corporation along with Roger Sant, T. Z. Chou, Michael Story, and William Fies. In early 1968, they delivered the first prototype quadrupole GC/MS instruments to Stanford and Purdue University. When Finnigan Instrument Corporation was acquired by Thermo Instrument Systems (later Thermo Fisher Scientific) in 1990, it was considered "the world's leading manufacturer of mass spectrometers".

In 1996 the top-of-the-line high-speed GC-MS units completed analysis of fire accelerants in less than 90 seconds, whereas first-generation GC-MS would have required at least 16 minutes. By the 2000s computerized GC/MS instruments using quadrupole technology had become both essential to chemical research and one of the foremost instruments used for organic analysis. Today computerized GC/MS instruments are widely used in environmental monitoring of water, air, and soil; in the regulation of agriculture and food safety; and in the discovery and production of medicine.

Instrumentation

The insides of the GC-MS, with the column of the gas chromatograph in the oven on the right.

The GC-MS is composed of two major building blocks: the gas chromatograph and the mass spectrometer. The gas chromatograph utilizes a capillary column which depends on the column's dimensions (length, diameter, film thickness) as well as the phase properties (e.g. 5% phenyl polysiloxane). The difference in the chemical properties between different molecules in a mixture and their relative affinity for the stationary phase of the column will promote separation of the molecules as the sample travels the length of the column. The molecules are retained by the column and then elute (come off) from the column at different times (called the retention time), and this allows the mass spectrometer downstream to capture, ionize, accelerate, deflect, and detect the ionized molecules separately. The mass spectrometer does this by breaking each molecule into ionized fragments and detecting these fragments using their mass-to-charge ratio.

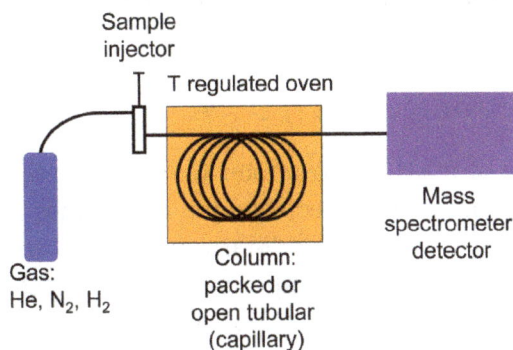

GC-MS schematic

These two components, used together, allow a much finer degree of substance identification than either unit used separately. It is not possible to make an accurate identification of a particular molecule by gas chromatography or mass spectrometry alone. The mass spectrometry process normally requires a very pure sample while gas chromatography using a traditional detector (e.g. Flame ionization detector) cannot differentiate between multiple molecules that happen to take the same amount of time to travel through the column (*i.e.* have the same retention time), which results in two or more molecules that co-elute. Sometimes two different molecules can also have a similar pattern of ionized fragments in a mass spectrometer (mass spectrum). Combining the two processes reduces the possibility of error, as it is extremely unlikely that two different molecules will behave in the same way in both a gas chromatograph and a mass spectrometer. Therefore, when an identifying mass spectrum appears at a characteristic retention time in a GC-MS analysis, it typically increases certainty that the analyte of interest is in the sample.

Purge and Trap GC-MS

For the analysis of volatile compounds, a purge and trap (P&T) concentrator system may be used to introduce samples. The target analytes are extracted and mixed with water and introduced into an airtight chamber. An inert gas such as Nitrogen (N_2) is bubbled through the water; this is known as purging or sparging. The volatile compounds move into the headspace above the water and are drawn along a pressure gradient (caused by the introduction of the purge gas) out of the chamber. The volatile compounds are drawn along a heated line onto a 'trap'. The trap is a column of adsorbent material at ambient temperature that holds the compounds by returning them to the liquid phase. The trap is then heated and the sample compounds are introduced to the GC-MS column via a volatiles interface, which is a split inlet system. P&T GC-MS is particularly suited to volatile organic compounds (VOCs) and BTEX compounds (aromatic compounds associated with petroleum).

A faster alternative is the "purge-closed loop" system. In this system the inert gas is bubbled through the water until the concentrations of organic compounds in the vapor phase are at equilibrium with concentrations in the aqueous phase. The gas phase is then analysed directly.

Types of Mass Spectrometer Detectors

The most common type of mass spectrometer (MS) associated with a gas chromatograph (GC) is the quadrupole mass spectrometer, sometimes referred to by the Hewlett-Packard (now Agilent) trade name "Mass Selective Detector" (MSD). Another relatively common detector is the ion trap mass spectrometer. Additionally one may find a magnetic sector mass spectrometer, however these particular instruments are expensive and bulky and not typically found in high-throughput service laboratories. Other detectors may be encountered such as time of flight (TOF), tandem quadrupoles

(MS-MS) , or in the case of an ion trap MS^n where n indicates the number mass spectrometry stages.

GC-tandem MS

When a second phase of mass fragmentation is added, for example using a second quadrupole in a quadrupole instrument, it is called tandem MS (MS/MS). MS/MS can sometimes be used to quantitate low levels of target compounds in the presence of a high sample matrix background.

The first quadrupole (Q1) is connected with a collision cell (Q2) and another quadrupole (Q3). Both quadrupoles can be used in scanning or static mode, depending on the type of MS/MS analysis being performed. Types of analysis include product ion scan, precursor ion scan, selected reaction monitoring (SRM) (sometimes referred to as multiple reaction monitoring (MRM)) and neutral loss scan. For example: When Q1 is in static mode (looking at one mass only as in SIM), and Q3 is in scanning mode, one obtains a so-called product ion spectrum (also called "daughter spectrum"). From this spectrum, one can select a prominent product ion which can be the product ion for the chosen precursor ion. The pair is called a "transition" and forms the basis for SRM. SRM is highly specific and virtually eliminates matrix background.

Ionization

After the molecules travel the length of the column, pass through the transfer line and enter into the mass spectrometer they are ionized by various methods with typically only one method being used at any given time. Once the sample is fragmented it will then be detected, usually by an electron multiplier diode, which essentially turns the ionized mass fragment into an electrical signal that is then detected.

The ionization technique chosen is independent of using full scan or SIM.

Electron Ionization

By far the most common and perhaps standard form of ionization is electron ionization (EI). The molecules enter into the MS (the source is a quadrupole or the ion trap itself in an ion trap MS) where they are bombarded with free electrons emitted from a filament, not unlike the filament one would find in a standard light bulb. The electrons bombard the molecules, causing the molecule to fragment in a characteristic and reproducible way. This "hard ionization" technique results in the creation of more fragments of low mass to charge ratio (m/z) and few, if any, molecules approaching the molecular mass unit. Hard ionization is considered by mass spectrometrists as the employ of molecular electron bombardment, whereas "soft ionization" is charge by molecular collision with an introduced gas. The molecular fragmentation pattern is dependent upon the electron energy applied to the system, typically 70 eV (electron Volts). The use of 70 eV facilitates comparison of generated spectra with library spectra using manufacturer-supplied software or software developed by the National Institute of Standards (NIST-USA). Spectral library searches employ matching algorithms such as Probability Based Matching and dot-product matching that are used with methods of analysis written by many method standardization agencies. Sources of libraries include NIST, Wiley, the AAFS, and instrument manufacturers.

Cold Electron Ionization

The "hard ionization" process of electron ionization can be softened by the cooling of the molecules before their ionization, resulting in mass spectra that are richer in information. In this method named cold electron ionization (Cold-EI) the molecules exit the GC column, mixed with added helium make up gas and expand into vacuum through a specially designed supersonic nozzle, forming a supersonic molecular beam (SMB). Collisions with the make up gas at the expanding supersonic jet reduce the internal vibrational (and rotational) energy of the analyte molecules, hence reducing the degree of fragmentation caused by the electrons during the ionization process. Cold-EI mass spectra are characterized by an abundant molecular ion while the usual fragmentation pattern is retained, thus making Cold-EI mass spectra compatible with library search identification techniques. The enhanced molecular ions increase the identification probabilities of both known and unknown compounds, amplify isomer mass spectral effects and enable the use of isotope abundance analysis for the elucidation of elemental formulae.

Chemical Ionization

In chemical ionization a reagent gas, typically methane or ammonia is introduced into the mass spectrometer. Depending on the technique (positive CI or negative CI) chosen, this reagent gas will interact with the electrons and analyte and cause a 'soft' ionization of the molecule of interest. A softer ionization fragments the molecule to a lower degree than the hard ionization of EI. One of the main benefits of using chemical ionization is that a mass fragment closely corresponding to the molecular weight of the analyte of interest is produced.

In positive chemical ionization (PCI) the reagent gas interacts with the target molecule, most often with a proton exchange. This produces the species in relatively high amounts.

In negative chemical ionization (NCI) the reagent gas decreases the impact of the free electrons on the target analyte. This decreased energy typically leaves the fragment in great supply.

Analysis

A mass spectrometer is typically utilized in one of two ways: full scan or selected ion monitoring (SIM). The typical GC-MS instrument is capable of performing both functions either individually or concomitantly, depending on the setup of the particular instrument.

The primary goal of instrument analysis is to quantify an amount of substance. This is done by comparing the relative concentrations among the atomic masses in the generated spectrum. Two kinds of analysis are possible, comparative and original. Comparative analysis essentially compares the given spectrum to a spectrum library to see if its characteristics are present for some sample in the library. This is best performed by a computer because there are a myriad of visual distortions that can take place due to variations in scale. Computers can also simultaneously correlate more data (such as the retention times identified by GC), to more accurately relate certain data.

Another method of analysis measures the peaks in relation to one another. In this method, the tallest peak is assigned 100% of the value, and the other peaks being assigned proportionate values. All values above 3% are assigned. The total mass of the unknown compound is normally indicated by the parent peak. The value of this parent peak can be used to fit with a chemical formula containing the various elements which are believed to be in the compound. The isotope pattern in the spectrum, which is unique for elements that have many isotopes, can also be used to identify the various elements present. Once a chemical formula has been matched to the spectrum, the molecular structure and bonding can be identified, and must be consistent with the characteristics recorded by GC-MS. Typically, this identification done automatically by programs which come with the instrument, given a list of the elements which could be present in the sample.

A "full spectrum" analysis considers all the "peaks" within a spectrum. Conversely, selective ion monitoring (SIM) only monitors selected ions associated with a specific substance. This is done on the assumption that at a given retention time, a set of ions is characteristic of a certain compound. This is a fast and efficient analysis, especially if the analyst has previous information about a sample or is only looking for a few specific substances. When the amount of information collected about the ions in a given gas chromatographic peak decreases, the sensitivity of the analysis increases. So, SIM analysis allows for a smaller quantity of a compound to be detected and measured, but the degree of certainty about the identity of that compound is reduced.

Full Scan MS

When collecting data in the full scan mode, a target range of mass fragments is determined and put into the instrument's method. An example of a typical broad range of mass fragments to monitor would be m/z 50 to m/z 400. The determination of what range to use is largely dictated by what one anticipates being in the sample while being cognizant of the solvent and other possible interferences. A MS should not be set to look for mass fragments too low or else one may detect air (found as m/z 28 due to nitrogen), carbon dioxide (m/z 44) or other possible interference. Additionally if one is to use a large scan range then sensitivity of the instrument is decreased due to performing fewer scans per second since each scan will have to detect a wide range of mass fragments.

Full scan is useful in determining unknown compounds in a sample. It provides more information than SIM when it comes to confirming or resolving compounds in a sample. During instrument method development it may be common to first analyze test solutions in full scan mode to determine the retention time and the mass fragment fingerprint before moving to a SIM instrument method.

Selected Ion Monitoring

In selected ion monitoring (SIM) certain ion fragments are entered into the instrument method and only those mass fragments are detected by the mass spectrometer. The advantages of SIM are that the detection limit is lower since the instrument is only looking at a small number of fragments (e.g. three fragments) during each scan. More scans can take place each second. Since only a few mass fragments of interest are being monitored, matrix interferences are typically lower. To additionally confirm the likelihood of a potentially positive result, it is relatively important to be sure that the ion ratios of the various mass fragments are comparable to a known reference standard.

Applications

Environmental Monitoring and Cleanup

GC-MS is becoming the tool of choice for tracking organic pollutants in the environment. The cost of GC-MS equipment has decreased significantly, and the reliability has increased at the same time, which has contributed to its increased adoption in environmental studies.

Criminal Forensics

GC-MS can analyze the particles from a human body in order to help link a criminal to a crime. The analysis of fire debris using GC-MS is well established, and there is even an established American Society for Testing and Materials (ASTM) standard for fire debris analysis. GCMS/MS is especially useful here as samples often contain very complex matrices and results, used in court, need to be highly accurate.

Law Enforcement

GC-MS is increasingly used for detection of illegal narcotics, and may eventually supplant drug-sniffing dogs. It is also commonly used in forensic toxicology to find drugs and/or poisons in biological specimens of suspects, victims, or the deceased.

Sports Anti-doping Analysis

GC-MS is the main tool used in sports anti-doping laboratories to test athletes' urine samples for prohibited performance-enhancing drugs, for example anabolic steroids.

Security

A post–September 11 development, explosive detection systems have become a part of all US airports. These systems run on a host of technologies, many of them based on GC-MS. There are only three manufacturers certified by the FAA to provide these systems,one of which is Thermo Detection (formerly Thermedics), which produces the EGIS, a GC-MS-based line of explosives detectors. The other two manufacturers are Barringer Technologies, now owned by Smith 's Detection Systems, and Ion Track Instruments, part of General Electric Infrastructure Security Systems.

Chemical Warfare Agent Detection

As part of the post-September 11 drive towards increased capability in homeland security and public health preparedness, traditional GC-MS units with transmission quadrupole mass spectrometers, as well as those with cylindrical ion trap (CIT-MS) and toroidal ion trap (T-ITMS) mass spectrometers have been modified for field portability and near real-time detection of chemical warfare agents (CWA) such as sarin, soman, and VX. These complex and large GC-MS systems have been modified and configured with resistively heated low thermal mass (LTM) gas chromatographs that reduce analysis time to less than ten percent of the time required in traditional laboratory systems. Additionally, the systems are smaller, and more mobile, including units that are mounted in mobile analytical laboratories (MAL), such as those used by the United States Marine Corps Chemical and Biological Incident Response Force MAL and other similar laboratories, and systems that are hand-carried by two-person teams or individuals, much ado to the smaller mass detectors. Depending on the system, the analytes can be introduced via liquid injection, desorbed from sorbent tubes through a thermal desorption process, or with solid-phase micro extraction (SPME).

Chemical Engineering

GC-MS is used for the analysis of unknown organic compound mixtures. One critical use of this technology is the use of GC-MS to determine the composition of bio-oils processed from raw biomass.

Food, Beverage and Perfume Analysis

Foods and beverages contain numerous aromatic compounds, some naturally present in the raw materials and some forming during processing. GC-MS is extensively used for the analysis of these compounds which include esters, fatty acids, alcohols, aldehydes, terpenes etc. It is also used to detect and measure contaminants from spoilage or adulteration which may be harmful and which is often controlled by governmental agencies, for example pesticides.

Astrochemistry

Several GC-MS have left earth. Two were brought to Mars by the Viking program. Venera 11 and 12 and Pioneer Venus analysed the atmosphere of Venus with GC-MS. The Huygens probe of the Cassini-Huygens mission landed one GC-MS on Saturn's largest moon, Titan. The material in the comet 67P/Churyumov-Gerasimenko will be analysed by the Rosetta mission with a chiral GC-MS in 2014.

Medicine

Dozens of congenital metabolic diseases also known as Inborn error of metabolism are now detectable by newborn screening tests, especially the testing using gas chromatography–mass spectrometry. GC-MS can determine compounds in urine even in minor concentration. These compounds are normally not present but appear in individuals suffering with metabolic disorders. This is increasingly becoming a common way to diagnose IEM for earlier diagnosis and institution of treatment eventually leading to a better outcome. It is now possible to test a newborn for over 100 genetic metabolic disorders by a urine test at birth based on GC-MS.

In combination with isotopic labeling of metabolic compounds, the GC-MS is used for determining metabolic activity. Most applications are based on the use of ^{13}C as the labeling and the measurement of ^{13}C-^{12}C ratios with an isotope ratio mass spectrometer (IRMS); an MS with a detector designed to measure a few select ions and return values as ratios.

Inverse Gas Chromatography

Analytical Gas Chromatography

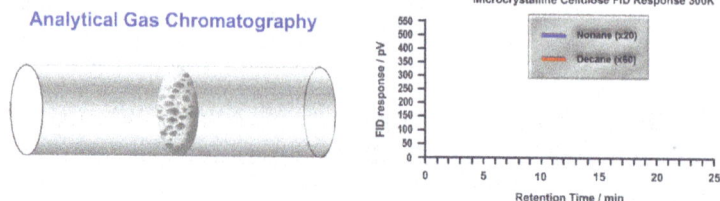

Microcrystalline Cellulose FID Response 300K

Nonane (x20)
Decane (x50)

FID response / pV
550
500
400
350
300
250
200
150
100
50
0

Retention Time / min
0   5   10   15   20   25

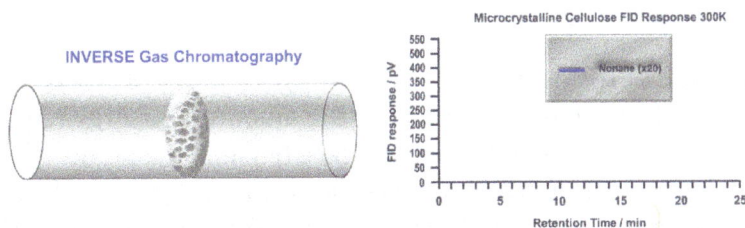

Analytical Gas Chromatography A (top) compared with Inverse Gas Chromatography B (bottom). While in gas chromatography a sample containing multiple species is separated into its components on a stationary phase, Inverse gas chromatography uses injection of a single species to probe the characteristics of a stationary phase sample.

Inverse gas chromatography is a physical characterization technique that is used in the analysis of the surfaces of solids. Traditional GC is an analytical technique.

Inverse gas chromatography or IGC is a highly sensitive and versatile gas phase technique developed over 40 years ago to study the surface and bulk properties of particulate and fibrous materials. In IGC the roles of the stationary (solid) and mobile (gas or vapor) phases are inverted from traditional analytical gas chromatography (GC). In GC, a standard column is used to separate and characterize several gases and/or vapors. In IGC, a single gas or vapor (probe molecule) is injected into a column packed with the solid sample under investigation. Instead of an analytical technique, IGC is considered a materials characterization technique.

During an IGC experiment a pulse or constant concentration of a known gas or vapor (probe molecule) is injected down the column at a fixed carrier gas flow rate. The retention time of the probe molecule is then measured by traditional GC detectors (i.e. flame ionization detector or thermal conductivity detector). Measuring how the retention time changes as a function of probe molecule chemistry, probe molecule size, probe molecule concentration, column temperature, or carrier gas flow rate can elucidate a wide range of physico-chemical properties of the solid under investigation. Several in depth reviews of IGC have been published previously.

IGC experiments are typically carried out at infinite dilution where only small amounts of probe molecule are injected. This region is also called Henry's law region or linear region of the sorption isotherm. At infinite dilution probe-probe interactions are assumed negligible and any retention is only due to probe-solid interactions. The resulting retention volume, V_R°, is given by the following equation:

$$V_R^\circ = \frac{j}{m} F(t_R - t_o) \frac{T}{273.15}$$

where j is the James–Martin pressure drop correction, m is the sample mass, F is the carrier gas flow rate at standard temperature and pressure, t_R is the gross retention time for the injected probe, t_o is the retention time for a non-interaction probe (i.e. dead-time), and T is the absolute temperature.

Surface Energy Determination

The main application of IGC is to measure the surface energy of solids (fibers, particulates, and films). Surface energy is defined as the amount of energy required to create a unit area of a solid surface; analogous to surface tension of a liquid. Also, the surface energy can be defined as the excess energy at the surface of a material compared to the bulk. The surface energy (γ) is directly related to the thermodynamic work of adhesion (W_{adh}) between two materials as given by the following equation:

$$W_{adh} = 2(\gamma_1\gamma_2)^{1/2}$$

where 1 and 2 represent the two components in the composite or blend. When determining if two materials will adhere it is common to compare the work of adhesion with the work of cohesion, $W_{coh} = 2\gamma$. If the work of adhesion is greater than the work of cohesion, then the two materials are thermodynamically favored to adhere.

Surface energies are commonly measured by contact angle methods. However, these methods are ideally designed for flat, uniform surfaces. For contact angle measurements on powders, they are typically compressed or adhered to a substrate which can effectively change the surface characteristics of the powder. Alternatively, the Washburn method can be used, but this has been shown to be affected by column packing, particle size, and pore geometry. IGC is a gas phase technique, thus is not subject to the above limitations of the liquid phase techniques.

To measure the solid surface energy by IGC a series of injections using different probe molecules is performed at defined column conditions. It is possible to ascertain both the dispersive component of the surface energy and acid-base properties via IGC. For the dispersive surface energy, the retention volumes for a series of n-alkane vapors (i.e. decane, nonane, octane, heptanes, etc.) are measured. The Dorris and Gray. or Schultz methods can then be used to calculate the dispersive surface energy. Retention volumes for polar probes (i.e. toluene, ethyl acetate, acetone, ethanol, acetonitrile, chloroform, dichloromethane, etc.) can then be used to determine the acid-base characteristics of the solid using either the Gutmann, or Good-van Oss theory.

Other parameters accessible by IGC include: heats of sorption , adsorption isotherms, energetic heterogeneity profiles, diffusion coefficients,glass transition temperatures , Hildebrand and Hansen solubility parameters, and crosslink densities.

Applications

IGC experiments have applications over a wide range of industries. Both surface and bulk properties obtained from IGC can yield vital information for materials ranging from pharmaceuticals to carbon nanotubes. Although surface energy experiments are most common, there are a wide range of experimental parameters that can be controlled in IGC, thus allowing the determination of a variety of sample parameters. The below sections highlight how IGC experiments are utilized in several industries.

Polymers and Coatings

IGC has been used extensively for the characterization of polymer films, beads, and powders. For instance, IGC was used to study surface properties and interactions amongst components in paint formulations. Also, IGC has been used to investigate the degree of crosslinking for ethylene propylene rubber using the Flory–Rehner equation . Additionally, IGC is a sensitive technique for the detection and determination of first and second order phase transitions like melting and glass transition temperatures of polymers. Although other techniques like dynamic scanning calorimetry are capable of measuring these transition temperatures, IGC has the capability of glass transition temperatures as a function of relative humidity.

Pharmaceuticals

The increasing sophistication of pharmaceutical materials has necessitated the use for more sensitive, thermodynamic based techniques for materials characterization. For these reasons, IGC, has seen increased use throughout the pharmaceutical industry. Applications include polymorph characterization, effect of processing steps like milling, and drug-carrier interactions for dry powder formulations. In other studies, IGC was used to relate surface energy and acid-base values with triboelectric charging and differentiate the crystalline and amorphous phases .

Fibers

Surface energy values obtained by IGC have been used extensively on fibrous materials including textiles, natural fibers, glass fibers, and carbon fibers. Most of these and other related studies investigating the surface energy of fibers are focusing on the use of these fibers in composites. Ultimately, the changes in surface energy can be related to composite performance via the works of adhesion and cohesion discussed previously.

Nanomaterials

Similar to fibers, nanomaterials like carbon nanotubes, nanoclays, and nanosilicas are being used as composite reinforcement agents. Therefore, the surface energy and surface treatment of these materials has been actively studied by IGC. For instance, IGC has been used to study the surface activity of nanosilica, nanohematite, and nanogeoethite. Further, IGC was used to characterize the surface of as received and modified carbon nanotubes.

Metakaolins

IGC was used to characterize the adsorption surface properties of calcined kaolin (metakaolin) and the grinding effect on this material.

Other

Other applications for IGC include paper-toner adhesion, wood composites, porous materials , and food materials.

Comprehensive Two-dimensional Gas Chromatography

Comprehensive Two-dimensional gas chromatography, or GCxGC was originally described in 1991 by Professor Phillips and his student Liu. Since then the GC × GC has been extensively applied to solve complex problems of separations. Started in the Oil and Gas Industry the technology was mainly used for the complex oil samples to determine the many different types of Hydrocarbons and its isomers. Nowadays in these types of samples it has been reported that over 30000 different compounds could be identified in a crude oil with this Comprehensive Chromatography Technology (CCT). The CCT evolved from a technology only used in academic R&D laboratories, into a more robust technology used in many different industrial labs. Comprehensive Chromatography is used in forensics, food and flavor, environmental, metabolomics, biomarkers and clinical applications. And almost daily new applications are being developed in laboratories all over the world. Some of the most well-established research groups in the world that are found in Australia,Italy,Holland, Canada,United States, and Brazil use this analytical technique.

Modulation: The Process

In GC × GC two columns are connected sequentially, typically the first dimension is a conventional column and the second dimension is a short fast GC type, with a modulator positioned between them. The function of the modulator can be divided into basically three processes:

1. continuously collect small fractions of the effluent from 1D, ensuring that the separation is maintained in this dimension;

2. focus or refocus the effluent of a narrow band;

3. to quickly transfer the 2D fraction collected and focused as a narrow pulse. Taken together, these three steps is called modulation cycle, which is repeated throughout the chromatographic run.

The most frequently used type of modulation is the thermal modulation (patent holder is ZOEX Corporation), where liquid nitrogen is used to (cryogenic) trap (immobilize) all the components eluting from the first dimension. After a fixed time interval a hot stream pulse is mobilizing a part of the compounds again. This hot pulse can be considered as the injection starting point into the second dimension column. The latest version is called a loop-type Thermal modulator where the released compounds are trapped and refocused for a second time (and released again) to have perfect peak shapes and maximum resolution in the second dimension. With the thermal modulator also very volatile compounds can be modulated. The thermal modulation in practice is a liquid nitrogen cooled loop system that provides the lowest temperature for thermal modulation, and modulates the widest range (C2 to C55) of organic compounds. The temperature at the jet is -189 °C. The maximum temperature of the hot jet is 475 °C. Even methane has been modulated with liquid nitrogen cooled gas jets like those in the this type of modulator. The second important modulator type is the so-called Closed Cycle Refrigerated Loop Modulation. This loop modulation system eliminates the need for liquid nitrogen for thermal modulation. The system employs a closed cycle refrigerator/heat exchanger to produce -90 °C at the jet. The cooling is done by indirect cooling of gaseous nitrogen and therefore this type modulates volatile and semi volatile compounds over the C6+ range. Another type of modulation is flow modulation. This is a valve-based approach, where differential flows are used to 'fill' and 'flush' a sample loop. Flow modulation does not suffer from the same volatility restrictions as thermal modulation, as it does not rely on trapping analytes using a cool jet - meaning volatiles <C5 can be efficiently modulated.

The time required to complete a cycle is called the period of modulation (modulation time) and is actually the time in between two hot pulses, which typically lasts between 2 and 10 seconds is related to the time needed for the compounds to eluted in 2D.

Another key aspect of GC x GC that can be highlighted is that the result from the refocusing in the 1D, which occurs during the modulation, causes a significant increase in sensitivity. The modulation process causes the chromatographic bands in GC × GC systems are 10-50 times closer than in 1D-GC, resulting in values for much better peak widths (FWHM Full Width Half Mass) between 50 ms to 500 ms, which requires detectors with fast response and small internal volumes.

Columnset

Regarding the set of columns, can be combined in various types. The Original column sets in the beginning were mainly poly(dimethylsiloxane) in the first dimension and poly(ethyleneglycol) in the second dimension. These so-called straight phase column sets are very suitable for Hydrocarbon analysis. Therefore, these are still used most frequent in Oil and Gas industry. If the application is to determine more polar compounds in a non-polar matrix. The reverse phase column set gives more resolution for the polar compounds. The first dimension column in this situation is a polar column, followed by a mid-polar second dimension column. For each application there can be an optimal column set. In this respective you can use also chiral columns for optical isomer separation or for example PLOT columns for volatiles and gas samples. To optimize the application is a bit more complex compared to 1D separations, as there are more parameters involved. Flow and oven temperature program are still very important but also hot jet pulse duration, length of the second dimension column and modulation time have a big influence on the final results. The outcome is also different. No longer a traditional chromatogram is produced, but now a 3 dimensional plot is produced by specially designed software packages. It is a different way of evaluate data, but you get much more information. With modern software you can perform group-type separation as well as automated peak identification (with Mass Spectrometry).

Detectors

Due to the small width of the peak in the second dimension suitable detectors are needed. For example, flame ionization detector (FID), (micro) electron capture detector (µECD) and mass spectrometry analyzers such as fast time of flight (TOF). Several authors have published work using quadrupole Mass Spectrometry (qMS), though some trade-offs have to be accepted as these are much slower.

Headspace Gas Chromatography for Dissolved Gas Measurement

Headspace Gas Chromatography uses headspace gas injected directly onto a gas chromatographic column. Chemists often use the phrase "standard temperature and pressure or STP" to convey that they are working at a temperature of 25 °C and one atmosphere of pressure. There are three states of matter under these conditions: solids, liquids and gases. Although all three are distinct states both solids and gases can dissolve (or disperse) in liquids. The most commonly occurring liquid in the biosphere is water. All components of the atmosphere are capable of dissolving in water to some degree. The bulk of the stable natural components of the atmosphere are nitrogen, oxygen, carbon dioxide, gaseous water, argon and other trace gases.

Materials that exist primarily in the gas phase at STP are referred to as "volatile." Many natural and man-made (anthropogenic) materials are stable in two states at STP, earning them the title "semi-volatile." A naturally occurring volatile that is sometimes found in aqueous solution is methane, water itself is semi-volatile. Man-made or anthropogenic chemicals also occur in these classes. Examples of volatile anthropogenic chemicals include the refrigerants chlorofluorocarbons (CFCs) and hydrofluorocarbons (HCFCs). Semivolatile anthropogenics can exist as mixtures, such as petroleum distillates or as pure chemicals like trichloroethylene (TCE).

There is a need to analyze the dissolved gas content of aqueous solutions. Dissolved gases can directly interact with aquatic organisms or can volatilize from solution (the latter described by Henry's Law). These processes can result in exposure that, depending on the nature of the dissolved material, can have negative health effects. There is natural occurrence of various dissolved gases in groundwater and can be a measure of health for lakes, streams and rivers. Dissolved gases also occur as a result of human contamination from fuel and chlorinated spill sites. This method can be used to determine if there is natural biodegradation processes occurring in contaminated aquifers. For example, fuel hydrocarbons will break down into methane. Chlorinated solvents such as trichloroethylene, break down into ethene and chloride. Detecting these compounds can determine if biodegradation processes are occurring and possibly at what rate.Natural gas extracted from the earth also contains many low molecular weight hydrocarbon compounds such as methane, ethane, propane and butane. For example, methane has been found in many water wells in West Virginia.

Chromatographic techniques are often useful when mixtures of analytes are present because they are capable of measuring multiple analytes during a single application. They require isolation of the analyte from the matrix they come in (the body of the sample). One of the simpler techniques is to simply trap the analytes in a bubble of air above the sample (headspace) and inject part of that bubble directly into the instrument. This procedure is called headspace analysis. One of the most widely used methods for headspace analysis is described by the United States Environmental Protection Agency (USEPA) and is called RSKSOP-175. This method is described below.

RSKSOP-175

RSKSOP-175 is an unofficial method employed by the USEPA to detect and quantify dissolved gases in water. This method has been used to quantify dissolved hydrogen, methane, ethylene, ethane, propane, butane, acetylene, nitrogen, nitrous oxide and oxygen. The method uses headspace gas injected into a gas chromatographic column (GC) to determine the original concentration in a water sample.

A sample of water is collected in the field in a vial without headspace and capped with a Teflon septum or crimp top to minimize the escape of volatile gases. It is beneficial to store the bottles upside down to further minimize loss of analytes. Before analysis begins, the sample is brought to room temperature and temperature is recorded. In the laboratory, a headspace is created by displacing water with high purity helium. The bottle is then shaken upside down for a minimum of five minutes in order to equilibrate the dissolved gases into the headspace. It's important to note that the bottle must be kept upside down for the remainder of analysis if manually injected. A known volume of headspace gas is then injected onto a gas chromatographic column. An automated process can also be utilized. Individual components (gases) are separated and detected by either a thermal conductivity detector (TCD), a flame ionization detector (FID) or an electron capture detector (ECD). Using the known temperature of the sample, the bottle volume, the concentrations of gas in the headspace (as determined by GC), and Henry's law constant, the concentration of the original water sample is calculated.

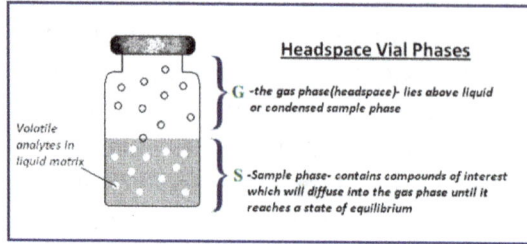

Phases of a headspace vial used in gas chromatography

Calculations

Using the known temperature of the sample, the bottle volume, the concentrations of gas in the headspace (as determined by GC), and Henry's law constant, the concentration of the original water sample is calculated. Total gas concentration (TC) in the original water sample is calculated by determining the concentration of headspace and converting this to the partial pressure and then solving for the aqueous concentration which partitioned in the gas phase (CAH) and the concentration remaining in the aqueous phase (CA). The total concentration of gas in original sample (TC) is the sum of the concentration partitioned in the gas phase (CAH) and the concentration remaining in the aqueous phase (CA).

TC = CAH + CA

Henry's law states that the mole fraction of a dissolved gas (Xg) is equal to the partial pressure of the gas (pg) at equilibrium divided by Henry's law constant (H). Gas solubility coefficients are used to calculate Henry's law constant.

Xg=Pg/H

After manipulating equations and substituting volumes of each phase, the molar concentration of water (55.5 mol/L) and the molecular weight of the gas analyte (MW), a final equation is solved for.

TC=[(55.5 mol/L)* Pg/H*(10)^3 mg/g]+ ([Vh/((Vb-Vh))]*Cg*MW(g/mol)/22.4(L/mol) *[273K/((T+273K)]* (10)^3 mg/g)

Where Vb is the bottle volume and Vh is the volume of headspace. Cg is the volumetric concentration of gas. For full calculation examples reference the RSK-175SOP.

Practical Considerations

One of the major concerns for this method is reproducibility. Due to the nature of calculations, this method is reliant on temperatures to be constant and volumes to be exact. When gases are spiked manually into the GC, the speed and technique in which an analyst does this plays a role in reproducibility. If one analyst is faster in removing gas from the vial and injecting it onto the instrument, then it is important to have the same analyst run on the calibration they prepped, otherwise error will more than likely be introduced. A headspace auto-sampler may remove some of this error, but constant heat and variable temperature on the instrument becomes an issue.

Methanizer

Methanizer is an appliance used in gas chromatography, which allows to detect very low concentrations of carbon monoxide and carbon dioxide. It consists of a flame ionization detector, preceded by a hydrogenating reactor, which converts CO_2 and CO into methane CH_4.

Chemical Reaction

On-line catalytic reduction of carbon monoxide to methane for detection by FID was described by Porter & Volman, who suggested that both carbon dioxide and carbon monoxide could also be converted to methane with the same nickelcatalyst. This was confirmed by Johns & Thompson, who determined optimum operating parameters for each of the gases.

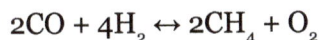

$$CO_2 + 2H_2 \leftrightarrow CH_4 + O_2$$

$$2CO + 4H_2 \leftrightarrow 2CH_4 + O_2$$

Typical Design

The catalyst usually consists of a 2% coating of Ni in the form of nickel nitrate deposited on a chromatographic packing material (e.g. Chromosorb G). A 1½" long bed is packed around the bend of an 8"×1/8" SS U-tube. The tube is clamped in a block so that the ends protrude down into the column oven for easy connection between column or TCD outlet and FID base. Heat is provided by a pair of cartridge heaters and controlled by a temperature controller.

Hydrogen for the reduction can be provided either by adding it via a tee at the inlet to the catalyst (preferred), or by using hydrogen as carrier gas.

Start-up

Since the raw catalyst is supplied in the form of nickel oxide, it is necessary to reduce it to metallic nickel before it will operate properly.

The following procedure is recommended:

- Condition all columns in the normal way. Never condition a column while connected to the catalyst.

- Connect columns to detectors(s) and catalyst as required by the application.

- Set normal carrier gas flow (25-30 mL/min for 1/8" columns). Either He or N_2 is satisfactory.

- Set H_2 flow to the catalyst at about 20 mL/min and H_2 to the FID at 10 mL/min. If H_2 is used as carrier, 20 mL/min N_2 or He make-up should be provided through the normal FID H_2 make-up line.

- When the detectors are up to operating temperature, set the column oven temperature as required, turn on the catalyst heater, and set to 400 °C.

- By the time the injector temperature reaches 400 °C, the catalyst will be reduced and ready for use.

Inject a sample containing known amounts of CH_4, CO and CO_2 to check conversion efficiency and peak shape. The retention times of these compounds should be known. If not, and light hydrocarbons are present in the sample, there might be some confusion in identification. The user should be aware, also, that the FID does respond slightly to O_2 so at high sensitivities an air peak might also be evident. As a very rough indication, 1% O_2 gives a signal similar to that of 1 ppm CO or CO_2.

If there is any doubt about retention times, the following pointers might be useful:

- On a Mol. Sieve 5A, CO retention time is about three times that of CH_4.

- On a Mol. Sieve 13X, CO has about 25% longer retention time than CH_4.

- On a porous polymers and silica gel, CO elutes with air just before CH_4, and CO_2 elutes between CH_4 and C_2H_6 except on Chromosorb 104, from which CO_2 elutes just after C_2H_6. It also elutes just after ethane from silica gel and the retention time is considerably longer than on porous polymers.

- For confirmation of CO_2, atmospheric air contains about 300 ppm and a breath sample 5-15%.

If necessary, adjust catalyst temperature to optimize conversion efficiency and peak symmetry. Also adjust the H_2 flow to optimize sensitivity. The H_2 flow through the catalyst and the ratio of H_2 to catalyst and H_2 to FID are not critical.

Operating Characteristics

Temperature

Conversion of both CO and CO_2 to CH_4 starts at a catalyst temperature below 300 °C, but the conversion is incomplete and peak tailing is evident. At around 340 °C, conversion is complete, as indicated by area measurements, but some tailing limits the peak height. At 360-380 °C, tailing is eliminated and there is little change in peak height up to 400 °C.

Although carbonization of CO has been reported at temperatures above 350°, it is rather a rare phenomenon.

Range

The conversion efficiency is essentially 100% from minimum detectable levels up to a flow of CO or CO_2 at the detector of about 5×10^{-5} g/s. These represent a detection limit of about 200 ppb and a maximum concentration of about 10% in a 0.5 mL sample. Both values are dependent upon peak width.

Catalyst Poisoning

Some elements and compounds can deactivate the catalyst:

- H_2S. Very small amounts of H_2S, SF_6, and probably any other sulfur containing gases, cause immediate and complete deactivation of the catalyst. It is not possible to regenerate a poisoned catalyst that has been deactivated by sulfur, by treating with either oxygen or hydrogen. If sulfur containing gases are present in the sample, a switching valve should be used either to bypass the catalyst, or to back-flush the column to vent after elution of CO_2.

- Air or O_2. Reports of oxygen poisoning seem to be rather rumors than real facts. Small amounts of air through a catalyst will not kill it but anything over about 5 cc/min will cause an immediate and continual degradation of the catalyst. This has been seen first hand on several systems over 30 years of personal experience with a catalytic FID designed for analysis of U.S. EPA Method 25 and 25-C samples.

- Unsaturated hydrocarbons. Samples of pure ethylene cause immediate, but partial, degradation of the catalyst, evidenced by slight tailing of CO and CO_2 peaks. The effect of 2 or 3 samples might be tolerable, but since it is cumulative, such gases should be back-flushed or by-passed. Low concentrations do not cause any degradation. Samples of pure acetylene affect the catalyst much more severely than does ethylene. Low concentrations have no effect. Probably some carbonization with high concentrations of unsaturates occurs, resulting in the deposit of soot on the catalyst surface. It is likely that aromatics would have the same effect.

- Other compounds. Water has no effect on the catalyst, as well as various Freons and NH_3. Here again, with NH_3, there is conflicting evidence from some users, who have seen a degradation after several injections, but other researchers were not able to confirm it. As with sulfur containing gases, NH_3 can be back-flushed to vent or by-passed if desired.

Troubleshooting

In general, the catalyst works perfectly unless it is degraded by sample components, possible minute amounts of sulfur gases at otherwise undetectable levels. The effect is always the same — the CO and CO_2 peaks start to tail. If only CO tails, it might well be a column effect, e.g., a Mol. Sieve 13X always causes slight tailing of CO. If the tailing is minimal, raising the catalyst temperature might provide enough improvement to permit further use.

With a newly packed catalyst, tailing usually indicates that part of the catalyst bed is not hot enough. This can happen if the bed extends too far up the arms of the U-tube. Possibly a longer bed will improve the upper conversion limit, but if this is the aim, the packing must not extend beyond the confines of the heater block.

Catalyst Preparation

Dissolve 1 g of nickel nitrate $Ni(NO_3)_2 \cdot 6H_2O$ in 4-5 mL of methanol. Add 10 g of Chromosorb G. A/W, 80-100 mesh. There should be just enough methanol to completely wet the support without excess. Mix the slurry, pour into a flat Pyrex pan and dry on a hot plate at about 80-90 °C with occasional gentle shaking or mixing. When dry, heat in air at about 400 °C to decompose the salt to NiO. Note that NO_2 is emitted during baking — provide adequate ventilation. About an hour at 400 °C, longer at lower temperatures, will be needed to complete the process. After baking, the material is dark gray, with no trace of the original green.

Pour the raw catalyst into both arms of an 8"×1/8" nickel U-tube, checking the depth in both with a wire. The final bed should extend 3/8" to 1/2" above the bottom of the U in both arms. Plug with glass wool and install in the injector block.

Disadvantages

A major limitation of methanizers is that they only enable the detection of carbon monoxide or carbon dioxide. A flame ionization detector is also insensitive to other highly oxygenated and functionalized compounds such as formic acid and formaldehyde.

Alternative Solution

In order to quantify these compounds a sequential reactor technique has been proposed to first oxidize the compounds and then reduce the resulting carbon dioxide to methane. This technique enables the accurate quantification of any number of compounds that contain carbon beyond just carbon monoxide and carbon dioxide. In addition to increasing the sensitivity of the FID to particular compounds, the response factors of all species become equivalent to that of methane, thereby eliminating the need for calibration curves and the standards they rely on. The sequential reactor is available exclusively from the Activated Research Company and is known as the Polyarc reactor.

References

- Pavia, Donald L., Gary M. Lampman, George S. Kritz, Randall G. Engel (2006). Introduction to Organic Laboratory Techniques (4th Ed.). Thomson Brooks/Cole. pp. 797–817. ISBN 978-0-495-28069-9.

- Harris, Daniel C. (1999). "24. Gas Chromatography". Quantitative chemical analysis (Chapter) (Fifth ed.). W. H. Freeman and Company. pp. 675–712. ISBN 0-7167-2881-8.

- Grob, Robert L.; Barry, Eugene F. (2004). Modern Practice of Gas Chromatography (4th Ed.). John Wiley & Sons. ISBN 0-471-22983-0.

- Grob, Konrad "Carrier Gases for GC" (1997) Restek Advantage, Restek Corporation http://www.restek.com/Technical-Resources/Technical-Library/Editorial/editorial_A017, Retrieved 09 March 2016

- JeromeJeyakumar, J.; et. al. (April 2013). "A Study of Phytochemical Constituents in Caralluma Umbellata By Gc – Ms Anaylsis [sic]" (PDF). International Journal of Pharmaceutical Science Invention: 37–41. Retrieved 23 January 2015.

- "Thermo Instrument Systems Inc. History". International Directory of Company Histories (Volume 11 ed.). St. James Press. 1995. pp. 513–514. Retrieved 23 January 2015.

6

Various Detectors in Chromatography

The device that is used in liquid and gas chromatography to detect various components is known as chromatography detector. These detectors can also be classified into destructive detectors and non-destructive detectors. Some of the destructive detectors are flame ionization detectors and mass spectrometry whereas the non-destructive detectors explained are thermal conductivity detector and electron capture detector. The following chapter helps the readers in developing an in-depth understanding on the various detectors in chromatography.

Flame Ionization Detector

A flame ionization detector (FID) is a scientific instrument that measures the concentration of organic species in a gas stream. It is frequently used as a detector in gas chromatography. Standalone FIDs can also be used in applications such as landfill gas monitoring, fugitive emissions monitoring and internal combustion engine emissions measurement in stationary or portable instruments.

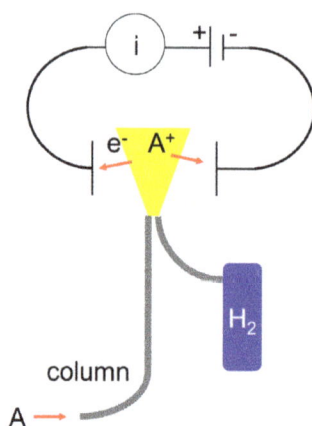

Schematic of a flame ionization detector for gas chromatography

History

The first flame ionization detectors were developed simultaneously and independently in 1957 by McWilliam and Dewar at Imperial Chemical Industries of Australia and New Zealand Central Research Laboratory, Ascot Vale, Melbourne,

Australia. and by Harley and Pretorius at the University of Pretoria in Pretoria, South Africa.

In 1959, Perkin Elmer Corp. included a flame ionization detector in its Vapor Fractometer

Operating Principle

The operation of the FID is based on the detection of ions formed during combustion of organic compounds in a hydrogenflame. The generation of these ions is proportional to the concentration of organic species in the sample gas stream. Hydrocarbons generally have molar response factors that are equal to the number of carbon atoms in their molecule, while oxygenates and other species that contain heteroatoms tend to have a lower response factor. Carbon monoxide and carbon dioxide are not detectable by FID.

Advantages and Disadvantages

Advantages

Flame ionization detectors are used very widely in gas chromatography because of a number of advantages.

- Cost: Flame ionization detectors are relatively inexpensive to acquire and operate.

- Low maintenance requirements: Apart from cleaning or replacing the FID jet, these detectors require little maintenance.

- Rugged construction: FIDs are relatively resistant to misuse.

- Linearity and detection ranges: FIDs can measure organic substance concentration at very low and very high levels, having a linear response of 10^7.

Disadvantages

Flame ionization detectors cannot detect inorganic substances and some highly oxygenated or functionalized species like infrared and laser technology can. In some systems, CO and CO_2 can be detected in the FID using a methanizer, which is a bed of Ni catalyst that reduces CO and CO_2 to methane, which can be in turn detected by the FID. The methanizer is limited by its inability to reduce compounds other than CO and CO_2 and its tendency to be poisoned by a number of chemicals commonly found in GC effluents.

Another important disadvantage is that the FID flame oxidizes all oxidizable compounds that pass through it; all hydrocarbons and oxygenates are oxidized to carbon dioxide and water and other heteroatoms are oxidized according to thermodynamics. For this reason, FIDs tend to be the last in a detector train and also cannot be used for preparatory work.

Alternative Solution

An improvement to the methanizer exists in the form of a sequential reactor that oxidizes compounds before reducing them to methane. This method can be used to improve the response of the FID and allow for the detection of many more carbon-containing compounds. The complete conversion of compounds to methane and the now equivalent response in the detector also eliminates the need for calibrations and standards because response factors are all equivalent to those of methane. This allows for the rapid analysis of complex mixtures that contain molecules where standards are not available. The sequential reactor is sold commercially as the Polyarc reactor, available online from the Activated Research Company.

Operation

To detect these ions, two electrodes are used to provide a potential difference. The positive electrode doubles as the nozzle head where the flame is produced. The other, negative electrode is positioned above the flame. When first designed, the negative electrode was either tear-drop shaped or angular piece of platinum. Today, the design has been modified into a tubular electrode, commonly referred to as a collector plate. The ions thus are attracted to the collector plate and upon hitting the plate, induce a current. This current is measured with a high-impedance picoammeter and fed into an integrator. The manner in which the final data is displayed is based on the computer and software. In general, a graph is displayed that has time on the x-axis and total ion on the y-axis.

The current measured corresponds roughly to the proportion of reduced carbon atoms in the flame. Specifically how the ions are produced is not necessarily understood, but the response of the detector is determined by the number of carbon atoms (ions) hitting the detector per unit time. This makes the detector sensitive to the mass rather than the concentration, which is useful because the response of the detector is not greatly affected by changes in the carrier gas flow rate.

Description of a Generic Detector

The design of the flame ionization detector varies from manufacturer to manufacturer, but the principles are the same. Most commonly, the FID is attached to a gas chromatography system.

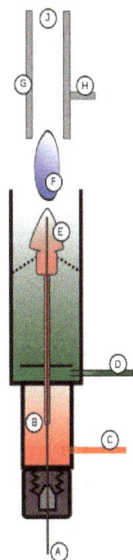

FID Schematic: A) Capillary tube; B) Platinum jet; C) Hydrogen; D) Air; E) Flame; F) Ions; G) Collector H) Gas outlet; J) Coaxial cable to Analog to Digital converter

The eluent exits the GC column (A) and enters the FID detector's oven (B). The oven is needed to make sure that as soon as the eluent exits the column, it does not come out of the gaseous phase and deposit on the interface between the column and FID. This deposition would result in loss of eluent and errors in detection. As the eluent travels up the FID, it is first mixed with the hydrogen fuel (C) and then with the oxidant (D). The eluent/fuel/oxidant mixture continues to travel up to the nozzle head where a positive bias voltage exists (E). This positive bias helps to repel the reduced carbon ions created by the flame (F) pyrolyzing the eluent. The ions are repelled up toward the collector plates (G) which are connected to a very sensitive ammeter, which detects the ions hitting the plates, then feeds that signal (H) to an amplifier, integrator, and display system. The products of the flame are finally vented out of the detector through the exhaust port (J).

Mass Spectrometry

SIMS mass spectrometer, model IMS 3f.

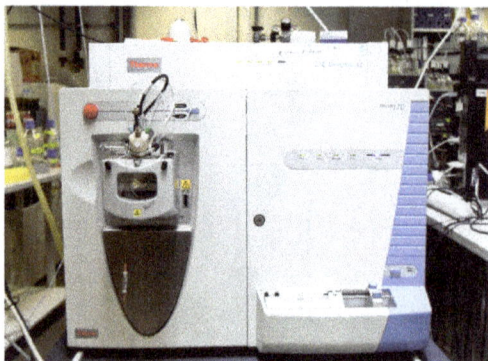

Orbitrap mass spectrometer.

Mass spectrometry (MS) is an analytical technique that ionizes chemical species and sorts the ions based on their mass to charge ratio. In simpler terms, a mass spectrum measures the masses within a sample. Mass spectrometry is used in many different fields and is applied to pure samples as well as complex mixtures.

A mass spectrum is a plot of the ion signal as a function of the mass-to-charge ratio. These spectra are used to determine the elemental or isotopic signature of a sample, the masses of particles and of molecules, and to elucidate the chemical structures of molecules, such as peptides and other chemical compounds.

In a typical MS procedure, a sample, which may be solid, liquid, or gas, is ionized, for example by bombarding it with electrons. This may cause some of the sample's molecules to break into charged fragments. These ions are then separated according to their mass-to-charge ratio, typically by accelerating them and subjecting them to an electric or magnetic field: ions of the same mass-to-charge ratio will undergo the same amount of deflection. The ions are detected by a mechanism capable of detecting charged particles, such as an electron multiplier. Results are displayed as spectra of the relative abundance of detected ions as a function of the mass-to-charge ratio. The atoms or molecules in the sample can be identified by correlating known masses to the identified masses or through a characteristic fragmentation pattern.

History

Replica of J.J. Thompson's third mass spectrometer.

In 1886, Eugen Goldstein observed rays in gas discharges under low pressure that traveled away from the anode and through channels in a perforated cathode, opposite to the direction of negatively charged cathode rays (which travel from cathode to anode). Goldstein called these positively charged anode rays "Kanalstrahlen"; the standard translation of this term into English is "canal rays". Wilhelm Wien found that strong electric or magnetic fields deflected the canal rays and, in 1899, constructed a device with parallel electric and magnetic fields that separated the positive rays according to their charge-to-mass ratio (Q/m). Wien found that the charge-to-mass ratio depended on the nature of the gas in the discharge tube. English scientist J.J. Thomson later improved on the work of Wien by reducing the pressure to create the mass spectrograph.

Calutron mass spectrometers were used in the Manhattan Project for uranium enrichment.

The word *spectrograph* had become part of the international scientific vocabulary by 1884. Early *spectrometry* devices that measured the mass-to-charge ratio of ions were called *mass spectrographs* which consisted of instruments that recorded a spectrum of mass values on a photographic plate. A *mass spectroscope* is similar to a *mass spectrograph* except that the beam of ions is directed onto a phosphor screen. A mass spectroscope configuration was used in early instruments when it was desired that the effects of adjustments be quickly observed. Once the instrument was properly adjusted, a photographic plate was inserted and exposed. The term mass spectroscope continued to be used even though the direct illumination of a phosphor screen was replaced by indirect measurements with an oscilloscope. The use of the term *mass spectroscopy* is now discouraged due to the possibility of confusion with light spectroscopy. Mass spectrometry is often abbreviated as *mass-spec* or simply as *MS*.

Modern techniques of mass spectrometry were devised by Arthur Jeffrey Dempster and F.W. Aston in 1918 and 1919 respectively.

Sector mass spectrometers known as calutrons were used for separating the isotopes of uranium developed by Ernest O. Lawrence during the Manhattan Project. Calutron mass spectrometers were used for uranium enrichment at the Oak Ridge, Tennessee Y-12 plant established during World War II.

In 1989, half of the Nobel Prize in Physics was awarded to Hans Dehmelt and Wolfgang Paul for the development of the ion trap technique in the 1950s and 1960s.

In 2002, the Nobel Prize in Chemistry was awarded to John Bennett Fenn for the development of electrospray ionization (ESI) and Koichi Tanaka for the development of soft laser desorption (SLD) and their application to the ionization of biological macromolecules, especially proteins.

Parts of a Mass Spectrometer

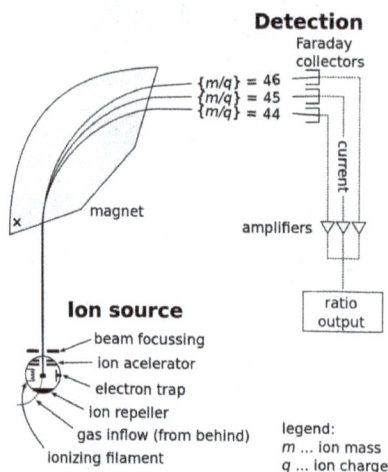

Schematics of a simple mass spectrometer with sector type mass analyzer. This one is for the measurement of carbon dioxide isotope ratios (IRMS) as in the carbon-13 urea breath test

A mass spectrometer consists of three components: an ion source, a mass analyzer, and a detector. The *ionizer* converts a portion of the sample into ions. There is a wide variety of ionization techniques, depending on the phase (solid, liquid, gas) of the sample and the efficiency of various ionization mechanisms for the unknown species. An extraction system removes ions from the sample, which are then targeted through the mass analyzer and onto the *detector*. The differences in masses of the fragments allows the mass analyzer to sort the ions by their mass-to-charge ratio. The detector measures the value of an indicator quantity and thus provides data for calculating the abundances of each ion present. Some detectors also give spatial information, e.g., a multichannel plate.

Theoretical Example

The following example describes the operation of a spectrometer mass analyzer, which is of the sector type. (Other analyzer types are treated below.) Consider a sample of sodium chloride (table salt). In the ion source, the sample is vaporized (turned into gas) and ionized (transformed into electrically charged particles) into sodium (Na^+) and chloride (Cl^-) ions. Sodium atoms and ions are monoisotopic, with a mass of about 23 u. Chloride atoms and ions come in two isotopes with masses of approximately 35 u (at a natural abundance of about 75 percent) and approximately 37 u (at a natural abundance of about 25 percent). The analyzer part of the spectrometer contains electric and magnetic fields, which exert forces on ions traveling through these fields. The speed of a charged particle may be increased or decreased while passing through the electric field, and its direction may be altered by the magnetic field. The magnitude of the deflection of the moving ion's trajectory depends on its mass-to-charge ratio. Lighter ions get deflected by the magnetic force more than heavier ions (based on Newton's second law of motion, $F = ma$). The streams of sorted ions pass from the analyzer to the detector, which records the relative abundance of each ion type. This information is used to determine the chemical element composition of the original sample (i.e. that both sodium and chlorine are present in the sample) and the isotopic composition of its constituents (the ratio of ^{35}Cl to ^{37}Cl).

Creating Ions

Surface ionization source at the Argonne National Laboratory linear accelerator

The ion source is the part of the mass spectrometer that ionizes the material under analysis (the analyte). The ions are then transported by magnetic or electric fields to the mass analyzer.

Techniques for ionization have been key to determining what types of samples can be analyzed by mass spectrometry. Electron ionization and chemical ionization are used for gases and vapors. In chemical ionization sources, the analyte is ionized by chemical ion-molecule reactions during collisions in the source. Two techniques often used with liquid and solid biological samples include electrospray ionization (invented by John Fenn) and matrix-assisted laser desorption/ionization (MALDI, initially developed as a similar technique "Soft Laser Desorption (SLD)" by K. Tanaka for which a Nobel Prize was awarded and as MALDI by M. Karas and F. Hillenkamp).

Hard Ionization and Soft Ionization

Quadrupole mass spectrometer and electrospray ion source used for Fenn's early work

In mass spectrometry, ionization refers to the production of gas phase ions suitable for resolution in the mass analyser or mass filter. Ionization occurs in the ion source. There are several ion sources available; each has advantages and disadvantages for particular applications. For example, electron ionization (EI) gives a high degree of fragmentation, yielding highly detailed mass spectra which when skilfully analysed can provide important information for structural elucidation/characterisation and facilitate identification of unknown compounds by comparison to mass spectral libraries obtained under identical operating conditions. However, EI is not suitable for coupling to HPLC, i.e. LC-MS, since at atmospheric pressure, the filaments used to generate electrons burn out rapidly. Thus EI is coupled predominantly with GC, i.e. GC-MS, where the entire system is under high vacuum.

Hard ionization techniques are processes which impart high quantities of residual energy in the subject molecule invoking large degrees of fragmentation (i.e. the systematic rupturing of bonds acts to remove the excess energy, restoring stability to the resulting ion). Resultant ions tend to have m/z lower than the molecular mass (other than in the case of proton transfer and not including isotope peaks). The most common example of hard ionization is electron ionization (EI).

Soft ionization refers to the processes which impart little residual energy onto the subject molecule and as such result in little fragmentation. Examples include fast atom bombardment (FAB), chemical ionization (CI), atmospheric-pressure chemical ionization (APCI), electrospray ionization (ESI), and matrix-assisted laser desorption/ionization (MALDI)

Inductively Coupled Plasma

Inductively coupled plasma (ICP) sources are used primarily for cation analysis of a wide array of sample types. In this source, a plasma that is electrically neutral overall, but that has had a substantial fraction of its atoms ionized by high temperature, is used to atomize introduced sample molecules and to further strip the outer electrons from those atoms. The plasma is usually generated from argon gas, since the first ionization energy of argon atoms is higher than the first of any other elements except He, O, F and Ne, but lower than the second ionization energy of all except the most electropositive metals. The heating is achieved by a radio-frequency current passed through a coil surrounding the plasma.

Inductively coupled plasma ion source

Other Ionization Techniques

Others include photoionization, glow discharge, field desorption (FD), fast atom bombardment (FAB), thermospray, desorption/ionization on silicon (DIOS), Direct Analysis in Real Time (DART), atmospheric pressure chemical ionization (APCI), secondary ion mass spectrometry (SIMS), spark ionization and thermal ionization (TIMS).

Mass Selection

Mass analyzers separate the ions according to their mass-to-charge ratio. The following two laws govern the dynamics of charged particles in electric and magnetic fields in vacuum:

$\mathbf{F} = Q(\mathbf{E} + \mathbf{v} \times \mathbf{B})$ (Lorentz force law);

$\mathbf{F} = m\mathbf{a}$ (Newton's second law of motion in non-relativistic case, i.e. valid only at ion velocity much lower than the speed of light).

Here F is the force applied to the ion, m is the mass of the ion, a is the acceleration, Q is the ion charge, E is the electric field, and v × B is the vector cross product of the ion velocity and the magnetic field

Equating the above expressions for the force applied to the ion yields:

$(m / Q)\mathbf{a} = \mathbf{E} + \mathbf{v} \times \mathbf{B}.$

This differential equation is the classic equation of motion for charged particles. Together with the particle's initial conditions, it completely determines the particle's motion in space and time in terms of m/Q. Thus mass spectrometers could be thought of as "mass-to-charge spectrometers". When presenting data, it is common to use the (officially) dimensionless m/z, where z is the number of elementary charges (e) on the ion (z=Q/e). This quantity, although it is informally called the mass-to-charge ratio, more accurately speaking represents the ratio of the mass number and the charge number, z.

There are many types of mass analyzers, using either static or dynamic fields, and magnetic or electric fields, but all operate according to the above differential equation. Each analyzer type has its strengths and weaknesses. Many mass spectrometers use two or more mass analyzers for tandem mass spectrometry (MS/MS). In addition to the more common mass analyzers listed below, there are others designed for special situations.

There are several important analyser characteristics. The mass resolving power is the measure of the ability to distinguish two peaks of slightly different m/z. The mass accuracy is the ratio of the m/z measurement error to the true m/z. Mass accuracy is usually measured in ppm or milli mass units. The mass range is the range of m/z amenable to analysis by a given analyzer. The linear dynamic range is the range over which ion signal is linear with analyte concentration. Speed refers to the time frame of the experiment and ultimately is used to determine the number of spectra per unit time that can be generated.

Sector Instruments

ThermoQuest AvantGarde sector mass spectrometer

A sector field mass analyzer uses an electric and/or magnetic field to affect the path and/or velocity of the charged particles in some way. As shown above, sector instruments bend the trajectories of the ions as they pass through the mass analyzer, according to their mass-to-charge ratios, deflecting the more charged and faster-moving, lighter ions more. The analyzer can be used to select a narrow range of m/z or to scan through a range of m/z to catalog the ions present.

Time-of-flight

The time-of-flight (TOF) analyzer uses an electric field to accelerate the ions through the same potential, and then measures the time they take to reach the detector. If the particles all have the same charge, the kinetic energies will be identical, and their velocities will depend only on their masses. Ions with a lower mass will reach the detector first.

Quadrupole Mass Filter

Quadrupole mass analyzers use oscillating electrical fields to selectively stabilize or destabilize the paths of ions passing through a radio frequency (RF) quadrupole field created between 4 parallel rods. Only the ions in a certain range of mass/charge ratio are passed through the system at any time, but changes to the potentials on the rods allow a wide range of m/z values to be swept rapidly, either continuously or in a succession of discrete hops. A quadrupole mass analyzer acts as a mass-selective filter and is closely related to the quadrupole ion trap, particularly the linear quadrupole ion trap except that it is designed to pass the untrapped ions rather than collect the trapped ones, and is for that reason referred to as a transmission quadrupole. A magnetically enhanced quadrupole mass analyzer includes the addition of a magnetic field, either applied axially or transversely. This novel type of instrument leads to an additional performance enhancement in terms of resolution and/or sensitivity depending upon the magnitude and orientation of the applied magnetic field. A common variation of the transmission quadrupole is the triple quadrupole mass spectrometer. The "triple quad" has three consecutive quadrupole stages, the first acting as a mass filter to transmit a particular incoming ion to the second quadrupole, a collision chamber, wherein that ion can be broken into fragments. The third quadrupole also acts as a mass filter, to transmit a particular fragment ion to the detector. If a quadrupole is made to rapidly and repetitively cycle through a range of mass filter settings, full spectra can be reported. Likewise, a triple quad can be made to perform various scan types characteristic of tandem mass spectrometry.

Ion Traps

Three-dimensional Quadrupole Ion Trap

The quadrupole ion trap works on the same physical principles as the quadrupole mass analyzer, but the ions are trapped and sequentially ejected. Ions are trapped in a mainly quadrupole RF field, in a space defined by a ring electrode (usually connected to the main RF potential) between two endcap electrodes (typically connected to DC or auxiliary AC potentials). The sample is ionized either internally (e.g. with an electron or laser beam), or externally, in which case the ions are often introduced through an aperture in an endcap electrode.

There are many mass/charge separation and isolation methods but the most commonly used is the mass instability mode in which the RF potential is ramped so that the orbit of ions with a mass $a > b$ are stable while ions with mass b become unstable and are ejected on the z-axis onto a detector. There are also non-destructive analysis methods.

Ions may also be ejected by the resonance excitation method, whereby a supplemental oscillatory excitation voltage is applied to the endcap electrodes, and the trapping voltage amplitude and/or excitation voltage frequency is varied to bring ions into a resonance condition in order of their mass/charge ratio.

Cylindrical Ion Trap

The cylindrical ion trap mass spectrometer (CIT) is a derivative of the quadrupole ion trap where the electrodes are formed from flat rings rather than hyperbolic shaped electrodes. The architecture lends itself well to miniaturization because as the size of a trap is reduced, the shape of the electric field near the center of the trap, the region where the ions are trapped, forms a shape similar to that of a hyperbolic trap.

Linear Quadrupole Ion Trap

A linear quadrupole ion trap is similar to a quadrupole ion trap, but it traps ions in a two dimensional quadrupole field, instead of a three-dimensional quadrupole field as in a 3D quadrupole ion trap. Thermo Fisher's LTQ ("linear trap quadrupole") is an example of the linear ion trap.

A toroidal ion trap can be visualized as a linear quadrupole curved around and connected at the ends or as a cross section of a 3D ion trap rotated on edge to form the toroid, donut shaped trap. The trap can store large volumes of ions by distributing them throughout the ring-like trap structure. This toroidal shaped trap is a configuration that allows the increased miniaturization of an ion trap mass analyzer. Additionally all ions are stored in the same trapping field and ejected together simplifying detection that can be complicated with array configurations due to variations in detector alignment and machining of the arrays.

As with the toroidal trap, linear traps and 3D quadrupole ion traps are the most commonly miniaturized mass analyzers due to their high sensitivity, tolerance for mTorr pressure, and capabilities for single analyzer tandem mass spectrometry (e.g. product ion scans).

Orbitrap

Orbitrap instruments are similar to Fourier transform ion cyclotron resonance mass spectrometers. Ions are electrostatically trapped in an orbit around a central, spindle shaped electrode. The electrode confines the ions so that they both orbit around the central electrode and oscillate back and forth along the central electrode's long axis. This oscillation generates an image current in the detector plates which is recorded by the instrument. The frequencies of these image currents depend on the mass to charge ratios of the ions. Mass spectra are obtained by Fourier transformation of the recorded image currents.

Orbitraps have a high mass accuracy, high sensitivity and a good dynamic range.

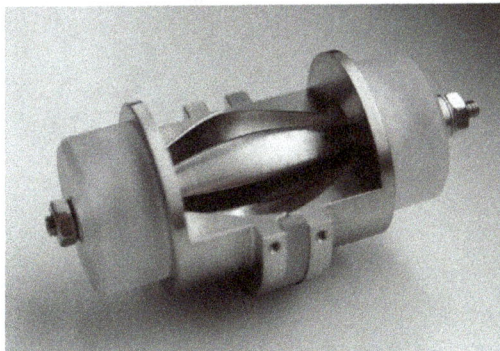

Orbitrap mass analyzer

Fourier Transform Ion Cyclotron Resonance

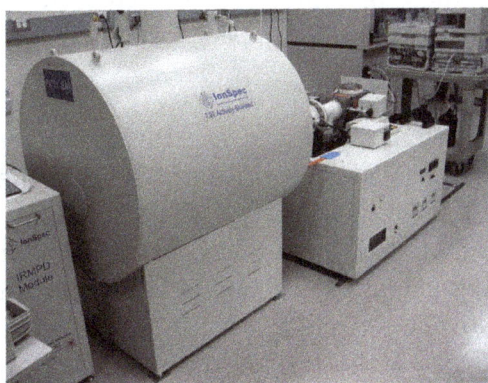

Fourier transform ion cyclotron resonance mass spectrometer

Fourier transform mass spectrometry (FTMS), or more precisely Fourier transform ion cyclotron resonance MS, measures mass by detecting the image current produced by ions cyclotroning in the presence of a magnetic field. Instead of measuring the deflection of ions with a detector such as an electron multiplier, the ions are injected into a Penning trap (a static electric/magnetic ion trap) where they effectively form part of a circuit. Detectors at fixed positions in space measure the electrical signal of ions which pass near them over time, producing a periodic signal. Since the frequency of an ion's cycling is determined by its mass to charge ratio, this can be deconvoluted by performing a Fourier transform on the signal. FTMS has the advantage of high sensitivity (since each ion is "counted" more than once) and much higher resolution and thus precision.

Ion cyclotron resonance (ICR) is an older mass analysis technique similar to FTMS except that ions are detected with a traditional detector. Ions trapped in a Penning trap are excited by an RF electric field until they impact the wall of the trap, where the detector is located. Ions of different mass are resolved according to impact time.

Detectors

A continuous dynode particle multiplier detector.

The final element of the mass spectrometer is the detector. The detector records either the charge induced or the current produced when an ion passes by or hits a surface. In a scanning instrument, the signal produced in the detector during the course of the scan versus where the instrument is in the scan (at what m/Q) will produce a mass spectrum, a record of ions as a function of m/Q.

Typically, some type of electron multiplier is used, though other detectors including Faraday cups and ion-to-photon detectors are also used. Because the number of ions leaving the mass analyzer at a particular instant is typically quite small, considerable amplification is often necessary to get a signal. Microchannel plate detectors are commonly used in modern commercial instruments. In FTMS and Orbitraps, the detector consists of a pair of metal surfaces within the mass analyzer/ion trap region which the ions only pass near as they oscillate. No direct current is produced, only a weak AC image current is produced in a circuit between the electrodes. Other inductive detectors have also been used.

Tandem Mass Spectrometry

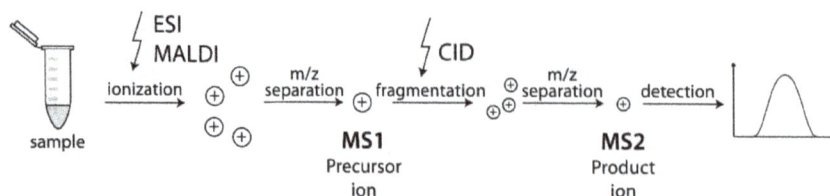

Tandem mass spectrometry for biological molecules using ESI or MALDI

A tandem mass spectrometer is one capable of multiple rounds of mass spectrometry, usually separated by some form of molecule fragmentation. For example, one mass analyzer can isolate one peptide from many entering a mass spectrometer. A second mass analyzer then stabilizes the peptide ions while they collide with a gas, causing them to fragment by collision-induced dissociation (CID). A third mass analyzer then sorts the fragments produced from the peptides. Tandem MS can also be done in a single mass analyzer over time, as in a quadrupole ion trap. There are various methods for fragmenting molecules for tandem MS, including collision-induced dissociation (CID), electron capture dissociation (ECD), electron transfer dissociation (ETD), infrared multiphoton dissociation (IRMPD), blackbody infrared radiative dissociation (BIRD), electron-detachment dissociation (EDD) and surface-induced dissociation (SID). An important application using tandem mass spectrometry is in protein identification.

Tandem mass spectrometry enables a variety of experimental sequences. Many commercial mass spectrometers are designed to expedite the execution of such routine sequences as selected reaction monitoring (SRM) and precursor ion scanning. In SRM, the first analyzer allows only a single mass through and the second analyzer monitors for multiple user-defined fragment ions. SRM is most often used with scanning instruments where the second mass analysis event is duty cycle limited. These experiments are used to increase specificity of detection of known molecules, notably in pharmacokinetic studies. Precursor ion scanning refers to monitoring for a specific loss from the precursor ion. The first and second mass analyzers scan across the spectrum as partitioned by a user-defined m/z value. This experiment is used to detect specific motifs within unknown molecules.

Another type of tandem mass spectrometry used for radiocarbon dating is accelerator mass spectrometry (AMS), which uses very high voltages, usually in the mega-volt range, to accelerate negative ions into a type of tandem mass spectrometer.

Common Mass Spectrometer Configurations and Techniques

When a specific configuration of source, analyzer, and detector becomes conventional in practice, often a compound acronym arises to designate it, and the compound acronym may be better known among nonspectrometrists than the component acronyms. The epitome of this is MALDI-TOF, which simply refers to combining a matrix-assisted laser desorption/ionization source with a time-of-flight mass analyzer. The MALDI-TOF moniker is more widely recognized by the non-mass spectrometrists than MALDI or TOF individually. Other examples include inductively coupled plasma-mass spectrometry (ICP-MS), accelerator mass spectrometry (AMS), thermal ionization-mass spectrometry (TIMS) and spark source mass spectrometry (SSMS). Sometimes the use of the generic "MS" actually connotes a very specific mass analyzer and detection system, as is the case with AMS, which is always sector based.

Certain applications of mass spectrometry have developed monikers that although strictly speaking would seem to refer to a broad application, in practice have come instead to connote a specific or a limited number of instrument configurations. An example of this is isotope ratio mass spectrometry (IRMS), which refers in practice to the use of a limited number of sector based mass analyzers; this name is used to refer to both the application and the instrument used for the application.

Separation Techniques Combined with Mass Spectrometry

An important enhancement to the mass resolving and mass determining capabilities of mass spectrometry is using it in tandem with chromatographic and other separation techniques.

Gas Chromatography

A gas chromatograph (right) directly coupled to a mass spectrometer (left)

A common combination is gas chromatography-mass spectrometry (GC/MS or GC-MS). In this technique, a gas chromatograph is used to separate different compounds. This stream of separated compounds is fed online into the ion source, a metallicfilament to which voltage is applied. This filament emits electrons which ionize the compounds. The ions can then further fragment, yielding predictable patterns. Intact ions and fragments pass into the mass spectrometer's analyzer and are eventually detected.

Liquid Chromatography

Indianapolis Museum of Art conservation scientist performing liquid chromatography–mass spectrometry.

Similar to gas chromatography MS (GC/MS), liquid chromatography-mass spectrometry (LC/MS or LC-MS) separates compounds chromatographically before they are introduced to the ion source and mass spectrometer. It differs from GC/MS in that the mobile phase is liquid, usually a mixture of water and organic solvents, instead of gas. Most commonly, an electrospray ionization source is used in LC/MS. Other popular and commercially available LC/MS ion sources are atmospheric pressure chemical ionization and atmospheric pressure photoionization. There are also some newly developed ionization techniques like laser spray.

Capillary Electrophoresis–mass Spectrometry

Capillary electrophoresis–mass spectrometry (CE-MS) is a technique that combines the liquid separation process of capillary electrophoresis with mass spectrometry. CE-MS is typically coupled to electrospray ionization.

Ion Mobility

Ion mobility spectrometry-mass spectrometry (IMS/MS or IMMS) is a technique where ions are first separated by drift time through some neutral gas under an applied electrical potential gradient before being introduced into a mass spectrometer. Drift time is a measure of the radius relative to the charge of the ion. The duty cycle of IMS (the time over which the experiment takes place) is longer than most mass spectrometric techniques, such that the mass spectrometer can sample along the course of the IMS separation. This produces data about the IMS separation and the mass-to-charge ratio of the ions in a manner similar to LC/MS.

The duty cycle of IMS is short relative to liquid chromatography or gas chromatography separations and can thus be coupled to such techniques, producing triple modalities such as LC/IMS/MS.

Data and Analysis

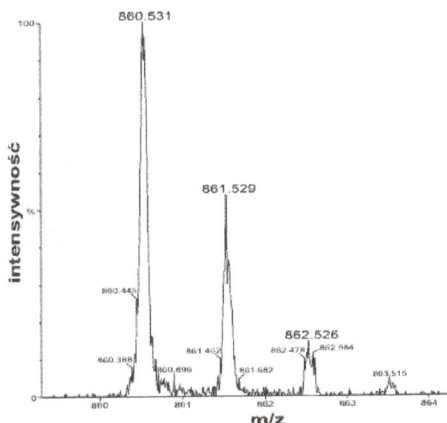

Mass spectrum of a peptide showing the isotopic distribution

Data Representations

Mass spectrometry produces various types of data. The most common data representation is the mass spectrum.

Certain types of mass spectrometry data are best represented as a mass chromatogram. Types of chromatograms include selected ion monitoring (SIM), total ion current (TIC), and selected reaction monitoring (SRM), among many others.

Other types of mass spectrometry data are well represented as a three-dimensional contour map. In this form, the mass-to-charge, m/z is on the x-axis, intensity the y-axis, and an additional experimental parameter, such as time, is recorded on the z-axis.

Data Analysis

Mass spectrometry data analysis is specific to the type of experiment producing the data. General subdivisions of data are fundamental to understanding any data.

Many mass spectrometers work in either *negative ion mode* or *positive ion mode*. It is very important to know whether the observed ions are negatively or positively charged. This is often important in determining the neutral mass but it also indicates something about the nature of the molecules.

Different types of ion source result in different arrays of fragments produced from the original molecules. An electron ionization source produces many fragments and mostly single-charged (1-) radicals (odd number of electrons), whereas an electrospray source usually produces non-radical quasimolecular ions that are frequently multiply charged. Tandem mass spectrometry purposely produces fragment ions post-source and can drastically change the sort of data achieved by an experiment.

Knowledge of the origin of a sample can provide insight into the component molecules of the sample and their fragmentations. A sample from a synthesis/manufacturing process will probably contain impurities chemically related to the target component. A crudely prepared biological sample will probably contain a certain amount of salt, which may form adducts with the analyte molecules in certain analyses.

Results can also depend heavily on sample preparation and how it was run/introduced. An important example is the issue of which matrix is used for MALDI spotting, since much of the energetics of the desorption/ionization event is controlled by the matrix rather than the laser power. Sometimes samples are spiked with sodium or another ion-carrying species to produce adducts rather than a protonated species.

Mass spectrometry can measure molar mass, molecular structure, and sample purity. Each of these questions requires a different experimental procedure; therefore, adequate definition of the experimental goal is a prerequisite for collecting the proper data and successfully interpreting it.

Interpretation of Mass Spectra

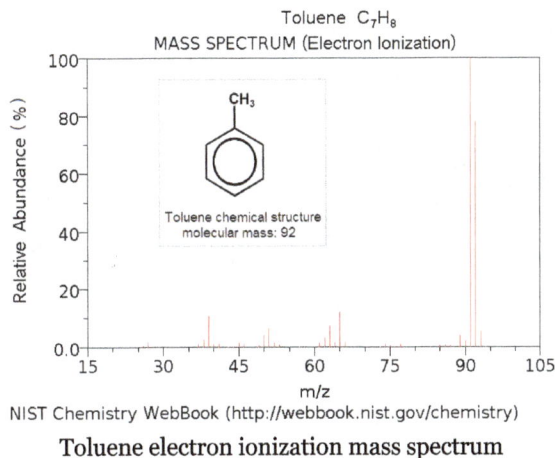

Toluene electron ionization mass spectrum

Since the precise structure or peptide sequence of a molecule is deciphered through the set of fragment masses, the interpretation of mass spectra requires combined use of various techniques. Usually the first strategy for identifying an unknown compound is to compare its experimental mass spectrum against a library of mass spectra. If no matches result from the search, then manual interpretation or software assisted interpretation of mass spectra must be performed. Computer simulation of ionization and fragmentation processes occurring in mass spectrometer is the primary tool for assigning structure or peptide sequence to a molecule. An *a priori* structural information is fragmented *in silico* and the resulting pattern is compared with observed spectrum. Such simulation is often supported by a fragmentation library that contains published patterns of known decomposition reactions. Software taking advantage of this idea has been developed for both small molecules and proteins.

Analysis of mass spectra can also be spectra with accurate mass. A mass-to-charge ratio value (m/z) with only integer precision can represent an immense number of theoretically possible ion structures; however, more precise mass figures significantly reduce the number of candidate molecular formulas. A computer algorithm called formula generator calculates all molecular formulas that theoretically fit a given mass with specified tolerance.

A recent technique for structure elucidation in mass spectrometry, called precursor ion fingerprinting, identifies individual pieces of structural information by conducting a search of the tandem spectra of the molecule under investigation against a library of the product-ion spectra of structurally characterized precursor ions.

Applications

Mass spectrometry has both qualitative and quantitative uses. These include identifying unknown compounds, determining the isotopic composition of elements in a molecule, and determining the structure of a compound by observing its fragmentation. Other uses include quantifying the amount of a compound in a sample or studying the fundamentals of gas phase ion chemistry (the chemistry of ions and neutrals in a vacuum). MS is now in very common use in analytical laboratories that study physical, chemical, or biological properties of a great variety of compounds.

NOAA Particle Analysis by Laser Mass Spectrometry aerosol mass spectrometer aboard a NASAWB-57 high-altitude research aircraft

As an analytical technique it possesses distinct advantages such as: Increased sensitivity over most other analytical techniques because the analyzer, as a mass-charge filter, reduces background interference, Excellent specificity from characteristic fragmentation patterns to identify unknowns or confirm the presence of suspected compounds, Information about molecular weight, Information about the isotopic abundance of elements, Temporally resolved chemical data.

A few of the disadvantages of the method is that often fails to distinguish between optical and geometrical isomers and the positions of substituent in o-, m- and p- positions in an aromatic ring. Also, its scope is limited in identifying hydrocarbons that produce similar fragmented ions.

Isotope Ratio MS: Isotope dating and tracing

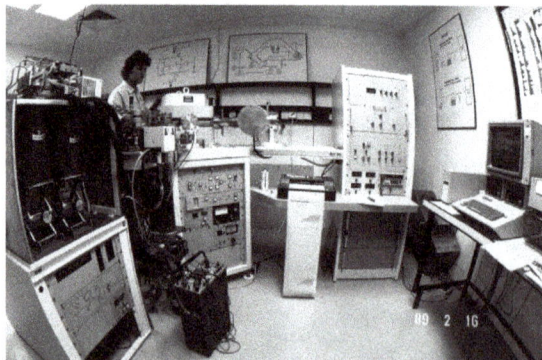

Mass spectrometer to determine the $^{16}O/^{18}O$ and $^{12}C/^{13}C$ isotope ratio on biogenous carbonate

Mass spectrometry is also used to determine the isotopic composition of elements within a sample. Differences in mass among isotopes of an element are very small, and the less abundant isotopes of an element are typically very rare, so a very sensitive instrument is required. These instruments, sometimes referred to as isotope ratio mass spectrometers (IR-MS), usually use a single magnet to bend a beam of ionized particles towards a series of Faraday cups which convert particle impacts to electric current. A fast on-line analysis of deuterium content of water can be done using flowing afterglow mass spectrometry, FA-MS. Probably the most sensitive and accurate mass spectrometer for this purpose is the accelerator mass spectrometer (AMS). This is because it provides ultimate sensitivity, capable of measuring individual atoms and measuring nuclides with a dynamic range of $\sim 10^{15}$ relative to the major stable isotope. Isotope ratios are important markers of a variety of processes. Some isotope ratios are used to determine the age of materials for example as in carbon dating. Labeling with stable isotopes is also used for protein quantification.

Trace Gas Analysis

Several techniques use ions created in a dedicated ion source injected into a flow tube or a drift tube: selected ion flow tube (SIFT-MS), and proton transfer reaction (PTR-MS), are variants of chemical ionization dedicated for trace gas analysis of air, breath or liquid headspace using well defined reaction time allowing calculations of analyte concentrations from the known reaction kinetics without the need for internal standard or calibration.

Atom Probe

An atom probe is an instrument that combines time-of-flight mass spectrometry and field-evaporation microscopy to map the location of individual atoms.

Pharmacokinetics

Pharmacokinetics is often studied using mass spectrometry because of the complex nature of the matrix (often blood or urine) and the need for high sensitivity to observe low dose and long time point data. The most common instrumentation used in this application is LC-MS with a triple quadrupole mass spectrometer. Tandem mass spectrometry is usually employed for added specificity. Standard curves and internal standards are used for quantitation of usually a single pharmaceutical in the samples. The samples represent different time points as a pharmaceutical is administered and then metabolized or cleared from the body. Blank or t=0 samples taken before administration are important in determining background and ensuring data integrity with such complex sample matrices. Much attention is paid to the linearity of the standard curve; however it is not uncommon to use curve fitting with more complex functions such as quadratics since the response of most mass spectrometers is less than linear across large concentration ranges.

There is currently considerable interest in the use of very high sensitivity mass spectrometry for microdosing studies, which are seen as a promising alternative to animal experimentation.

Protein Characterization

Mass spectrometry is an important method for the characterization and sequencing of proteins. The two primary methods for ionization of whole proteins are electrospray ionization (ESI) and matrix-assisted laser desorption/ionization (MALDI). In keeping with the performance and mass range of available mass spectrometers, two approaches are used for characterizing proteins. In the first, intact proteins are ionized by either of the two techniques described above, and then introduced to a mass analyzer. This approach is referred to as "top-down" strategy of protein analysis. In the second, proteins are enzymatically digested into smaller peptides using proteases such as trypsin or pepsin, either in solution or in gel after electrophoretic separation. Other proteolytic agents are also used. The collection of peptide products are then introduced to the mass analyzer. When the characteristic pattern of peptides is used for the identification of the protein the method is called peptide mass fingerprinting (PMF), if the identification is performed using the sequence data determined in tandem MS analysis it is called de novo peptide sequencing. These procedures of protein analysis are also referred to as the "bottom-up" approach.

Glycan Analysis

Mass spectrometry (MS), with its low sample requirement and high sensitivity, has been predominantly used in glycobiology for characterization and elucidation of glycan structures. Mass spectrometry provides a complementary method to HPLC for the analysis of glycans. Intact glycans may be detected directly as singly charged ions by matrix-assisted laser desorption/ionization mass spectrometry (MALDI-MS) or, following permethylation or peracetylation, by fast atom bombardment mass spectrometry (FAB-MS).Electrospray ionization mass spectrometry (ESI-MS) also gives good signals for the smaller glycans. Various free and commercial software are now available which interpret MS data and aid in Glycan structure characterization.

Space Exploration

As a standard method for analysis, mass spectrometers have reached other planets and moons. Two were taken to Mars by the Viking program. In early 2005 the Cassini–Huygens mission delivered a specialized GC-MS instrument aboard the Huygens probe through the atmosphere of Titan, the largest moon of the planet Saturn. This instrument analyzed atmospheric samples along its descent trajectory and was able to vaporize and analyze samples of Titan's frozen, hydrocarbon covered surface once the probe had landed. These measurements compare the abundance of isotope(s) of each particle comparatively to earth's natural abundance. Also on board the Cassini–Huygens spacecraft is an ion and neutral mass spectrometer which has been taking measurements of Titan's atmospheric composition as well as the composition of Enceladus' plumes. A Thermal and Evolved Gas Analyzer mass spectrometer was carried by the Mars Phoenix Lander launched in 2007.

NASA's Phoenix Mars Lander analyzing a soil sample from the "Rosy Red" trench with the TEGA mass spectrometer

Mass spectrometers are also widely used in space missions to measure the composition of plasmas. For example, the Cassini spacecraft carries the Cassini Plasma Spectrometer (CAPS), which measures the mass of ions in Saturn's magnetosphere.

Respired Gas Monitor

Mass spectrometers were used in hospitals for respiratory gas analysis beginning around 1975 through the end of the century. Some are probably still in use but none are currently being manufactured.

Found mostly in the operating room, they were a part of a complex system, in which respired gas samples from patients undergoing anesthesia were drawn into the instrument through a valve mechanism designed to sequentially connect up to 32 rooms to the mass spectrometer. A computer directed all operations of the system. The data collected from the mass spectrometer was delivered to the individual rooms for the anesthesiologist to use.

The uniqueness of this magnetic sector mass spectrometer may have been the fact that a plane of detectors, each purposely positioned to collect all of the ion species expected to be in the samples, allowed the instrument to simultaneously report all of the gases respired by the patient. Although the mass range was limited to slightly over 120 u, fragmentation of some of the heavier molecules negated the need for a higher detection limit.

Preparative Mass Spectrometry

The primary function of mass spectrometry is as a tool for chemical analyses based on detection and quantification of ions according to their mass-to-charge ratio. However, mass spectrometry also shows promise for material synthesis. Ion soft landing is characterized by deposition of intact species on surfaces at low kinetic energies which precludes the fragmentation of the incident species. The soft landing technique was first reported in 1977 for the reaction of low energy sulfur containing ions on a lead surface.

Thermal Conductivity Detector

The thermal conductivity detector (TCD), also known as a Katharometer, is a bulk property detector and a chemical specific detector commonly used in gas chromatography. This detector senses changes in the thermal conductivity of the column effluent and compares it to a reference flow of carrier gas. Since most compounds have a thermal conductivity much less than that of the common carrier gases of helium or hydrogen, when an analyte elutes from the column the effluent thermal conductivity is reduced, and a detectable signal is produced.

Operation

The TCD consists of an electrically heated filament in a temperature-controlled cell. Under normal conditions there is a stable heat flow from the filament to the detector body. When an analyte elutes and the thermal conductivity of the column effluent is reduced, the filament heats up and changes resistance. This resistance change is often sensed by a Wheatstone bridge circuit which produces a measurable voltage change. The column effluent flows over one of the resistors while the reference flow is over a second resistor in the four-resistor circuit.

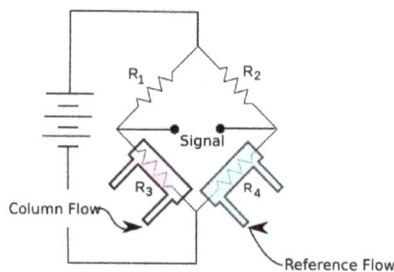

TCD Schematic

A schematic of a classic thermal conductivity detector design utilizing a wheatstone bridge circuit is shown. The reference flow across resistor 4 of the circuit compensates for drift due to flow or temperature fluctuations. Changes in the thermal conductivity of the column effluent flow across resistor 3 will result in a temperature change of the resistor and therefore a resistance change which can be measured as a signal.

Since all compounds, organic and inorganic, have a thermal conductivity different from helium, all compounds can be detected by this detector. The TCD is often called a universal detector because it responds to all compounds. Also, since the thermal conductivity of organic compounds are similar and very different from helium, a TCD will respond similarly to similar concentrations of analyte. Therefore the TCD can be used without calibration and the concentration of a sample component can be estimated by the ratio of the analyte peak area to all components (peaks) in the sample.

The TCD is a good general purpose detector for initial investigations with an unknown sample. Since the TCD is less sensitive than the flame ionization detector and has a larger dead volume it will not provide as good resolution as the FID. However, in combination with thick film columns and correspondingly larger sample volumes, the overall detection limit can be similar to that of an FID. The TCD is not as sensitive as other detectors but it is non-specific and non-destructive.

The TCD is also used in the analysis of permanent gases (argon, oxygen, nitrogen, carbon dioxide) because it responds to all these pure substances unlike the FID which cannot detect compounds which do not contain carbon-hydrogen bonds.

Process Description

It functions by having two parallel tubes both containing gas and heating coils. The gases are examined by comparing the rate of loss of heat from the heating coils into the gas. The coils are arranged in a bridge circuit so that resistance changes due to unequal cooling can be measured. One channel normally holds a reference gas and the mixture to be tested is passed through the other channel.

Applications

In the oil industry katharometers have been used for a long time for hydrocarbon detection but have a history of unstable calibrations in non-stationary oil-related applications. In normal drilling practice, 5 hydrocarbon gases, plus a couple of non-hydrocarbon gases, are expected in normal samples resulting in cross-talk between the methane absorption line and the ethane. Hence the current use of flame ionization detectors.

Katharometers are used medically in lung function testing equipment and in gas chromatography. The results are slower to obtain compared to a mass spectrometer, but the device is inexpensive, and has good accuracy when the gases in question are known, and it is only the proportion that must be determined.

Monitoring of hydrogen purity in hydrogen-cooled turbogenerators.

Detection of helium loss from the helium vessel of an MRI superconducting magnet.

Electron Capture Detector

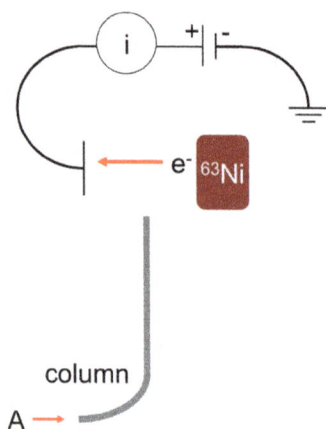

Schematic of an electron capture detector for a gas chromatograph with a ^{63}Ni source.

An electron capture detector (ECD) is a device for detecting atoms and molecules in a gas through the attachment of electrons via electron capture ionization. The device was invented in 1957 by James Lovelock and is used in gas chromatography to detect trace amounts of chemical compounds in a sample.

Gas Chromatograph Detector

Electron capture detector developed by James Lovelock in the Science Museum, London

Electron capture detector, Chemical Heritage Foundation

The electron capture detector is used for detecting electron-absorbing components (high electronegativity) such as halogenated compounds in the output stream of a gas chromatograph. The ECD uses a radioactive beta particle (electron) emitter in conjunction with a so-called makeup gas flowing through the detector chamber. The electron emitter typically consists of a metal foil holding 10 millicuries (370 MBq) of the radionuclide[63]Ni. Usually, nitrogen is used as makeup gas, because it exhibits a low excitation energy, so it is easy to remove an electron from a nitrogen molecule. The electrons emitted from the electron emitter collide with the molecules of the makeup gas, resulting in many more free electrons. The electrons are accelerated towards a positively charged anode, generating a current. There is therefore always a background signal present in the chromatogram. As the sample is carried into the detector by the carrier gas, electron-ab-sorbing analyte molecules capture electrons and thereby reduce the current between the collector anode and a cathode. The analyte concentration is thus proportional to the degree of electron capture. ECD detectors are particularly sensitive to halogens, organometallic compounds, nitriles, or nitro compounds.

Sensitivity

Depending on the analyte, an ECD can be 10-1000 times more sensitive than a flame ionization detector (FID), and one million times more sensitive than a thermal conductivity detector (TCD). An ECD has a limited dynamic range and finds its greatest application in analysis of halogenated compounds. The detection limit for electron capture detectors is 5 femtograms per second (fg/s), and the detector commonly exhibits a 10,000-fold linear range. This made it possible to detect halogenated compounds such as pesticides and CFCs, even at levels of only one part per trillion (ppt), thus revolutionizing our understanding of the atmosphere and pollutants.

Photoionization Detector

A photoionization detector or PID is a type of gas detector.

Typical photoionization detectors measure volatile organic compounds and other gases in concentrations from sub parts per billion to 10 000 parts per million (ppm). The photoionization detector is an efficient and inexpensive detector for many gas and vapor analytes. PIDs produce instantaneous readings, operate continuously, and are commonly used as detectors for gas chromatography or as hand-held portable instruments. Hand-held, battery-operated versions are widely used in military, industrial, and confined working facilities for health and safety. Their primary use is for monitoring possible worker exposure to volatile organic compounds (VOCs) such as solvents, fuels, degreasers, plastics & their precursors, heat transfer fluids, lubricants, etc. during manufacturing processes and waste handling.

Portable PIDs are used as monitoring solutions for:

- Industrial hygiene and safety

- Environmental contamination and remediation

- Hazardous materials handling

- Ammonia detection

- Lower explosive limit measurements

- Arson investigation

- Indoor air quality

- Cleanroom facility maintenance

Principle

In a photoionization detector high-energy photons, typically in the vacuum ultraviolet (VUV) range, break molecules into positively chargedions. As compounds enter the detector they are bombarded by high-energy UV photons and are ionized when they absorb the UV light, resulting in ejection of electrons and the formation of positively charged ions. The ions produce an electric current, which is the signal output of the detector. The greater the concentration of the component, the more ions are produced, and the greater the current. The current is amplified and displayed on an ammeter or digital concentration display. The ions can undergo numerous reactions including reaction with oxygen or water vapor, rearrangement, and fragmentation. A few of them may recapture an electron within the detector to reform their original molecules; however only a small portion of the airborne analytes are ionized to begin with so the practical impact of this (if it occurs) is usually negligible. Thus, PIDs are non-destructive and can be used before other sensors in multiple-detector configurations.

The PID will only respond to components that have ionization energies similar to or lower than the energy of the photons produced by the PID lamp. As stand alone detectors, PIDs are broad band and not selective, as these may ionize everything with an ionization energy less than or equal to the lamp photon energy. The more common commercial lamps have photons energy upper limits of approximately 8.4 eV, 10.0 eV, 10.6 eV, and 11.7 eV. The major and minor components of clean air all have ionization energies above 12.0 eV and thus do not interfere significantly in the measurement of VOCs, which typically have ionization energies below 12.0 eV.

Lamp Types and Detectable Compounds

PID lamp photon emissions depend on the type of fill gas (which defines the light energy produced) and the lamp window, which affects the energy of photons that can exit the lamp:

Main Photon Energy Fill Gas Window Material

- 11.7 eV Ar LiF Short-lived
- 10.6 eV Kr MgF_2 Most robust
- 10.2 eV H2 MgF_2
- 10.0 eV Kr CaF_2
- 9.6 eV Xe BaF_2
- 8.4 eV Xe Al_2O_3

The 10.6 eV lamp is the most common because it has strong output, has the longest life and responds to many compounds. In approximate order from most sensitive to least sensitive, these compounds include:

- Aromatics
- Olefins
- Bromides & Iodides
- Sulfides & Mercaptans
- Organic Amines
- Ketones
- Ethers
- Esters & Acrylates
- Aldehydes
- Alcohols
- Alkanes
- Some Inorganics, including NH3, H2S, and PH3

Applications

The first commercial application of photoionization detection was in 1973 as a hand-held instrument for the purpose of detecting leaks of VOCs, specifically vinyl chloride monomer (VCM), at a chemical manufacturing facility. The photoionization detector was applied to gas chromatography (GC) three years later, in 1976. A PID is highly selective when coupled with a chromatographic technique or a pre-treatment tube such as a Benzene specific tube. Broader cuts of selectivity for easily ionized compounds can be obtained by using a lower energy UV lamp. This selectivity can be useful when analyzing mixtures in which only some of the components are of interest.

The PID is usually calibrated using isobutylene, and other analytes may produce a relatively greater or lesser response on a concentration basis. Although many PID manufacturers provide the ability to program an instrument with a correction factor for quantitative detection of a specific chemical, the broad selectivity of the PID means that the user must know the identity of the gas or vapor species to be measured with high certainty. If a correction factor for benzene is entered into the instrument, but hexane vapor is measured instead, the lower relative detector response (higher correction factor) for hexane would lead to underestimation of the actual airborne concentration of hexane.

Matrix Gas Effects

With a gas chromatograph, filter tube, or other separation technique upstream of the PID, matrix effects are generally avoided because the analyte enters the detector isolated from interfering compounds.

Response to stand-alone PIDs is generally linear from the ppb range up to at least a few thousand ppm. In this range, response to mixtures of components is also linearly additive. At the higher concentrations, response gradually deviates from linearity because of recombination of oppositely charged ions formed in close proximity and/ or 2) absorption of UV light without ionization. The signal produced by a PID may be quenched when measuring in high humidity environments, or when a compound such as methane is present in high concentrations of $\geq 1\%$ by volume This attenuation is due to the ability of water, methane, and other compounds with high ionization energies to absorb the photons emitted by the UV lamp without leading to the production of an ion current. This reduces the number of energetic photons available to ionize target analytes.

Helium Ionization Detector

A helium ionization detector (HID) is a type of detector used in gas chromatography.

Principle

HID connected to a GC has the great advantage to use helium as both the carrier gas and the ionization gas. HID is a ion detector which uses a radioactive source, typically β-emitters, to create metastable helium species.The radioactive source ionizes helium atoms by bombarding them with emissions. The metastable helium species have an energy of up to 19.8 eV. These metastable helium species can ionize all compounds with the exception of neon which has a bigger ionization potential of 21.56 eV. As components elute from the GC's column they collide with the metastable helium ions, which then ionize the components. The ions produce an electric current, which is the signal output of the detector. The greater the concentration of the component, the more ions are produced, and the greater the current.

Application

HIDs are sensitive to a broad range of components. They must use helium as a carrier gas.

HID is classified as a mass sensitive detector, which means that the analyte is destroyed during reaction. Therefore, it is considered a destructive detector.

The drawback to HIDs are that they contain a radioactive source. In the United States, this means they fall under a number of federal regulations concerning their use in the workplace, shipping, disposal, etc. Discharge ionization detectors have generally supplanted them.

Pulsed Discharge Ionization Detector

A pulsed discharge ionization detector (pulsed discharge detector) is a detector for gas chromatography that utilizes a stable, low powered, pulsed DC discharge in helium as an ionization source.

Eluants from the GC column, flowing counter to the flow of helium from the discharge zone, are ionized by photons from the helium discharge. Bias electrode(s) focus the resulting electrons toward the collector electrode, where they cause changes in the standing current which are quantified as the detector output.

Electron Capture Mode

In the electron capture mode, the PDD is a selective detector for monitoring high electron affinity compounds such as freons, chlorinated pesticides, and other halogen compounds. For this type of compound, the minimum detectable quantity (MDQ) is at the femtogram ($10-15$) or picogram ($10-12$) level. The PDD is similar in sensitivity and response characteristics to a conventional radioactive ECD, and can be operated at temperatures up to 400°C. For operation in this mode, He and $CH4$ are introduced just upstream from the column exit.

Helium Photoionization Mode

In the helium photoionization mode, the PDD is a universal, non-destructive, high sensitivity detector. The response to both inorganic and organic compounds is linear over a wide range. Response to fixed gases is positive (increase in standing current), with an MDQ in the low ppb range. The PDD in helium photoionization mode is an excellent replacement for flame ionization detectors in petrochemical or refinery environments, where the flame and use of hydrogen can be problematic. In addition, when the helium discharge gas is doped with a suitable noble gas, such as argon, krypton, or xenon (depending on the desired cutoff point), the PDD can function as a specific photoionization detector for selective determination of aliphatics, aromatics, amines, as well as other species.

Gas Chromatography Ion Detector

Many gas chromatograph detectors are ion detectors with varying methods of ionizing the components eluting from the gas chromatograph's column.

An ion detector is analogous to a capacitor or vacuum tube. It can be envisioned as two metal grids separated by air with inverse charges placed on them. An electric potential difference (voltage) exist between the two grids. After components are ionized in the detector, they enter the region between the two grids, causing current to pass from one to the other. This current is amplified and is the signal generated by the detector. The higher the concentration of the component, the more ions are generated, and the greater the current.

Some early FIDs actually used two metal grids as their ion detectors. However, more efficient designs have been developed, so few current ion-type detectors use two metal grids. But the principle is the same, and it can be easiest to think of the detector in this manner.

Types of Ion Detectors

- Flame ionization detector (FID) -- uses a flame to produce ions
- Electron capture detector (ECD) -- uses bta radiation
- Photo-ionization detector (PID) -- uses UV light to produce ions
- Helium ionization detector (HID) -- uses a radioactive source to produce helium ions, which in turn ionize the components
- Discharge ionization detector (DID) -- uses an electric spark source to produce helium ions, which in turn ionize the components
- Pulsed discharge ionization detector (PDD) -- similar to a Discharge ionization detector (DID), but uses a different sort of spark

References

- Downard, Kevin (2004). "Mass Spectrometry - A Foundation Course". Royal Society of Chemistry. doi:10.1039/9781847551306. ISBN 978-0-85404-609-6.

- Eiceman, G.A. (2000). Gas Chromatography. In R.A. Meyers (Ed.), Encyclopedia of Analytical Chemistry: Applications, Theory, and Instrumentation, pp. 10627. Chichester: Wiley. ISBN 0-471-97670-9

- Tureček, František; McLafferty, Fred W. (1993). Interpretation of mass spectra. Sausalito: University Science Books. ISBN 0-935702-25-3.

- Various. "Advisory Opinion". Innovative Technology: Gas Chromatography Field Analysis. NEW-MOA Technology Review Committee. Retrieved 2011-04-21.

Chromatography Software: An Integrated Study

Chromatography software is a software that is used in collecting and analyzing chromatographic results. They are usually provided by manufacturers and help in analyzing data. Agilent ChemStation, empower and OpenChrom are the topics that have been explained in the section. The following chapter helps the readers in developing an in-depth understanding of chromatography software.

Chromatography Software

A chromatography software, also known as a Chromatography data system (CDS), collects and analyzes chromatographic results delivered by chromatography detectors.

Many chromatography software packages are provided by manufacturers, and many of them only provide a simple interface to acquire data. They also provide different tools to analyze this data.

The most common tool is "integration": It computes simple areas to delimit peaks by adding valleys automatically or not. Areas are computed between two valleys, a signal, and a baseline.

The following is a list of software and the (unexplained) tools that each provides:

Chromatography software					
Name	**Publisher**	**Control**	**Analysis**	**Type**	**Comments**
Malcom	Schlumberger	No	Yes	GC/GC-MS	
ChemStation	Agilent	Yes	Yes	LC/GC/LC-MS/GC-MS	XML, Macros
EZChrom Elite	Agilent	Yes	Yes	LC/GC	
OpenLAB CDS	Agilent	Yes	Yes	LC/GC/LC-MS/GC-MS	
Chromeleon	Thermo Scientific	Yes	Yes	GC/LC/IEX/MS	
Empower	Waters	Yes	Yes	GC/LC	

OpenChrom	Philip Wenig		Yes	GC/LC	open-source EPL
Atlas CDS	Thermo Scientific	Yes	Yes	GC/LC	
Clarity	DataApex	Yes	Yes	GC/LC/MS	
Mass Frontier	HighChem		Yes	GC/LC/MS	

Agilent ChemStation

Agilent ChemStation is a software package to control Agilentliquid chromatography and gas chromatography systems such as the 1050, 1100 and 1200 Series HPLC system. It is an evolution of the Hewlett-Packard ChemStation System.

Two versions are available: one ("online") in connection with the modules of the HPLC chain is designed to control instruments and run experiments, and the other ("offline"), without a connection with the HPLC chain, is designed to analyze data.

ChemStation is structured around a number of registers. Two of the more important registers are CHROMREG and CHROMRES, the chromatographic data registers.

ChemStation has a command line interpreter and can run macros. Those macros are files grouping a set of commands. These files possess a .mac extension.

ChemStation can import analysis lists and export result files in XML by adding new lines to the ChemStation.ini configuration file. This is a feature to implement the connection with a Laboratory information management system (LIMS

Empower (Software)

Empower is a software produced by Waters Corporation allowing the user to control chromatography instruments and to process data (for use with HPLC, UPLC, and Gas Chromatography systems)

History

The impetuses for the development of the chromatographysoftware that would ultimately become Empower were both internally and externally driven. Regulatory discussions led by the U.S. Food and Drug Administration (FDA) within the 1988 time frame centered on the need for an audit trail to track process related procedures in chromatography data systems. The FDA ultimately did not create a regulation addressing this particular issue, but many of these ideas were later solidified in the 21 CFR Part

11 regulations. Concurrent to theses regulatory focused discussions, Waters sought to simplify its chromatography software development efforts. At the time, Waters was developing and supporting three software packages: Expert, Maxima, and ExpertEase. Based on discussions with customers for better management of chromatography data and suggestions by the FDA, Waters undertook a software development project to combine its chromatography data systems into one solution.

A key component of the new CDS platform was an integrated relational database that improved overall data management as well as data traceability. The data traceability functionality would later be instrumental in providing 21 CFR Part 11 safe guards such as audit trails. Empower was the first commercially available CDS to integrate a relational database. Inclusion of a database enabled Millennium integrated reporting and later provided a more robust interface for integrating with other laboratory software packages such as LIMS.

Versions

Expert – 1984

Maxima – 1986 (workstation only)

Expert Ease -1986-87 (Network only)

Millennium V1 1992- Workstation only First commercially available CDS to include integrated reporting. Prior to Millennium, chromatography results were exported to a third party tool for report generation. Functionality included in version 1: custom calculations, reporting, photo diode array acquisition and processing, and system suitability determination. The integrated relational database was a key technology enabler of the custom calculations and integration report generation. This was also the first CDS to operate in a Microsoft Windows environment.

Millennium V2 1994 – The second version of Millennium. Offered a client/server network configuration in addition to the workstation only offering. Version 2 was the first version of Millennium offered as a client/server configuration. Whether deployed as a workstation or client/server network, the installed Millennium shared the same code base, dramatically reducing the software training effort as customer migrated from workstation to client/server network .

Millennium32 V3.0 – 1998-Millennium migrated from 16-bit to 32-bit to provide increase memory access.

Millennium32 V3.2 -1999- Millennium V3.2 offered a complete set of 21 CFR Part 11 Compliant-ready safeguards for the first time.

Millennium32 V4.0 - Released in September 2001—First version of Millennium to support other vendor hardware (Agilent). With the release of version 4, Millennium adopted the Open Instrument Portal (OIP) model for supporting both 3rd party hardware and Waters hardware. The OIP module acted similar to a printer driver and allowed addition of new instruments without the need for rewriting and recompiling the entire code base. All that was required was an updated OIP module.

Empower (Build 1154) - Released in May 2002 – Millennium received a major update and facelift. Empower provided multiple user interfaces, e.g., expert user interface, open access interface, etc. It also added supported for Microsoft Windows 2000. The peak detection algorithms were updated by the inclusion of ApexTrack Integration. ApexTrack integration allowed more automated peak detection and was an important requirement for process chromatography adoption.

Empower 2 (Build 2154) - Released in May 2005 – Empower 2 provided support for Microsoft Windows XP. First CDS network to support mass spectrometry quadrupole detection. Instrument control through the OIP module was dramatically expanded to include xxx instruments. An optional module, called Method Validation Manager, for validating chromatography methods according to regulatory guidelines was added.

Empower 3 (Build 3471) - Released in November 2010 – Latest version of Empower. Designed for the Enterprise and offers certification for virtualization by VMWare and remote desktop access through Citrix. A second aspect of Designed for the Enterprise includes more wide area network support, greater redundancy and uptime capabilities. – Empower 3 provided support for Microsoft Windows 7.

OpenChrom

OpenChrom is an open source software for the analysis and visualization of mass spectrometric and chromatographic data. Its focus is to handle native data files from several mass spectrometry systems (e.g. GC/MS, LC/MS, Py-GC/MS, HPLC-MS), vendors like Agilent Technologies, Varian, Shimadzu, Thermo Fisher, PerkinElmer and others. But also data formats from other detector types are supported recently.

OpenChrom supports only the analysis and representation of chromatographic and mass spectrometric data. It has no capabilities for data acquisition or control of vendor hardware. OpenChrom is built on the Eclipse Rich Client Platform (RCP), hence it is available for various operating systems, e.g. Microsoft Windows, Mac OS X and Linux. It is distributed under the Eclipse Public License 1.0 (EPL). Third-party libraries are separated into single bundles and are released under various OSI compatible licenses.

History

OpenChrom was developed by Philip Wenig (SCJP, LPIC-1) as part of his PhD thesis at the University of Hamburg, Germany. The focus of the thesis was to apply pattern recognition techniques on datasets recorded by analytical pyrolysis coupled with chromatography and mass spectrometry (Py-GC/MS).

OpenChrom won the Thomas Krenn Open Source Award 2010 as well as the Eclipse Community Award 2011.

Supported Data Formats

Each system vendor stores the recorded analysis data in its own data format. These mass spectrometry data formats are most commonly proprietary. That makes it difficult to compare data sets from different systems and vendors. Furthermore, it's a big drawback for interlaboratory tests. The aim of OpenChrom is to support a wide range of different data formats natively. OpenChrom takes care that the raw data files can't be modified. Hence, the raw data converters are not published under an open source license. The appropriate vendor converters can be installed from the OpenChrom marketplace. Their usage is free of charge. Moreover, OpenChrom supports several open formats to save the analysis results. In addition, OpenChrom offers its own open source format (*.ocb) that makes it possible to save the edited chromatogram as well as the peaks and identification results.

Mass Selective Detector

- Agilent *.D (DATA.MS and MSD1.MS)

- AMDIS Library (*.msl)

- Bruker Flex MALDI-MS (*.fid)

- Chromtech (*.dat)

- CSV (*.csv)

- Finnigan (*.RAW)

- Finnigan MAT95 (*.dat)

- Finnigan ITDS (*.DAT)

- Finnigan ITS40 (*.MS)

- Finnigan Element II (*.dat)

- JCAMP-DX (*.JDX)

- Microsoft Excel (*.xlsx)

- mzXML (*.mzXML)

- NetCDF (*.CDF)

- NIST Text (*.msp)

- OpenChrom XML (*.chrom)

- OpenChrom (*.ocb)

- Peak Loadings (*.mpl)

- PerkinElmer (*.raw)

- Varian SMS (*.SMS)

- Varian XMS (*.XMS)

- VG MassLab (*.DAT_001;1)

- Shimadzu (*.qgd)

- Shimadzu (*.spc)

- Waters (*.RAW)

- ZIP (*.zip)

- Agilent ICP-MS (*.icp)

- Finnigan ICIS (*.dat)

- mzML (*.mzML)

- SVG (*.svg)

- MassHunter (*.D)

- Finnigan ICIS (*.dat)

- MassLynx (*.RAW)

- Galactic Grams (*.cgm)

Flame Ionization Detector

- Agilent FID (*.D/*.ch)

- FID Text (*.xy)

- NetCDF (*.cdf)

- PerkinElmer (*.raw)

- Varian (*.run)

- Finnigan FID (*.dat)

- Finnigan FID (*.raw)

- Shimadzu (*.gcd)

- Arw (*.arw)

Other Formats

- Peak Loadings (*.mpl)

- NIST-DB (*.msp)

- AMDIS (*.msl)

- AMDIS (*.cal)

- AMDIS (*.ELU)

- MassBank (*.txt)

Major Features

OpenChrom offers a variety of features to analyze chromatographic data:

- Native handling of chromatographic data (MSD and FID)

- Batch process support

- Baseline detector support

- Peak detector, integrator support

- Peaks and mass spectrum identifier support

- Quantitation support

- Filter support (e.g. Mass Fragment and Scan Removal, noise reduction, Savitz-ky-Golay smoothing, CODA, backfolding)

- Retention time shift support

- Retention index support

- Chromatogram overlay mode

- Support for principal component analysis (PCA)

- Do/undo/redo support

- Integration of OpenOffice/LibreOffice and Microsoft Office

- Extensible by plug-ins

- Chromatogram and peak database support

- Update support

- Subtract mass spectra support

Releases

The software was first released in 2010. Each release is named after a famous scientist.

Codename	Eponym	Date	Version
Thomson	Joseph John Thomson	April 2010	0.1.0
Goldstein	Eugen Goldstein	May 2010	0.2.0
Wien	Wilhelm Wien	October 2010	0.3.0
Tswett	Mikhail Semyonovich Tswett	April 2011	0.4.0
Martin	Archer John Porter Martin	October 2011	0.5.0
Synge	Richard L. M. Synge	April 2012	0.6.0
Nernst	Walther Nernst	October 2012	0.7.0
Dempster	Arthur Jeffrey Dempster	July 2013	0.8.0
Mattauch	Josef Mattauch	July 2014	0.9.0
Aston	Francis William Aston	July 2015	1.0.0
Diels	Otto Diels	July 2016	**1.1.0**
Alder	Kurt Alder	July 2017	1.2.0

Old version

Older version, still supported

Latest version

Future release

Allied Fields of Chromatography

Chromatography is an interdisciplinary subject. The allied fields associated with chromatography are analytical chemistry, biochemistry and chemical engineering. This chapter will provide a glimpse of the allied fields of chromatography.

Analytical Chemistry

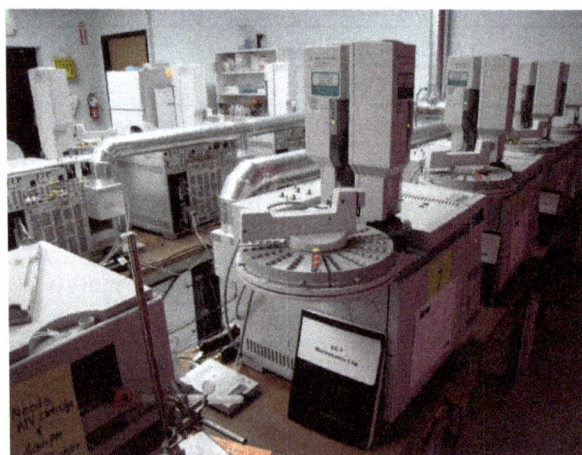

Gas chromatography laboratory

Analytical chemistry studies and uses instruments and methods used to separate, identify, and quantify matter. In practice separation, identification or quantification may constitute the entire analysis or be combined with another method. Separation isolates analytes. Qualitative analysis identifies analytes, while quantitative analysis determines the numerical amount or concentration.

Analytical chemistry consists of classical, wet chemical methods and modern, instrumental methods. Classical qualitative methods use separations such as precipitation, extraction, and distillation. Identification may be based on differences in color, odor, melting point, boiling point, radioactivity or reactivity. Classical quantitative analysis uses mass or volume changes to quantify amount. Instrumental methods may be used to separate samples using chromatography, electrophoresis or field flow fractionation. Then qualitative and quantitative analysis can be performed, often with the same instrument and may use light interaction, heat interaction, electric fields or magnetic fields . Often the same instrument can separate, identify and quantify an analyte.

Analytical chemistry is also focused on improvements in experimental design, chemometrics, and the creation of new measurement tools. Analytical chemistry has broad applications to forensics, medicine, science and engineering.

History

Gustav Kirchhoff (left) and Robert Bunsen (right)

Analytical chemistry has been important since the early days of chemistry, providing methods for determining which elements and chemicals are present in the object in question. During this period significant contributions to analytical chemistry include the development of systematic elemental analysis by Justus von Liebig and systematized organic analysis based on the specific reactions of functional groups.

The first instrumental analysis was flame emissive spectrometry developed by Robert Bunsen and Gustav Kirchhoff who discovered rubidium (Rb) and caesium (Cs) in 1860.

Most of the major developments in analytical chemistry take place after 1900. During this period instrumental analysis becomes progressively dominant in the field. In particular many of the basic spectroscopic and spectrometric techniques were discovered in the early 20th century and refined in the late 20th century.

The separation sciences follow a similar time line of development and also become increasingly transformed into high performance instruments. In the 1970s many of these techniques began to be used together as hybrid techniques to achieve a complete characterization of samples.

Starting in approximately the 1970s into the present day analytical chemistry has progressively become more inclusive of biological questions (bioanalytical chemistry), whereas it had previously been largely focused on inorganic or small organic molecules. Lasers have been increasingly used in chemistry as probes and even to initiate and influence a wide variety of reactions. The late 20th century also saw an expansion of the application of analytical chemistry from somewhat academic chemical questions to forensic, environmental, industrial and medical questions, such as in histology.

Modern analytical chemistry is dominated by instrumental analysis. Many analytical chemists focus on a single type of instrument. Academics tend to either focus on new applications and discoveries or on new methods of analysis. The discovery of a chemical present in blood that increases the risk of cancer would be a discovery that an analytical chemist might be involved in. An effort to develop a new method might involve the use of a tunable laser to increase the specificity and sensitivity of a spectrometric method. Many methods, once developed, are kept purposely static so that data can be compared over long periods of time. This is particularly true in industrial quality assurance (QA), forensic and environmental applications. Analytical chemistry plays an increasingly important role in the pharmaceutical industry where, aside from QA, it is used in discovery of new drug candidates and in clinical applications where understanding the interactions between the drug and the patient are critical.

Classical Methods

The presence of copper in this qualitative analysis is indicated by the bluish-green color of the flame

Although modern analytical chemistry is dominated by sophisticated instrumentation, the roots of analytical chemistry and some of the principles used in modern instruments are from traditional techniques many of which are still used today. These techniques also tend to form the backbone of most undergraduate analytical chemistry educational labs.

Qualitative Analysis

A qualitative analysis determines the presence or absence of a particular compound, but not the mass or concentration. By definition, qualitative analyses do not measure quantity.

Chemical Tests

There are numerous qualitative chemical tests, for example, the acid test for gold and the Kastle-Meyer test for the presence of blood.

Flame Test

Inorganic qualitative analysis generally refers to a systematic scheme to confirm the presence of certain, usually aqueous, ions or elements by performing a series of reactions that eliminate ranges of possibilities and then confirms suspected ions with a confirming test. Sometimes small carbon containing ions are included in such schemes. With modern instrumentation these tests are rarely used but can be useful for educational purposes and in field work or other situations where access to state-of-the-art instruments are not available or expedient.

Quantitative Analysis

Gravimetric Analysis

Gravimetric analysis involves determining the amount of material present by weighing the sample before and/or after some transformation. A common example used in undergraduate education is the determination of the amount of water in a hydrate by heating the sample to remove the water such that the difference in weight is due to the loss of water.

Volumetric Analysis

Titration involves the addition of a reactant to a solution being analyzed until some equivalence point is reached. Often the amount of material in the solution being analyzed may be determined. Most familiar to those who have taken chemistry during secondary education is the acid-base titration involving a color changing indicator. There are many other types of titrations, for example potentiometric titrations. These titrations may use different types of indicators to reach some equivalence point.

Instrumental Methods

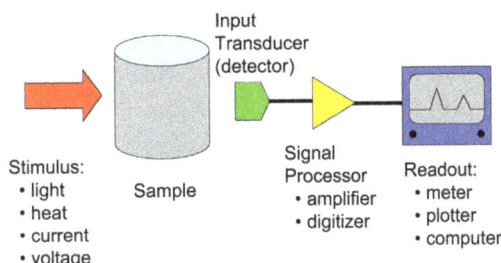

Block diagram of an analytical instrument showing the stimulus and measurement of response

Spectroscopy

Spectroscopy measures the interaction of the molecules with electromagnetic radiation. Spectroscopy consists of many different applications such as atomic absorption spectroscopy, atomic emission spectroscopy, ultraviolet-visible spectroscopy, x-ray fluorescence spectroscopy, infrared spectroscopy, Raman spectroscopy, dual polarization interferometry, nuclear magnetic resonance spectroscopy, photoemission spectroscopy, Mössbauer spectroscopy and so on.

Mass Spectrometry

An accelerator mass spectrometer used for radiocarbon dating and other analysis

Mass spectrometry measures mass-to-charge ratio of molecules using electric and magnetic fields. There are several ionization methods: electron impact, chemical ionization, electrospray, fast atom bombardment, matrix assisted laser desorption ionization, and others. Also, mass spectrometry is categorized by approaches of mass analyzers: magnetic-sector, quadrupole mass analyzer, quadrupole ion trap, time-of-flight, Fourier transform ion cyclotron resonance, and so on.

Electrochemical Analysis

Electroanalytical methods measure the potential (volts) and/or current (amps) in an electrochemical cell containing the analyte. These methods can be categorized according to which aspects of the cell are controlled and which are measured. The three main categories are potentiometry (the difference in electrode potentials is measured), coulometry (the transferred charge is measured over time), amperometry (the cell's current is measured over time), and voltammetry (the cell's current is measured while actively altering the cell's potential).

Thermal Analysis

Calorimetry and thermogravimetric analysis measure the interaction of a material and heat.

Separation

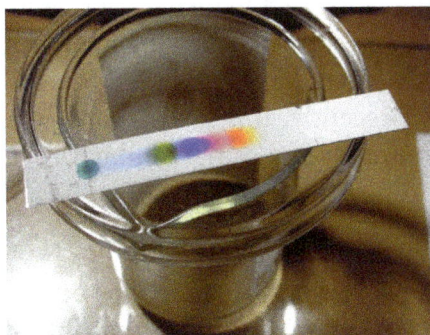

Separation of black ink on a thin layer chromatography plate

Separation processes are used to decrease the complexity of material mixtures. Chromatography, electrophoresis and Field Flow Fractionation are representative of this field.

Hybrid Techniques

Combinations of the above techniques produce a "hybrid" or "hyphenated" technique. Several examples are in popular use today and new hybrid techniques are under development. For example, gas chromatography-mass spectrometry, gas chromatography-infrared spectroscopy, liquid chromatography-mass spectrometry, liquid chromatography-NMR spectroscopy. liquid chromagraphy-infrared spectroscopy and capillary electrophoresis-mass spectrometry.

Hyphenated separation techniques refers to a combination of two (or more) techniques to detect and separate chemicals from solutions. Most often the other technique is some form of chromatography. Hyphenated techniques are widely used in chemistry and biochemistry. A slash is sometimes used instead of hyphen, especially if the name of one of the methods contains a hyphen itself.

Microscopy

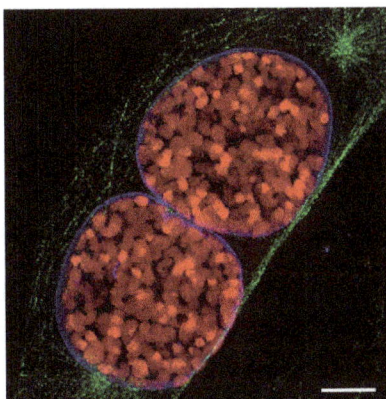

Fluorescence microscope image of two mouse cell nuclei in prophase (scale bar is 5 μm)

The visualization of single molecules, single cells, biological tissues and nanomaterials is an important and attractive approach in analytical science. Also, hybridization with other traditional analytical tools is revolutionizing analytical science. Microscopy can be categorized into three different fields: optical microscopy, electron microscopy, and scanning probe microscopy. Recently, this field is rapidly progressing because of the rapid development of the computer and camera industries.

Lab-on-a-chip

Devices that integrate (multiple) laboratory functions on a single chip of only millimeters to a few square centimeters in size and that are capable of handling extremely small fluid volumes down to less than picoliters.

Errors

Error can be defined as numerical difference between observed value and true value.

In error the true value and observed value in chemical analysis can be related with each other by the equation

$$E = O - T$$

where

E = absolute error,

O = observed value,

T = true value.

Error of a measurement is an inverse measure of accurate measurement i.e. smaller the error greater the accuracy of the measurement. Errors are expressed relatively as:

$$\frac{E}{T} \times 100 = \% \text{ error,}$$

$$\frac{E}{T} \times 1000 = \text{per thousand error}$$

Standards

Standard Curve

A general method for analysis of concentration involves the creation of a calibration curve. This allows for determination of the amount of a chemical in a material by comparing the results of unknown sample to those of a series of known standards. If the concentration of element or compound in a sample is too high for the detection range of the technique, it can simply be diluted in a pure solvent. If the amount in the sample is below an instrument's range of measurement, the method of addition can be used. In this method a known quantity of the element or compound under study is added, and the difference between the concentration added, and the concentration observed is the amount actually in the sample.

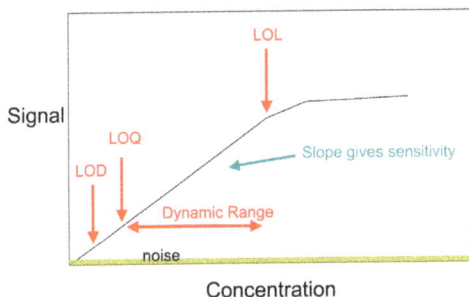

A calibration curve plot showing limit of detection (LOD), limit of quantification (LOQ), dynamic range, and limit of linearity (LOL)

Internal Standards

Sometimes an internal standard is added at a known concentration directly to an analytical sample to aid in quantitation. The amount of analyte present is then determined relative to the internal standard as a calibrant. An ideal internal standard is isotopically-enriched analyte which gives rise to the method of isotope dilution.

Standard Addition

The method of standard addition is used in instrumental analysis to determine concentration of a substance (analyte) in an unknown sample by comparison to a set of samples of known concentration, similar to using a calibration curve. Standard addition can be applied to most analytical techniques and is used instead of a calibration curve to solve the matrix effect problem.

Signals and Noise

One of the most important components of analytical chemistry is maximizing the desired signal while minimizing the associated noise. The analytical figure of merit is known as the signal-to-noise ratio (S/N or SNR).

Noise can arise from environmental factors as well as from fundamental physical processes.

Thermal Noise

Thermal noise results from the motion of charge carriers (usually electrons) in an electrical circuit generated by their thermal motion. Thermal noise is white noise meaning that the power spectral density is constant throughout the frequency spectrum.

The root mean square value of the thermal noise in a resistor is given by

$$v_{RMS} = \sqrt{4k_B TR\Delta f},$$

where k_B is Boltzmann's constant, T is the temperature, R is the resistance, and is the bandwidth of the frequency .

Shot Noise

Shot noise is a type of electronic noise that occurs when the finite number of particles (such as electrons in an electronic circuit or photons in an optical device) is small enough to give rise to statistical fluctuations in a signal.

Shot noise is a Poisson process and the charge carriers that make up the current follow a Poisson distribution. The root mean square current fluctuation is given by

$$i_{RMS} = \sqrt{2eI\Delta f}$$

where e is the elementary charge and I is the average current. Shot noise is white noise.

Flicker Noise

Flicker noise is electronic noise with a $1/f$ frequency spectrum; as f increases, the noise decreases. Flicker noise arises from a variety of sources, such as impurities in a conductive channel, generation and recombination noise in a transistor due to base current, and so on. This noise can be avoided by modulation of the signal at a higher frequency, for example through the use of a lock-in amplifier.

Environmental Noise

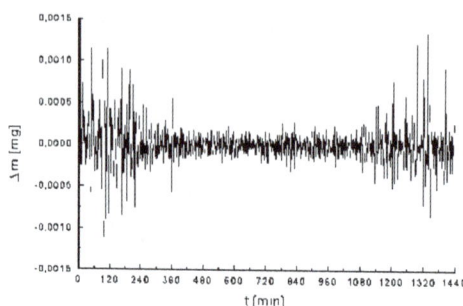

Noise in a thermogravimetric analysis; lower noise in the middle of the plot results from less human activity (and environmental noise) at night

Environmental noise arises from the surroundings of the analytical instrument. Sources of electromagnetic noise are power lines, radio and television stations, wireless devices, Compact fluorescent lamps and electric motors. Many of these noise sources are narrow bandwidth and therefore can be avoided. Temperature and vibration isolation may be required for some instruments.

Noise Reduction

Noise reduction can be accomplished either in computer hardware or software. Examples of hardware noise reduction are the use of shielded cable, analog filtering, and signal modulation. Examples of software noise reduction are digital filtering, ensemble average, boxcar average, and correlation methods.

Applications

Analytical chemistry has applications including in forensic science, bioanalysis, clinical analysis, environmental analysis, and materials analysis. Analytical chemistry research is largely driven by performance (sensitivity, detection limit, selectivity, robustness, dynamic range, linear range, accuracy, precision, and speed), and cost (purchase, operation, training, time, and space). Among the main branches of contemporary analytical atomic spectrometry, the most widespread and universal are optical and mass spectrometry. In the direct elemental analysis of solid samples, the new leaders are laser-induced breakdown and laser ablation mass spectrometry, and the related techniques with transfer of the laser ablation products into inductively coupled plasma. Advances in design of diode lasers and optical parametric oscillators promote developments in fluorescence and ionization spectrometry and also in absorption techniques where uses of optical cavities for increased effective absorption pathlength are expected to expand. The use of plasma- and laser-based methods is increasing. An interest towards absolute (standardless) analysis has revived, particularly in emission spectrometry.

Great effort is being put in shrinking the analysis techniques to chip size. Although there are few examples of such systems competitive with traditional analysis techniques, potential advantages include size/portability, speed, and cost. (micro total analysis system (μTAS) or lab-on-a-chip). Microscale chemistry reduces the amounts of chemicals used.

Many developments improve the analysis of biological systems. Examples of rapidly expanding fields in this area are genomics, DNA sequencing and related research in genetic fingerprinting and DNA microarray; proteomics, the analysis of protein concentrations and modifications, especially in response to various stressors, at various developmental stages, or in various parts of the body, metabolomics, which deals with metabolites; transcriptomics, including mRNA and associated fields; lipidomics - lipids and its associated fields; peptidomics - peptides and its associated fields; and metalomics, dealing with metal concentrations and especially with their binding to proteins and other molecules.

Analytical chemistry has played critical roles in the understanding of basic science to a variety of practical applications, such as biomedical applications, environmental monitoring, quality control of industrial manufacturing, forensic science and so on.

The recent developments of computer automation and information technologies have extended analytical chemistry into a number of new biological fields. For example, automated DNA sequencing machines were the basis to complete human genome projects leading to the birth of genomics. Protein identification and peptide sequencing by mass spectrometry opened a new field of proteomics.

Analytical chemistry has been an indispensable area in the development of nanotechnology. Surface characterization instruments, electron microscopes and scanning probe microscopes enables scientists to visualize atomic structures with chemical characterizations.

Biochemistry

Biochemistry, sometimes called biological chemistry, is the study of chemical processes within and relating to living organisms. By controlling information flow through biochemical signaling and the flow of chemical energy through metabolism, biochemical processes give rise to the complexity of life. Over the last decades of the 20th century, biochemistry has become so successful at explaining living processes that now almost all areas of the life sciences from botany to medicine to genetics are engaged in biochemical research. Today, the main focus of pure biochemistry is on understanding how biological molecules give rise to the processes that occur within living cells, which in turn relates greatly to the study and understanding of tissues, organs, and whole organisms—that is, all of biology.

Biochemistry is closely related to molecular biology, the study of the molecular mechanisms by which genetic information encoded in DNA is able to result in the processes of life. Depending on the exact definition of the terms used, molecular biology can be thought of as a branch of biochemistry, or biochemistry as a tool with which to investigate and study molecular biology.

Much of biochemistry deals with the structures, functions and interactions of biological macromolecules, such as proteins, nucleic acids, carbohydrates and lipids, which provide the structure of cells and perform many of the functions associated with life. The chemistry of the cell also depends on the reactions of smaller molecules and ions. These can be inorganic, for example water and metal ions, or organic, for example the amino acids, which are used to synthesize proteins. The mechanisms by which cells harness energy from their environment via chemical reactions are known as metabolism. The findings of biochemistry are applied primarily in medicine, nutrition, and agriculture. In medicine, biochemists investigate the causes and cures of diseases. In nutrition, they study how to maintain health and study the effects of nutritional deficiencies. In agriculture, biochemists investigate soil and fertilizers, and try to discover ways to improve crop cultivation, crop storage and pest control.

History

Gerty Cori and Carl Cori jointly won the Nobel Prize in 1947 for their discovery of the Cori cycle at RPMI.

At its broadest definition, biochemistry can be seen as a study of the components and composition of living things and how they come together to become life, and the history of biochemistry may therefore go back as far as the ancient Greeks. However, biochemistry as a specific scientific discipline has its beginning sometime in the 19th century, or a little earlier, depending on which aspect of biochemistry is being focused on. Some argued that the beginning of biochemistry may have been the discovery of the first enzyme, diastase (today called amylase), in 1833 by Anselme Payen, while others considered Eduard Buchner's first demonstration of a complex biochemical process alcoholic fermentation in cell-free extracts in 1897 to be the birth of biochemistry. Some might also point as its beginning to the influential 1842 work by Justus von Liebig, *Animal chemistry, or, Organic chemistry in its applications to physiology and pathology*, which presented a chemical theory of metabolism, or even earlier to the 18th century studies on fermentation and respiration by Antoine Lavoisier. Many other pioneers in the field who helped to uncover the layers of complexity of biochemistry have been proclaimed founders of modern biochemistry, for example Emil Fischer for his work on the chemistry of proteins, and F. Gowland Hopkins on enzymes and the dynamic nature of biochemistry.

The term "biochemistry" itself is derived from a combination of biology and chemistry. In 1877, Felix Hoppe-Seyler used the term (*biochemie* in German) as a synonym for physiological chemistry in the foreword to the first issue of *Zeitschrift für Physiologische Chemie* where he argued for the setting up of institutes dedicated to this field of study. The German chemistCarl Neuberg however is often cited to have coined the word in 1903, while some credited it to Franz Hofmeister.

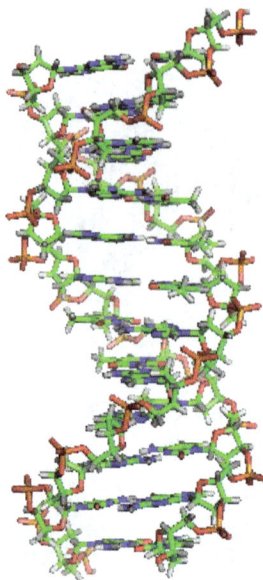

DNA structure (1D65)

It was once generally believed that life and its materials had some essential property or substance (often referred to as the "vital principle") distinct from any found in non-living matter, and it was thought that only living beings could produce the molecules of life. Then, in 1828, Friedrich Wöhler published a paper on the synthesis of urea, proving that organic compounds can be created artificially. Since then, biochemistry has advanced, especially since the mid-20th century, with the development of new techniques such as chromatography, X-ray diffraction, dual polarisation interferometry, NMR spectroscopy, radioisotopic labeling, electron microscopy, and molecular dynamics simulations. These techniques allowed for the discovery and detailed analysis of many molecules and metabolic pathways of the cell, such as glycolysis and the Krebs cycle (citric acid cycle).

Another significant historic event in biochemistry is the discovery of the gene and its role in the transfer of information in the cell. This part of biochemistry is often called molecular biology. In the 1950s, James D. Watson, Francis Crick, Rosalind Franklin, and Maurice Wilkins were instrumental in solving DNA structure and suggesting its relationship with genetic transfer of information. In 1958, George Beadle and Edward Tatum received the Nobel Prize for work in fungi showing that one gene produces one enzyme. In 1988, Colin Pitchfork was the first person convicted of murder with DNA evidence, which led to the growth of forensic science. More recently, Andrew Z. Fire and Craig C. Mello received the 2006 Nobel Prize for discovering the role of RNA interference (RNAi), in the silencing of gene expression.

Starting Materials: The Chemical Elements of Life

Around two dozen of the 92 naturally occurring chemical elements are essential to various kinds of biological life. Most rare elements on Earth are not needed by life (exceptions being selenium and iodine), while a few common ones (aluminum and titanium) are not used. Most organisms share element needs, but there are a few differences between plants and animals. For example, ocean algae use bromine, but land plants and animals seem to need none. All animals require sodium, but some plants do not. Plants need boron and silicon, but animals may not (or may need ultra-small amounts).

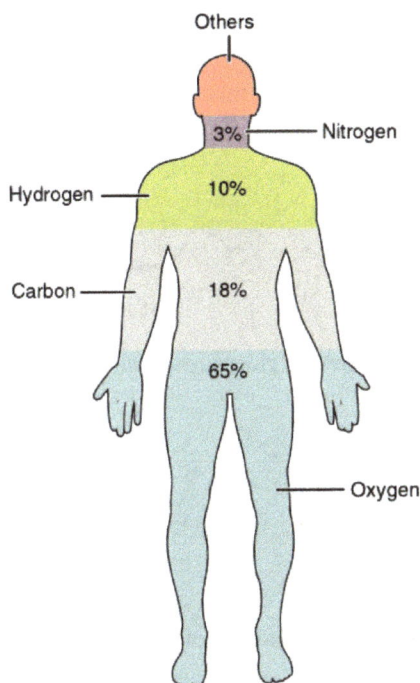

The main elements that compose the human body are shown from most abundant (by mass) to least abundant.

Just six elements—carbon, hydrogen, nitrogen, oxygen, calcium, and phosphorus—make up almost 99% of the mass of living cells, including those in the human body. In addition to the six major elements that compose most of the human body, humans require smaller amounts of possibly 18 more.

Biomolecules

The four main classes of molecules in biochemistry (often called biomolecules) are carbohydrates, lipids, proteins, and nucleic acids. Many biological molecules are polymers: in this terminology, *monomers* are relatively small micromolecules that are linked together to create large macromolecules known as *polymers*. When monomers are linked together to synthesize a biological polymer, they undergo a process called dehydration synthesis. Different macromolecules can assemble in larger complexes, often needed for biological activity.

Carbohydrates

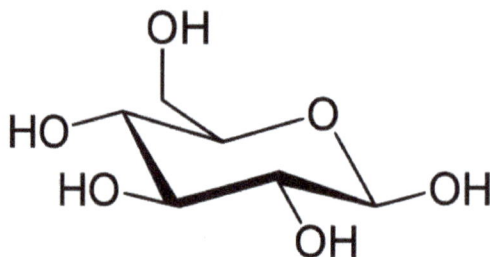

Glucose, a monosaccharide

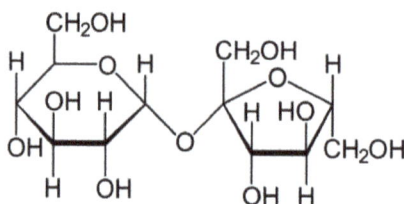

A molecule of sucrose (glucose + fructose), a disaccharide

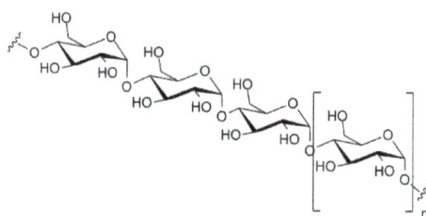

Amylose, a polysaccharide made up of several thousand glucose units

The function of carbohydrates includes energy storage and providing structure. Sugars are carbohydrates, but not all carbohydrates are sugars. There are more carbohydrates on Earth than any other known type of biomolecule; they are used to store energy and genetic information, as well as play important roles in cell to cell interactions and communications.

The simplest type of carbohydrate is a monosaccharide, which among other properties contains carbon, hydrogen, and oxygen, mostly in a ratio of 1:2:1 (generalized formula $C_nH_{2n}O_n$, where n is at least 3). Glucose ($C_6H_{12}O_6$) is one of the most important carbohydrates, others include fructose ($C_6H_{12}O_6$), the sugar commonly associated with the sweet taste of fruits,[a] and deoxyribose ($C_5H_{10}O_4$).

A monosaccharide can switch from the acyclic (open-chain) form to a cyclic form, through a nucleophilic addition reaction between the carbonyl group and one of the hydroxyls of the same molecule. The reaction creates a ring of carbon atoms closed by one bridging oxygen atom. The resulting molecule has an hemiacetal or hemiketal group, depending on whether the linear form was an aldose or a ketose. The reaction is easily reversed, yielding the original open-chain form.

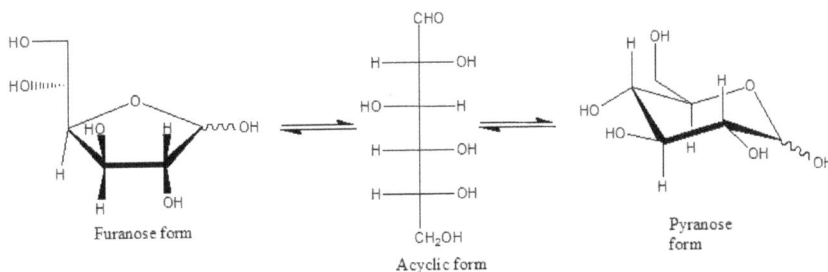

Conversion between the furanose, acyclic, and pyranose forms of D-glucose.

In these cyclic forms, the ring usually has 5 or 6 atoms. These forms are called fura-noses and pyranoses, respectively — by analogy with furan and pyran, the simplest compounds with the same carbon-oxygen ring (although they lack the double bonds of these two molecules). For example, the aldohexose glucose may form a hemiacetal linkage between the hydroxyl on carbon 1 and the oxygen on carbon 4, yielding a mol-ecule with a 5-membered ring, called glucofuranose. The same reaction can take place between carbons 1 and 5 to form a molecule with a 6-membered ring, called glucopy-ranose. Cyclic forms with a 7-atom ring (the same of oxepane), rarely encountered, are called heptoses.

When two monosaccharides undergo dehydration synthesis whereby a molecule of water is released, as two hydrogen atoms and one oxygen atom are lost from the two monosaccharides. The new molecule, consisting of two monosaccharides, is called a *disaccharide* and is conjoined together by a glycosidic or ether bond. The reverse re-action can also occur, using a molecule of water to split up a disaccharide and break the glycosidic bond; this is termed *hydrolysis*. The most well-known disaccharide is sucrose, ordinary sugar (in scientific contexts, called *table sugar* or *cane sugar* to dif-ferentiate it from other sugars). Sucrose consists of a glucose molecule and a fructose molecule joined together. Another important disaccharide is lactose, consisting of a glucose molecule and a galactose molecule. As most humans age, the production of lactase, the enzyme that hydrolyzes lactose back into glucose and galactose, typically decreases. This results in lactase deficiency, also called *lactose intolerance.*

When a few (around three to six) monosaccharides are joined, it is called an *oligosac-charide* (*oligo-* meaning "few"). These molecules tend to be used as markers and sig-nals, as well as having some other uses. Many monosaccharides joined together make a polysaccharide. They can be joined together in one long linear chain, or they may be branched. Two of the most common polysaccharides are cellulose and glycogen, both consisting of repeating glucose monomers. Examples are *Cellulose* which is an import-ant structural component of plant's cell walls, and *glycogen*, used as a form of energy storage in animals.

Sugar can be characterized by having reducing or non-reducing ends. A reducing end of a carbohydrate is a carbon atom that can be in equilibrium with the open-chain al-dehyde (aldose) or keto form (ketose). If the joining of monomers takes place at such a carbon atom, the free hydroxy group of the pyranose or furanose form is exchanged with an OH-side-chain of another sugar, yielding a full acetal. This prevents opening of the chain to the aldehyde or keto form and renders the modified residue non-reducing. Lactose contains a reducing end at its glucose moiety, whereas the galactose moiety form a full acetal with the C4-OH group of glucose. Saccharose does not have a reducing end because of full acetal formation between the aldehyde carbon of glucose (C1) and the keto carbon of fructose (C2).

Lipids

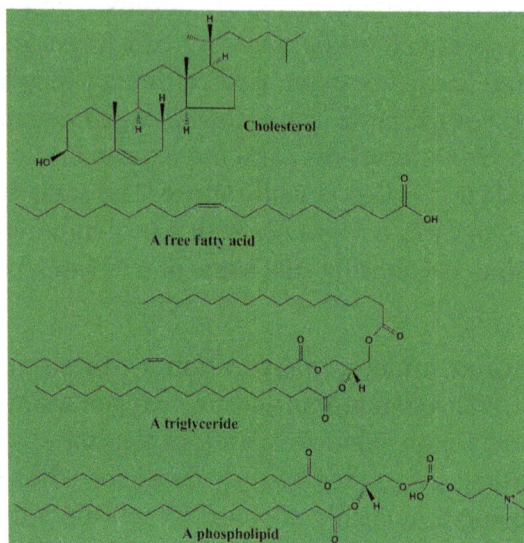

Structures of some common lipids. At the top are cholesterol and oleic acid. The middle structure is a triglyceride composed of oleoyl, stearoyl, and palmitoyl chains attached to a glycerol backbone. At the bottom is the common phospholipid, phosphatidylcholine.

Lipids comprises a diverse range of molecules and to some extent is a catchall for relatively water-insoluble or nonpolar compounds of biological origin, including waxes, fatty acids, fatty-acid derived phospholipids, sphingolipids, glycolipids, and terpenoids (e.g., retinoids and steroids). Some lipids are linear aliphatic molecules, while others have ring structures. Some are aromatic, while others are not. Some are flexible, while others are rigid.

Lipids are usually made from one molecule of glycerol combined with other molecules. In triglycerides, the main group of bulk lipids, there is one molecule of glycerol and three fatty acids. Fatty acids are considered the monomer in that case, and may be saturated (no double bonds in the carbon chain) or unsaturated (one or more double bonds in the carbon chain).

Most lipids have some polar character in addition to being largely nonpolar. In general, the bulk of their structure is nonpolar or hydrophobic ("water-fearing"), meaning that it does not interact well with polar solvents like water. Another part of their structure is polar or hydrophilic ("water-loving") and will tend to associate with polar solvents like water. This makes them amphiphilic molecules (having both hydrophobic and hydrophilic portions). In the case of cholesterol, the polar group is a mere -OH (hydroxyl or alcohol). In the case of phospholipids, the polar groups are considerably larger and more polar, as described below.

Lipids are an integral part of our daily diet. Most oils and milk products that we use for cooking and eating like butter, cheese, ghee etc., are composed of fats. Vegetable oils are rich in various polyunsaturated fatty acids (PUFA). Lipid-containing foods undergo digestion within the body and are broken into fatty acids and glycerol, which are the final degradation products of fats and lipids. Lipids, especially phospholipids, are also used in various pharmaceutical products, either as co-solubilisers (e.g., in parenteral infusions) or else as drug carrier components (e.g., in a liposome or transfersome).

Proteins

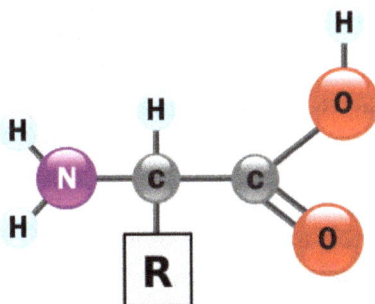

The general structure of an α-amino acid, with the amino group on the left and the carboxyl group on the right.

Proteins are very large molecules – macro-biopolymers – made from monomers called amino acids. An amino acid consists of a carbon atom bound to four groups. One is an amino group, $-NH_2$, and one is a carboxylic acid group, $-COOH$ (although these exist as $-NH_3^+$ and $-COO^-$ under physiologic conditions). The third is a simple hydrogen atom. The fourth is commonly denoted "—R" and is different for each amino acid. There are 20 standard amino acids, each containing a carboxyl group, an amino group, and a side-chain (known as an "R" group). The "R" group is what makes each amino acid different, and the properties of the side-chains greatly influence the overall three-dimensional conformation of a protein. Some amino acids have functions by themselves or in a modified form; for instance, glutamate functions as an important neurotransmitter. Amino acids can be joined via a peptide bond. In this dehydration synthesis, a water molecule is removed and the peptide bond connects the nitrogen of one amino acid's amino group to the carbon of the other's carboxylic acid group. The resulting molecule is called a *dipeptide*, and short stretches of amino acids (usually, fewer than thirty) are called *peptides* or polypeptides. Longer stretches merit the title *proteins*. As an example, the important blood serum protein albumin contains 585 amino acid residues.

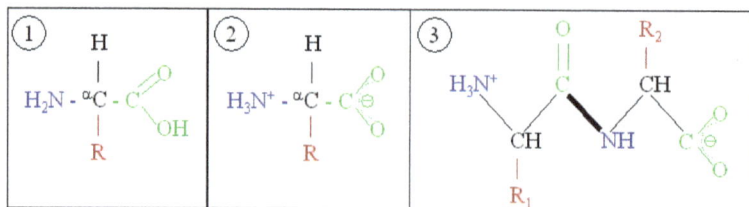

Generic amino acids (1) in neutral form, (2) as they exist physiologically, and (3) joined together as a dipeptide.

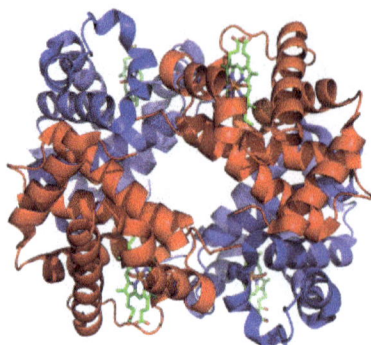

A schematic of hemoglobin. The red and blue ribbons represent the protein globin; the green structures are the heme groups.

Some proteins perform largely structural roles. For instance, movements of the proteins actin and myosin ultimately are responsible for the contraction of skeletal muscle. One property many proteins have is that they specifically bind to a certain molecule or class of molecules—they may be *extremely* selective in what they bind. Antibodies are an example of proteins that attach to one specific type of molecule. In fact, the enzyme-linked immunosorbent assay (ELISA), which uses antibodies, is one of the most sensitive tests modern medicine uses to detect various biomolecules. Probably the most important proteins, however, are the enzymes. Virtually every reaction in a living cell requires an enzyme to lower the activation energy of the reaction. These molecules recognize specific reactant molecules called *substrates*; they then catalyze the reaction between them. By lowering the activation energy, the enzyme speeds up that reaction by a rate of 10^{11} or more; a reaction that would normally take over 3,000 years to complete spontaneously might take less than a second with an enzyme. The enzyme itself is not used up in the process, and is free to catalyze the same reaction with a new set of substrates. Using various modifiers, the activity of the enzyme can be regulated, enabling control of the biochemistry of the cell as a whole.

The structure of proteins is traditionally described in a hierarchy of four levels. The primary structure of a protein simply consists of its linear sequence of amino acids; for instance,"alanine-glycine-tryptophan-serine-glutamate-asparagine-glycine-lysine-...". Secondary structure is concerned with local morphology (morphology being the study of structure). Some combinations of amino acids will tend to curl up in a coil called an α-helix or into a sheet called a β-sheet; some α-helixes can be seen in the hemoglobin schematic above. Tertiary structure is the entire three-dimensional shape of the protein. This shape is determined by the sequence of amino acids. In fact, a single change can change the entire structure. The alpha chain of hemoglobin contains 146 amino acid residues; substitution of the glutamate residue at position 6 with a valine residue changes the behavior of hemoglobin so much that it results in sickle-cell disease. Finally, quaternary structure is concerned with the structure of a protein with multiple peptide subunits, like hemoglobin with its four subunits. Not all proteins have more than one subunit.

Examples of protein structures from the Protein Data Bank

Members of a protein family, as represented by the structures of the isomerasedomains.

Ingested proteins are usually broken up into single amino acids or dipeptides in the small intestine, and then absorbed. They can then be joined to make new proteins. Intermediate products of glycolysis, the citric acid cycle, and the pentose phosphate pathway can be used to make all twenty amino acids, and most bacteria and plants possess all the necessary enzymes to synthesize them. Humans and other mammals, however, can synthesize only half of them. They cannot synthesize isoleucine, leucine, lysine, methionine, phenylalanine, threonine, tryptophan, and valine. These are the essential amino acids, since it is essential to ingest them. Mammals do possess the enzymes to synthesize alanine, asparagine, aspartate, cysteine, glutamate, glutamine, glycine, proline, serine, and tyrosine, the nonessential amino acids. While they can synthesize arginine and histidine, they cannot produce it in sufficient amounts for young, growing animals, and so these are often considered essential amino acids.

If the amino group is removed from an amino acid, it leaves behind a carbon skeleton called an α-keto acid. Enzymes called transaminases can easily transfer the amino group from one amino acid (making it an α-keto acid) to another α-keto acid (making it an amino acid). This is important in the biosynthesis of amino acids, as for many of the pathways, intermediates from other biochemical pathways are converted to the α-keto acid skeleton, and then an amino group is added, often via transamination. The amino acids may then be linked together to make a protein.

A similar process is used to break down proteins. It is first hydrolyzed into its component amino acids. Free ammonia (NH_3), existing as the ammonium ion (NH_4^+) in blood, is toxic to life forms. A suitable method for excreting it must therefore exist. Different tactics have evolved in different animals, depending on the animals' needs. Unicellular organisms, of course, simply release the ammonia into the environment. Likewise, bony fish can release the ammonia into the water where it is quickly diluted. In general, mammals convert the ammonia into urea, via the urea cycle.

In order to determine whether two proteins are related, or in other words to decide whether they are homologous or not, scientists use sequence-comparison methods. Methods like sequence alignments and structural alignments are powerful tools that help scientists identify homologies between related molecules. The relevance of finding homologies among proteins goes beyond forming an evolutionary pattern of protein families. By finding how similar two protein sequences are, we acquire knowledge about their structure and therefore their function.

Nucleic Acids

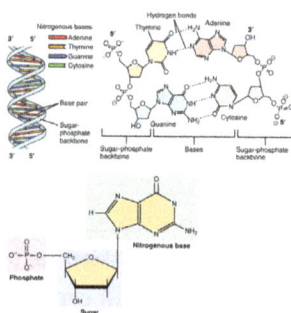

The structure of deoxyribonucleic acid (DNA), the picture shows the monomers being put together.

Nucleic acids, so called because of its prevalence in cellular nuclei, is the generic name of the family of biopolymers. They are complex, high-molecular-weight biochemical macromolecules that can convey genetic information in all living cells and viruses. The monomers are called nucleotides, and each consists of three components: a nitrogenous heterocyclic base (either a purine or a pyrimidine), a pentose sugar, and a phosphate group.

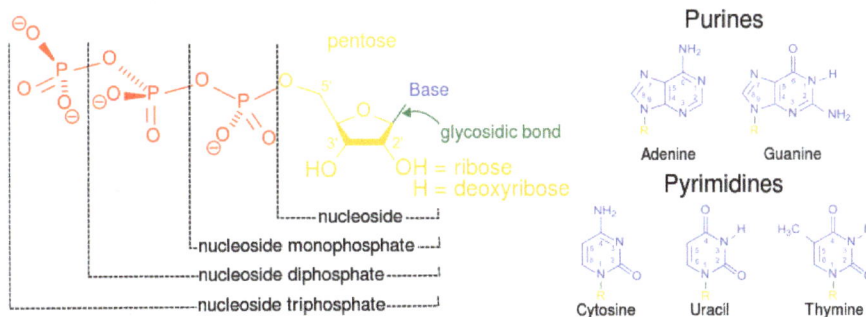

Structural elements of common nucleic acid constituents. Because they contain at least one phosphate group, the compounds marked *nucleoside monophosphate*, *nucleoside diphosphate* and *nucleoside triphosphate* are all nucleotides (not simply phosphate-lacking nucleosides).

The most common nucleic acids are deoxyribonucleic acid (DNA) and ribonucleic acid (RNA). The phosphate group and the sugar of each nucleotide bond with each other to form the backbone of the nucleic acid, while the sequence of nitrogenous bases stores the information. The most common nitrogenous bases are adenine, cytosine, guanine, thymine, and uracil. The nitrogenous bases of each strand of a nucleic acid will form hydrogen bonds with certain other nitrogenous bases in a complementary strand of nucleic acid (similar to a zipper). Adenine binds with thymine and uracil; Thymine binds only with adenine; and cytosine and guanine can bind only with one another.

Aside from the genetic material of the cell, nucleic acids often play a role as second messengers, as well as forming the base molecule for adenosine triphosphate (ATP), the primary energy-carrier molecule found in all living organisms. Also, the nitrogenous bases possible in the two nucleic acids are different: adenine, cytosine, and guanine occur in both RNA and DNA, while thymine occurs only in DNA and uracil occurs in RNA.

Metabolism

Carbohydrates as Energy Source

Glucose is the major energy source in most life forms. For instance, polysaccharides are broken down into their monomers (glycogen phosphorylase removes glucose residues from glycogen). Disaccharides like lactose or sucrose are cleaved into their two component monosaccharides.

Glycolysis (Anaerobic)

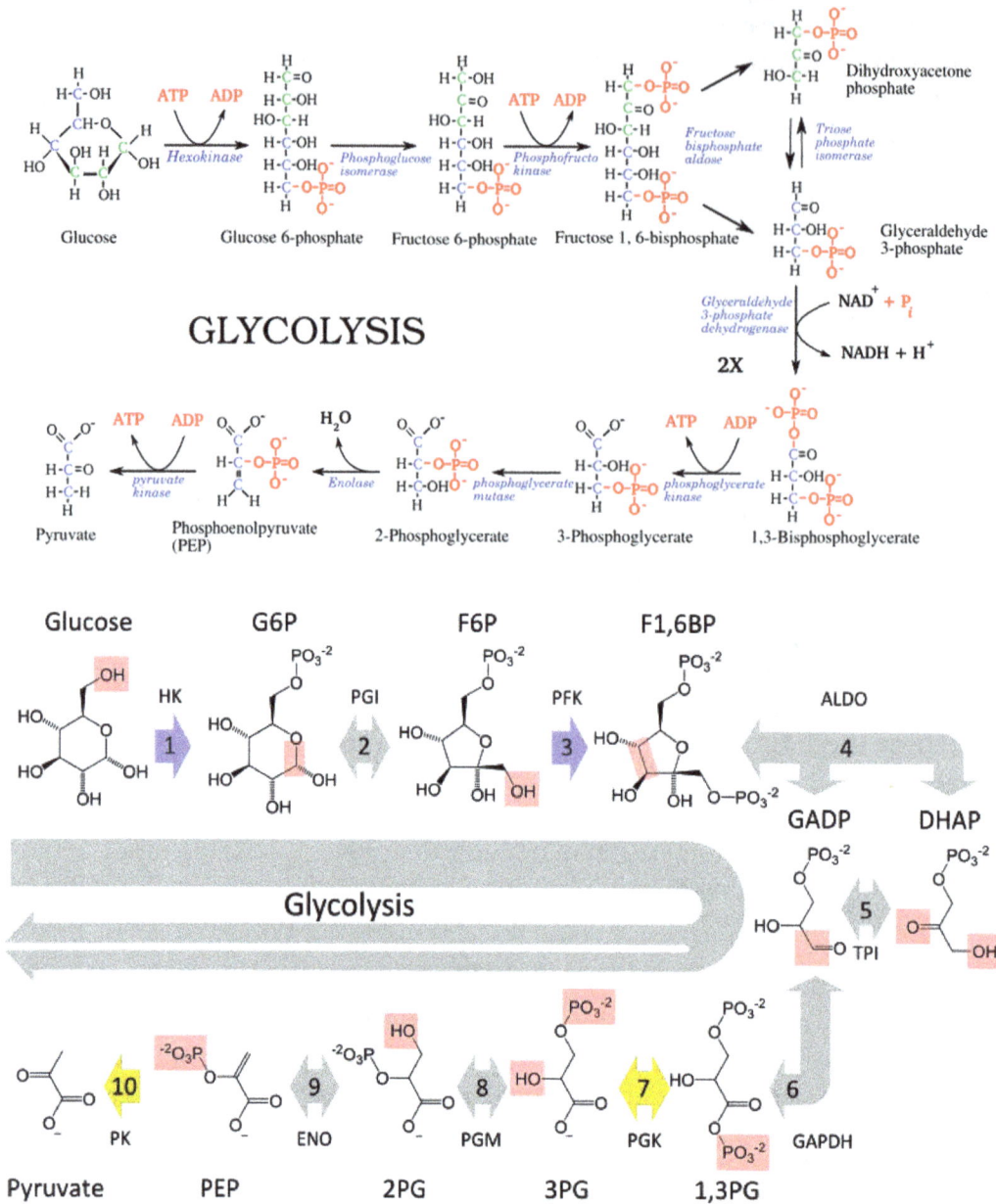

The metabolic pathway of glycolysis converts glucose to pyruvate by via a series of intermediate metabolites. Each chemical modification (red box) is performed by a different enzyme. Steps 1 and 3 consume ATP (blue) and steps 7 and 10 produce ATP (yellow). Since steps 6-10 occur twice per glucose molecule, this leads to a net production of ATP.

Glucose is mainly metabolized by a very important ten-step pathway called glycolysis, the net result of which is to break down one molecule of glucose into two molecules of pyruvate. This also produces a net two molecules of ATP, the energy currency of cells, along with two reducing equivalents of converting NAD^+ (nicotinamide adenine dinucleotide:oxidised form) to NADH (nicotinamide adenine dinucleotide:reduced form). This does not require oxygen; if no oxygen is available (or the cell cannot use oxygen), the NAD is restored by converting the pyruvate to lactate (lactic acid) (e.g., in humans) or to ethanol plus carbon dioxide (e.g., in yeast). Other monosaccharides like galactose and fructose can be converted into intermediates of the glycolytic pathway.

Aerobic

In aerobic cells with sufficient oxygen, as in most human cells, the pyruvate is further metabolized. It is irreversibly converted to acetyl-CoA, giving off one carbon atom as the waste product carbon dioxide, generating another reducing equivalent as NADH. The two molecules acetyl-CoA (from one molecule of glucose) then enter the citric acid cycle, producing two more molecules of ATP, six more NADH molecules and two reduced (ubi)quinones (via $FADH_2$ as enzyme-bound cofactor), and releasing the remaining carbon atoms as carbon dioxide. The produced NADH and quinol molecules then feed into the enzyme complexes of the respiratory chain, an electron transport system transferring the electrons ultimately to oxygen and conserving the released energy in the form of a proton gradient over a membrane (inner mitochondrial membrane in eukaryotes). Thus, oxygen is reduced to water and the original electron acceptors NAD^+ and quinone are regenerated. This is why humans breathe in oxygen and breathe out carbon dioxide. The energy released from transferring the electrons from high-energy states in NADH and quinol is conserved first as proton gradient and converted to ATP via ATP synthase. This generates an additional *28* molecules of ATP (24 from the 8 NADH + 4 from the 2 quinols), totaling to 32 molecules of ATP conserved per degraded glucose (two from glycolysis + two from the citrate cycle). It is clear that using oxygen to completely oxidize glucose provides an organism with far more energy than any oxygen-independent metabolic feature, and this is thought to be the reason why complex life appeared only after Earth's atmosphere accumulated large amounts of oxygen.

Gluconeogenesis

In vertebrates, vigorously contracting skeletal muscles (during weightlifting or sprinting, for example) do not receive enough oxygen to meet the energy demand, and so they shift to anaerobic metabolism, converting glucose to lactate. The liver regenerates the glucose, using a process called gluconeogenesis. This process is not quite the opposite of glycolysis, and actually requires three times the amount of energy gained from glycolysis (six molecules of ATP are used, compared to the two gained in glycolysis). Analogous to the above reactions, the glucose produced can then undergo glycolysis in tissues that need energy, be stored as glycogen (or starch in plants), or be converted to other monosaccharides or joined into di- or oligosaccharides. The combined pathways of glycolysis during exercise, lactate's crossing via the bloodstream to the liver, subsequent gluconeogenesis and release of glucose into the bloodstream is called the Cori cycle.

Relationship to Other "Molecular-scale" Biological Sciences

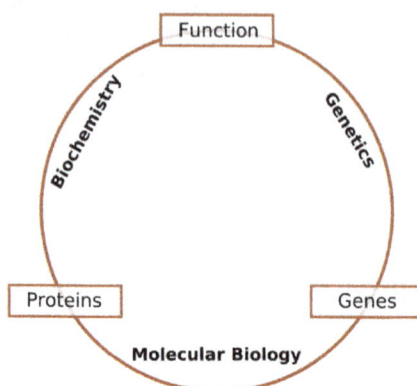

Schematic relationship between biochemistry, genetics, and molecular biology.

Researchers in biochemistry use specific techniques native to biochemistry, but increasingly combine these with techniques and ideas developed in the fields of genetics, molecular biology and biophysics. There has never been a hard-line among these disciplines in terms of content and technique. Today, the terms *molecular biology* and *biochemistry* are nearly interchangeable. The following figure is a schematic that depicts one possible view of the relationship between the fields:

- *Biochemistry* is the study of the chemical substances and vital processes occurring in living organisms. Biochemists focus heavily on the role, function, and structure of biomolecules. The study of the chemistry behind biological processes and the synthesis of biologically active molecules are examples of biochemistry.

- *Genetics* is the study of the effect of genetic differences on organisms. Often this can be inferred by the absence of a normal component (e.g., one gene). The study of "mutants" – organisms with a changed gene that leads to the organism being different with respect to the so-called "wild type" or normal phenotype. Genetic interactions (epistasis) can often confound simple interpretations of such "knock-out" or "knock-in" studies.

- *Molecular biology* is the study of molecular underpinnings of the process of replication, transcription and translation of the genetic material. The central dogma of molecular biology where genetic material is transcribed into RNA and then translated into protein, despite being an oversimplified picture of molecular biology, still provides a good starting point for understanding the field. This picture, however, is undergoing revision in light of emerging novel roles for RNA.

- *Chemical biology* seeks to develop new tools based on small molecules that allow minimal perturbation of biological systems while providing detailed information about their function. Further, chemical biology employs biological systems to create non-natural hybrids between biomolecules and synthetic devices (for example emptied viral capsids that can deliver gene therapy or drug molecules).

Chemical Engineering

Chemical engineers design, construct and operate process plants (distillation columns pictured)

Chemical engineering is a branch of engineering that applies physical sciences (physics and chemistry), life sciences (microbiology and biochemistry), together with applied mathematics and economics to produce, transform, transport, and properly use chemicals, materials and energy. Essentially, chemical engineers design large-scale processes that convert chemicals, raw materials, living cells, microorganisms and energy into useful forms and products.

Etymology

George E. Davis

A 1996 *British Journal for the History of Science* article cites James F. Donnelly for mentioning an 1839 reference to chemical engineering in relation to the production of sulfuric acid. In the same paper however, George E. Davis, an English consultant, was credited for having coined the term. The *History of Science in United States: An Encyclopedia* puts this at around 1890. "Chemical engineering", describing the use of mechanical equipment in the chemical industry, became common vocabulary in England after 1850. By 1910, the profession, "chemical engineer," was already in common use in Britain and the United States.

History

Chemical engineering emerged upon the development of unit operations, a fundamental concept of the discipline of chemical engineering. Most authors agree that Davis invented unit operations if not substantially developed it. He gave a series of lectures on unit operations at the Manchester Technical School (later part of the University of Manchester) in 1887, considered to be one of the earliest such about chemical engineering. Three years before Davis' lectures, Henry Edward Armstrong taught a degree course in chemical engineering at the City and Guilds of London Institute. Armstrong's course "failed simply because its graduates … were not especially attractive to employers." Employers of the time would have rather hired chemists and mechanical engineers. Courses in chemical engineering offered by Massachusetts Institute of Technology (MIT) in the United States, Owens College in Manchester, England, and University College London suffered under similar circumstances.

Students inside an industrial chemistry laboratory at MIT

Starting from 1888,Lewis M. Norton taught at MIT the first chemical engineering course in the United States. Norton's course was contemporaneous and essentially similar with Armstrong's course. Both courses, however, simply merged chemistry and engineering subjects. "Its practitioners had difficulty convincing engineers that they were engineers and chemists that they were not simply chemists." Unit operations was introduced into the course by William Hultz Walker in 1905. By the early 1920s, unit operations became an important aspect of chemical engineering at MIT and other US universities, as well as at Imperial College London. The American Institute of Chemical Engineers (AIChE), established in 1908, played a key role in making chemical engineering considered an independent science, and unit operations central to chemical engineering. For instance, it defined chemical engineering to be a "science of itself, the basis of which is … unit operations" in a 1922 report; and with which principle, it had published a list of academic institutions which offered "satisfactory" chemical engineering courses. Meanwhile, promoting chemical engineering as a distinct science in Britain lead to the establishment of the Institution of Chemical Engineers (IChemE) in 1922. IChemE likewise helped make unit operations considered essential to the discipline.

New Concepts and Innovations

By the 1940s, it became clear that unit operations alone was insufficient in developing chemical reactors. While the predominance of unit operations in chemical engineering courses in Britain and the United States continued until the 1960s, transport phenomena started to experience greater focus. Along with other novel concepts, such process systems engineering (PSE), a "second paradigm" was defined. Transport phenomena gave an analytical approach to chemical engineering while PSE focused on its synthetic elements, such as control system and process design. Developments in chemical engineering before and after World War II were mainly incited by the petrochemical industry, however, advances in other fields were made as well. Advancements in biochemical engineering in the 1940s, for example, found application in the pharmaceutical industry, and allowed for the mass production of various antibiotics, including penicillin and streptomycin. Meanwhile, progress in polymer science in the 1950s paved way for the "age of plastics".

Safety and Hazard Developments

Concerns regarding the safety and environmental impact of large-scale chemical manufacturing facilities were also raised during this period. *Silent Spring*, published in 1962, alerted its readers to the harmful effects of DDT, a potent insecticide. The 1974 Flixborough disaster in the United Kingdom resulted in 28 deaths, as well as damage to a chemical plant and three nearby villages. The 1984 Bhopal disaster in India resulted in almost 4,000 deaths. These incidents, along with other incidents, affected the reputation of the trade as industrial safety and environmental protection were given more focus. In response, the IChemE required safety to be part of every degree course that it accredited after 1982. By the 1970s, legislation and monitoring agencies were instituted in various countries, such as France, Germany, and the United States.

Recent Progress

Advancements in computer science found applications designing and managing plants, simplifying calculations and drawings that previously had to be done manually. The completion of the Human Genome Project is also seen as a major development, not only advancing chemical engineering but genetic engineering and genomics as well. Chemical engineering principles were used to produce DNA sequences in large quantities.

Concepts

Chemical engineering involves the application of several principles. Key concepts are presented below.

Chemical Reaction Engineering

Chemical engineering involves managing plant processes and conditions to ensure optimal plant operation. Chemical reaction engineers construct models for reactor analysis and design using laboratory data and physical parameters, such as chemical thermodynamics, to solve problems and predict reactor performance.

Plant Design and Construction

Chemical engineering design concerns the creation of plans, specification, and economic analyses for pilot plants, new plants or plant modifications. Design engineers often work in a consulting role, designing plants to meet clients' needs. Design is limited by a number of factors, including funding, government regulations and safety standards. These constraints dictate a plant's choice of process, materials and equipment.

Plant construction is coordinated by project engineers and project managers depending on the size of the investment. A chemical engineer may do the job of project engineer full-time or part of the time, which requires additional training and job skills, or act as a consultant to the project group. In USA the education of chemical engineering graduates from the Baccalaureate programs accredited by ABET do not usually stress project engineering education, which can be obtained by specialized training, as electives, or from graduate programs. Project engineering jobs are some of the largest employers for chemical engineers.

Process Design and Analysis

A unit operation is a physical step in an individual chemical engineering process. Unit operations (such as crystallization, filtration, drying and evaporation) are used to prepare reactants, purifying and separating its products, recycling unspent reactants, and controlling energy transfer in reactors. On the other hand, a unit process is the chemical equivalent of a unit operation. Along with unit operations, unit processes constitute a process operation. Unit processes (such as nitration and oxidation) involve the conversion of material by biochemical, thermochemical and other means. Chemical engineers responsible for these are called process engineers.

Process design requires the definition of equipment types and sizes as well as how they are connected together and the materials of construction. Details are often printed on a Process Flow Diagram which is used to control the capacity and reliability of a new or modified chemical factory.

Education for chemical engineers in the first college degree 3 or 4 years of study stresses the principles and practices of process design. The same skills are used in existing chemical plants to evaluate the efficiency and make recommendations for improvements.

Transport Phenomena

Modeling and analysis of transport phenomena is essential for many industrial applications. Transport phenomena involve fluid dynamics, heat transfer and mass transfer, which are governed mainly by momentum transfer, energy transfer and transport of chemical species respectively. Models often involve separate considerations for macroscopic, microscopic and molecular level phenomena. Modeling of transport phenomena requires therefore requires an understanding of applied mathematics.

Applications and Practice

Chemical engineers use computers to control automated systems in plants.

Operators in a chemical plant using an older analog control board, seen in East-Germany, 1986.

Chemical engineers "develop economic ways of using materials and energy". Chemical engineers use chemistry and engineering to turn raw materials into usable products, such as medicine, petrochemicals and plastics on a large-scale, industrial setting. They are also involved in waste management and research. Both applied and research facets could make extensive use of computers.

Chemical engineers may be involved in industry or university research where they are tasked with designing and performing experiments to create better and safer methods for production, pollution control, and resource conservation. They may be involved in designing and constructing plants as a project engineer. Chemical engineers serving as project engineers use their knowledge in selecting optimal production methods and plant equipment to minimize costs and maximize safety and profitability. After plant construction, chemical engineering project managers may be involved in equipment upgrades, process changes, troubleshooting, and daily operations in either full-time or consulting roles.

Related Fields and Topics

Today, the field of chemical engineering is a diverse one, covering areas from biotechnology and nanotechnology to mineral processing.

- Biochemical engineering
- Bioprocess engineering
- Bioinformatics
- Biomedical engineering
- Biomolecular engineering
- Biotechnology
- Biotechnology engineering
- Catalysts
- Ceramics
- Chemical process modeling
- Chemical Technologist
- Chemical reactor
- Chemical weapons
- Cheminformatics
- Computational fluid dynamics
- Corrosion engineering
- Cost estimation
- Electrochemistry
- Environmental engineering
- Earthquake engineering
- Fischer Tropsch synthesis

- Fluid dynamics
- Food engineering
- Fuel cell
- Gasification
- Heat transfer
- Industrial gas
- Industrial catalysts
- Mass transfer
- Materials science
- Metallurgy
- Microfluidics
- Mineral processing
- Molecular engineering
- Nanotechnology
- Natural environment
- Natural gas processing
- Nuclear reprocessing
- Oil exploration
- Oil refinery
- Petroleum engineering
- Pharmaceutical engineering

- Plastics engineering
- Polymers
- Process control
- Process design
- Process development
- Process engineering
- Process miniaturization
- Paper engineering
- Safety engineering
- Semiconductor device fabrication
- Separation processes
  - Crystallization processes
  - Distillation processes
  - Membrane processes
- Syngas production
- Textile engineering
- Thermodynamics
- Transport phenomena
- Unit operations
- Water technology

References

- Skoog, Douglas A.; West, Donald M.; Holler, F. James; Crouch, Stanley R. (2014). Fundamentals of Analytical Chemistry. Belmont: Brooks/Cole, Cengage Learning. p. 1. ISBN 0-495-55832-X.

- Skoog, Douglas A.; Holler, F. James; Crouch, Stanley R. (2007). Principles of Instrumental Analysis. Belmont, CA: Brooks/Cole, Thomson. p. 1. ISBN 0-495-01201-7.

- Crouch, Stanley; Skoog, Douglas A. (2007). Principles of instrumental analysis. Australia: Thomson Brooks/Cole. ISBN 0-495-01201-7.

- Arikawa, Yoshiko (2001). "Basic Education in Analytical Chemistry"(pdf). Analytical Sciences. The Japan Society for Analytical Chemistry. 17 (Supplement): i571–i573. Retrieved 10 January 2014.

- "Health Concerns associated with Energy Efficient Lighting and their Electromagnetic Emissions"(PDF). Trent University, Peterborough, ON, Canada. Retrieved 2011-11-12.

Evolution of Chromatography

The history of chromatography explains to the readers the growth of chromatography over the years. Chromatography was initially used to separate plant pigments. New forms of chromatography have been developed in the recent past. The section on evolution of chromatography offers an insightful focus, keeping in mind the subject matter.

The history of chromatography spans from the mid-19th century to the 21st. Chromatography, literally "color writing", was used—and named— in the first decade of the 20th century, primarily for the separation of plant pigments such as chlorophyll (which is green) and carotenoids (which are orange and yellow). New forms of chromatography developed in the 1930s and 1940s made the technique useful for a wide range of separation processes and chemical analysis tasks, especially in biochemistry.

Precursors

The earliest use of chromatography—passing a mixture through an inert material to create separation of the solution components based on differential adsorption—is sometimes attributed to German chemist Friedlieb Ferdinand Runge, who in 1855 described the use of paper to analyze dyes. Runge dropped spots of different inorganic chemicals onto circles of filter paper already impregnated with another chemical, and reactions between the different chemicals created unique color patterns. According to historical analysis of L. S. Ettre, however, Runge's work had "nothing to do with chromatography" (and instead should be considered a precursor of chemical spot tests such as the Schiff test).

In the 1860s, Christian Friedrich Schönbein and his student Friedrich Goppelsroeder published the first attempts to study the different rates at which different substances move through filter paper. Schönbein, who thought capillary action (rather than adsorption) was responsible for the movement, called the technique capillary analysis, and Goppelsroeder spent much of his career using capillary analysis to test the movement rates of a wide variety of substances. Unlike modern paper chromatography, capillary analysis used reservoirs of the substance being analyzed, creating overlapping zones of the solution components rather than separate points or bands.

Work on capillary analysis continued, but without much technical development, well into the 20th century. The first significant advances over Goppelsroeder's methods came with the work of Raphael E. Liesegang: in 1927, he placed filter strips in closed containers with atmospheres saturated by solvents, and in 1943 he began using discrete spots of sample adsorbed to filter paper, dipped in pure solvent to achieve separation. This method, essentially identical to modern paper chromatography, was published just before the independent—and far more influential—work of Archer Martin and his collaborators that inaugurated the widespread use of paper chromatography.

In 1897, the American chemist David Talbot Day (1859-1915), then serving with the U.S. Geological Survey, observed that crude petroleum generated bands of color as it seeped upwards through finely divided clay or limestone. In 1900, he reported his findings at the First International Petroleum Congress in Paris, where they created a sensation.

Tsvet and Column Chromatography

The first true chromatography is usually attributed to the Russian-Italian botanist Mikhail Tsvet. Tsvet applied his observations with filter paper extraction to the new methods of column fractionation that had been developed in the 1890s for separating the components of petroleum. He used a liquid-adsorption column containing calcium carbonate to separate yellow, orange, and green plant pigments (what are known today as xanthophylls, carotenes, and chlorophylls, respectively). The method was described on December 30, 1901 at the 11th Congress of Naturalists and Doctors (XI съезд естествоиспытателей и врачей) in Saint Petersburg. The first printed description was in 1903, in the Proceedings of the Warsaw Society of Naturalists, section of biology. He first used the term *chromatography* in print in 1906 in his two papers about chlorophyll in the German botanical journal, *Berichte der Deutschen Botanischen Gesellschaft*. In 1907 he demonstrated his chromatograph for the German Botanical Society. Interestingly, Mikhail's surname "Цвет" means "color" in Russian, so there is the possibility that his naming the procedure chromatography (literally "color writing") was a way that he could make sure that he, a commoner in Tsarist Russia, could be immortalized.

In a 1903 lecture (published in 1905), Tsvet also described using filter paper to approximate the properties of living plant fibers in his experiments on plant pigments—a precursor to paper chromatography. He found that he could extract some pigments (such as orange carotenes and yellow xanthophylls) from leaves with non-polar solvents, but others (such as chlorophyll) required polar solvents. He reasoned that chlorophyll was held to the plant tissue by adsorption, and that stronger solvents were necessary to overcome the adsorption. To test this, he applied dissolved pigments to filter paper, allowed the solvent to evaporate, then applied different solvents to see which could extract the pigments from the filter paper. He found the same pattern as from leaf extractions: carotene could be extracted from filter paper using non-polar solvents, but chlorophyll required polar solvents.

Tsvet's work saw little use until the 1930s.

Martin and Synge and Partition Chromatography

Chromatography methods changed little after Tsvet's work until the explosion of mid-20th century research in new techniques, particularly thanks to the work of Archer John Porter Martin and Richard Laurence Millington Synge. By "the marrying of two techniques, that of chromatography and that of countercurrent solvent extraction", Martin and Synge developed partition chromatography to separate chemicals with only slight differences in partition coefficients between two liquid solvents. Martin, who had previously been working in vitamin chemistry (including attempts to purify vitamin E), began collaborating with Synge in 1938, brought his experience with equipment design to Synge's project of separating amino acids. After unsuccessful experiments with complex countercurrent extraction machines and liquid-liquid chromatography methods where the liquids move in opposite directions, Martin hit on the idea of using silica gel in columns to hold water stationary while an organic solvent flows through the column. Martin and Synge demonstrated the potential of the methods by separating amino acids marked in the column by the addition of methyl red. In a series of publications beginning in 1941, they described increasingly powerful methods of separating amino acids and other organic chemicals.

In pursuit of better and easier methods of identifying the amino acid constituents of peptides, Martin and Synge turned to other chromatography media as well. A short abstract in 1943 followed by a detailed article in 1944 described the use of filter paper as the stationary phase for performing chromatography on amino acids: paper chromatography. By 1947, Martin, Synge and their collaborators had applied this method (along with Fred Sanger's reagent for identifying N-terminal residues) to determine the pentapeptide sequence of Gramicidin S. These and related paper chromatography methods were also foundational to Fred Sanger's effort to determine the amino acid sequence of insulin.

Refining the Techniques

Martin, in collaboration with Anthony T. James, went on to develop gas chromatography (the principles of which Martin and Synge had laid out in their landmark 1941 paper) beginning in 1949. In 1952, during his lecture for the Nobel Prize in Chemistry (shared with Synge, for their earlier chromatography work) Martin announced the successful separation of a wide variety of natural compounds by gas chromatography. (Austrian chemist Fritz Prior had achieved limited success with gas chromatography, separating oxygen and carbon dioxide, in 1947 during his Ph.D. research. The method of Martin and James, however, became the basis for subsequent developments in gas chromatography.)

The ease and efficiency of gas chromatography for separating organic chemicals spurred the rapid adoption of the method, as well as the rapid development of new detection methods for analyzing the output. The thermal conductivity detector, described in 1954 by N. H. Ray, was the foundation for several other methods: the flame ionization detector was described by J. Harley, W. Nel, and V. Pretorius in 1958, and James Lovelock introduced the electron capture detector that year as well. Others introduced mass spectrometers to gas chromatography in the late 1950s.

The work of Martin and Synge also set the stage for high performance liquid chromatography, suggesting that small sorbent particles and pressure could produce fast liquid chromatography techniques. This became widely practical by the late 1960s (and the method was used to separate amino acids as early as 1960).

Thin Layer Chromatography

The first developments in thin layer chromatography occurred in the 1940s, and techniques advanced rapidly in the 1950s after the introduction of relatively large plates and relatively stable materials for sorbent layers.

Later Developments

In 1987 Pedro Cuatrecasas and Meir Wilchek were awarded the Wolf Prize in Medicine for the invention and development of affinity chromatography and its applications to biomedical sciences.

Permissions

All chapters in this book are published with permission under the Creative Commons Attribution Share Alike License or equivalent. Every chapter published in this book has been scrutinized by our experts. Their significance has been extensively debated. The topics covered herein carry significant information for a comprehensive understanding. They may even be implemented as practical applications or may be referred to as a beginning point for further studies.

We would like to thank the editorial team for lending their expertise to make the book truly unique. They have played a crucial role in the development of this book. Without their invaluable contributions this book wouldn't have been possible. They have made vital efforts to compile up to date information on the varied aspects of this subject to make this book a valuable addition to the collection of many professionals and students.

This book was conceptualized with the vision of imparting up-to-date and integrated information in this field. To ensure the same, a matchless editorial board was set up. Every individual on the board went through rigorous rounds of assessment to prove their worth. After which they invested a large part of their time researching and compiling the most relevant data for our readers.

The editorial board has been involved in producing this book since its inception. They have spent rigorous hours researching and exploring the diverse topics which have resulted in the successful publishing of this book. They have passed on their knowledge of decades through this book. To expedite this challenging task, the publisher supported the team at every step. A small team of assistant editors was also appointed to further simplify the editing procedure and attain best results for the readers.

Apart from the editorial board, the designing team has also invested a significant amount of their time in understanding the subject and creating the most relevant covers. They scrutinized every image to scout for the most suitable representation of the subject and create an appropriate cover for the book.

The publishing team has been an ardent support to the editorial, designing and production team. Their endless efforts to recruit the best for this project, has resulted in the accomplishment of this book. They are a veteran in the field of academics and their pool of knowledge is as vast as their experience in printing. Their expertise and guidance has proved useful at every step. Their uncompromising quality standards have made this book an exceptional effort. Their encouragement from time to time has been an inspiration for everyone.

The publisher and the editorial board hope that this book will prove to be a valuable piece of knowledge for students, practitioners and scholars across the globe.

Index